Making God

MAKING GOD

A New Materialist Theory of the Person

Ann Long

ia

imprint-academic.com

Published in the UK by Imprint Academic
PO Box 200, Exeter EX5 5YX, UK

Published in the USA by Imprint Academic
Philosophy Documentation Center
PO Box 7147, Charlottesville, VA 22906-7147, USA

ISBN 9 781845 400909

A CIP catalogue record for this book is available from the
British Library and US Library of Congress

Cover Illustration:

Max Golding (1987), *The Bard*, a piece in oak and twine

CONTENTS

PART THREE

PREFACE

In this book I am trying to mount a complicated argument. At one and the same time, I am trying to challenge both personic exceptionalism, and personic reductionism. It is more usual to attempt either the one or the other.

By 'personic exceptionalism' I mean the argument that, different from all other entities in the universe, human persons (entities other than human animals, qualitatively different and special) escape cause-and-effect physical determination to be free-willing agents whose mental characteristics are other than physical: a stance which results in dualistic and religious-spiritual interpretations of the human condition; and in the view that there *cannot* therefore be a scientific study of persons. I think personic exceptionalism is wrong.

By 'personic reductionism' I mean the argument that, just like all other entities in the biosphere, human persons (entities really just human animals, nothing qualitatively different or special) cannot escape cause-and-effect biological determination, and are no more free-willing agents than is any other organism: a stance which results in selfish-gene and naturally-hierarchical interpretations of the human condition; and in the view that there *can* therefore be a scientific study of persons. I think personic reductionism is wrong as well.

Against the exceptionalists, with the reductionists, I *do* think that persons are as cause-and-effect determined as any other entity—but not by the biological. Against the reductionists, with the exceptionalists, I *do* think that persons are other than human animals, are qualitatively different and special—but not because they are free-willing agents. This is the stance which the book aims to defend.

I am especially concerned that young people should read it: maybe particularly, those who as sixth formers are making decisions about what they would like to study at university; or those who as undergraduates or graduates are studying the personic sciences, and are about to begin their research work. I would love them to take these ideas and fly with them:

sophisticate them; correct them; but above all, use them to help to make a better world.

Across the many years during which I have been thinking about the issues which this book addresses, many people have given me support and advice: not least my late partner, **Max Golding**; my three offspring; and indeed each and every member of my widely extended and close family. In addition, friends and colleagues too numerous to name individually have engaged me in that 'philosophical conversation' which eventually became its major object of study. A little more formally, so too have members of the **Oxford Philosophical Society** and the **Manchester Philosophy Group**, who have been variously sparring partners and/or philosophical allies across now several years. When the book was in its penultimate form, **Stephen Houseman**, among the former, and **Michael Kennedy**, among the latter, read it in part, and commented. And each of **Alan Bailey**, **Charles Brown**, **Don Cupitt**, **David Green**, **Andrew Shanks** and **Richard Smith** provided a close reading of the whole thing, and commented on it both wisely and generously. The two indexes, modified only slightly, were produced by **Judith Reading**. To all these, my thanks.

PART ONE

Chapter One

INTRODUCTION
The Mad Caprice

Just a few short years ago people all over the world were celebrating. These were millennial times. And as dates shifted from nineteens to twenties, a gale of for the most part hopeful applause, rejoicing and fireworks saluted the occasion. It seems to me that the author of almost any book addressing grand issues concerning 'the human condition', and written in the near wake of such things, might well be tempted to give her or his text at the very least a millennial feel, perhaps even some millennial reference. By quite deliberate choice, this author is not tempted but resolute. I want my book to have such feel and such reference. I want it to be a book which is at one and the same time both a product of, and a challenge to, the celebrations which then we made. And that being the case, let me begin at what therefore becomes my beginning: at the source, in the history of one of the world's three major monotheisms, of the millennial story.

The God of the Old Testament, we are told, created the world in six days and rested on the seventh. But what, to a God, might a day be? A clue comes from the book of Psalms. It is there that we discover (Psalm 90, verse 4) that 'a thousand years in thy sight are but as yesterday': a phrase which became the source of a powerful myth which grew up among some of the early Christians. Associating a day in the life of God with a thousand years, a millennium, in the lives of human beings, one version of the story ran as follows: that the less than perfect world, the world as we know it, would last for six thousand years; that at the end of that time Jesus would return; that his second coming would usher in a seventh thousand years, but this time of perfect bliss; and that at the end of that seventh and last millennium, the day of judgement would bring to an end the whole human enterprise. Thus the reference in Revelation (20.1–3) to 'that old serpent, which is the Devil' being cast into hell 'till the thousand years should be fulfilled' is one version of the tale.

This is a story with interesting implications. And not the least among them is the fact that if we could know when the world began, we could also know that six millennia later Jesus would come for the second time. We could also know that one millennium after that would see the end (or maybe the beginning) of everything. Skipping now many centuries, it was a seventeenth century prelate, Archbishop Ussher of Armagh, who made a subsequently notorious attempt actually to date the beginning of the world. Basing himself on the lists of the generations found in Genesis, he came up with the notion that it began in the year 4004 BCE. Add six millennia, that is six thousand years, to such a figure, and arrive in 1996. Add just four years to that, and arrive at the end of the second millennium CE. With such pregnant dates as these so recently around, perhaps it is no wonder that the embers of 'millenarianism' should have been stirred. Millenarianism is the belief that as a millennium closes so the end of the world is nigh. So perhaps it was only to be expected that, anyway in more suggestible places, a pre-millennial tension (PMT) associated with the years immediately preceding 2000 should have been succeeded by a post-millennial trauma (PMT) of the years immediately following 2000. Dates with lots of zeros at the end have such propensity to faze: a fact which makes it just nicely ironic that it was those very same zeros which, because the people of the Middle Ages had such an uncertain understanding of their status, led to all the debate a few years back as to whether the third millennium began on 1 January 2000 or 1 January 2001.

But if it was only to be expected, it was also surely somewhat odd. We have lived now with four centuries of modern science. Numerology, it might have been thought, is just dead. At the very least, most of us surely must now be aware, if not of the detail then certainly of the generality, of calendar adjustment. The earth rotates on its axis once every 23.9345 hours. The moon revolves around the earth once every 27.3217 days. The earth revolves around the sun once every 365.5306 days. Such untidy numbers just guarantee that from time to time some such adjustment will be necessary. Thus our calendar had 90 days added to it in 46 BCE; leap years omitted from it between 10 BCE and 4 CE; and 10 days removed from it in 1582. With the introduction, in 1752, of the modern Gregorian version, another eleven days were removed: this last of course the event which led to the famous 'give-us-back-our-eleven-days' campaign, which involved rather less the ignorant demand of folk who thought their lives had been shortened by the government (as sometimes it is portrayed) and rather more the totally reasonable demand of poor tenants trying to avoid giving their landlords eleven days rent for nothing. So if only in the light of such technicalities, was it not, in retrospect, somewhat odd to have treated midnight on either 31 December 1999 or 31 December 2000 as a time of any particular significance? What anyway is a 'new' year? Until 800 CE the French took March 1 to be New Year's Day. They then changed it to March 25, where it remained until 996, when for 55 years it coincided with Easter. Between 600 CE and 1338 CE the English took Christmas Day to be the start of the new year. But in 1339 they too shifted it to March 25. It is the Gregorian calendar which, from 1752 onwards, has taken January 1 to be New Year's Day: a fact which, nonetheless, did not stop it from being the case that in 1996, Gregorian style, the Chinese New Year was on 19 February; the Burmese New Year was on 15 April; the Islamic New Year was on 19 May; and the Jewish New Year was on 14 September. And we all know, or anyway half know, these and other like things.

Neither does it end there. Even if we consider the year 2000 as a date appropriate only to straight Christian celebrations—as Christ's 2000th birthday, as the posters had it—then still the dating is anything but precise. Christians did not start to celebrate his birth until some 350 years after it had happened. And the choice of December—not the 31st but the 25th—as the appropriate date probably had rather more to do with the winter solstice and the Roman feast of Saturnalia than with anything else. It was not until nearly two centuries later still, at a date which thereby became 525 AD, that the monk Dionysius Exiguus set up the system basing itself on a numbering of years backwards and forwards from 'the birth of our Lord'. It is this system, modified, which still we use today: the modification consisting in BC (before Christ) having now become BCE (before the common era) and AD (Anno Domini) having now become CE (the common era), which change appropriately respects the modern use of the system by people of all religions and of none. But just one very well recognised error that Dionysius made muddies the waters still further. By forgetting to include in his calculations at least four years of the rule of the emperor Octavian, he places the very nativity itself at least four years BCE. This, interestingly, gives us a date with which other calculations might tally: as members of a society of already some 50 million people which had been recording observations of the stars for hundreds of years, Chinese astronomers can offer us Halley's comet at 12 BCE and a broom star at 5 BCE as possible contenders for the star of Bethlehem. All of which means that if Jesus' birth took place perhaps some time between 12 BCE and 5 BCE, the rest of the anyway meagre dates of his life shift also: probably baptised by John somewhere around 28 CE; active then in public life for between two and five years; executed somewhere between 30 and 33 CE.

In the light of such things, while it might have been appropriate to choose 'Time' as the theme for Britain's millennial celebrations, it was perhaps less than appropriate to choose Greenwich as their site. Greenwich, if ever there was one, has been home to attempts at some sort of precision in our measuring of time. And it is surely just abundantly clear that the precise timing of our millennial festivities was of no consequence whatsoever: a fact which, despite every attempt to pretend to the contrary, almost everyone knew full well.

But then maybe it doesn't matter. Celebrations being things that we make, and then remake and make again to suit ourselves, maybe it doesn't matter that most of our millennial junketing was actually in celebration of nothing very much. After all, we do seem to need celebrations. And the television coverage of these particular festivities certainly provided some sort of image of one world. That was nice. That was good, surely?

Yes, it was good. But as anyway I will argue, not good enough. For while it is of no great consequence whether what we label 'the third millennium' began on 1 January 2000 or 1 January 2001, it is of considerable consequence, or so I will suggest, that what we were actually celebrating should have been appropriately characterised. It is my claim that it was not. It is my claim that what we should have been remembering and celebrating, but for the most part weren't, was the contribution which, 2000 or so years ago, our forebears made to the human conversation about how to live. It was, after all, a veritably huge such contribution. And in their for-the-most-part frothy nonsenses, I don't think our shindigs did any sort of justice to that fact. But all is not lost. Given the essential daftness of any precise dating of such things, we can even now re-think just what it was that was really the significance of the millennium, just what it was that we would do better to celebrate. And then celebrate it. That is the sort of millennial feel, of millennial reference, that I want this book to have.

I want to suggest that what we should have been celebrating was a moment in a very great making. In *The Origin and Goal of History* (1949), the German philosopher Karl Jaspers delineated the period 800 BCE to 200 BCE as the Axial Period: a period during which a transition was made between two distinctly different types of religion. And in *An Interpretation of Religion* (1989), the philosopher of religion John Hick suggested that this pre-Axial to post-Axial transition gives us the second of only two absolutely central and more or less universally accepted 'broad interpretative concepts' in the phenomenology of religion: the first, that recognisably religious practices are to be found in all human societies; this second, that in the Axial period these religious practices come to be concerned, for the first time in the history of the world, with salvation and liberation.

For it was during the period of Axial transition that the systems of ideas which have shaped the development of all the great faiths, of Hinduism, Judaism, Buddhism, Taoism, Confucianism, Christianity and Islam, were constructed. Many great teachers lived and worked during those times. In India, Gautama the Buddha taught non-violence and selflessness. In China, Lao Tzu the person (maybe) and Lao Tzu the text (certainly) taught that we should avoid striving for worldly goals, and Confucius advocated 'ren', the negative form of Golden Rule, whereby we were exhorted not to do to others what we would not wish for ourselves. In Persia, the ideas of Zoroaster, who proclaimed the existence of the one good god Ahura Mazda, gained ground. In the bible lands, Old Testament prophets such as Amos and Hosea began to insist with the people of Israel that what Yahweh wanted from them was compassion rather than sacrifice, with the resulting abandonment of human sacrifice somewhere between 600 BCE and 500 BCE marking a momentous shift in our conception of the nature of the human. And Isaiah became arguably the first person in the Judeo-Christian tradition clearly to conceive of full-blown monotheism. After that, Jesus taught that we should love our neighbours as ourselves; Mohammed laid down the fundamentally similar principles which became the basis of the third and last of the three great monotheisms; and post-Axial religion, that religion which is concerned with salvation and liberation, that religion which in Hick's words insists upon there being 'a limitlessly better reality in which we can come to participate', which participation involves 'our own moral and spiritual transformation', was firmly on the agenda.

Out of an idea which had begun its glimmering emergence some 12,000 years before in Central Asia, I will argue that the most important thing they did, these thinkers and prophets and teachers of the Axial Period of whom Jesus was just one, was to begin the making of the modern God. And that in doing so, they began the making of the single most important concept we have ever come up with. But I will also argue that the best way to celebrate them now, to remember them and love them now, is to stand on their giant shoulders and make again: to take their brilliant 'God' idea, examine it for the work it has done over now more than two millennia, and then, as our own contribution to what J. B. S. Haldane called the 'god-making tendency' of humankind, remake and make again. It is a task —or so I will suggest—which warrants no nostalgia. The way to celebrate their contribution, far from being to try to hang on to its creations, is precisely not to try to do so. After all, in making their God they were certainly not trying to hang on to the creations of their past. They were precisely challenging, contradicting and triumphantly superseding such creations. And in celebratory emulation, we for whom they are now our past need to do the same. Such an enterprise can never of course be undertaken without a sense of loss: the sort of loss that Milton gives voice to when, in his *Ode on the Morning of Christ's Nativity*, even as he knows (and celebrates that he knows) that the Jesus he lovingly calls 'our babe' is considerable progress upon the old gods, he nonetheless feels the pain of their passing, and bids them only a tender farewell. Such losses are indeed losses. But they are good losses: losses to celebrate, not to lament. So that where Hick suggested that the recognition of a pre-Axial to post-Axial transition is the second basic interpretative concept in the phenomenology of religion, I will be suggesting that we're now in need of a third. They made their personal God, I will argue, in their own image of what they were as persons. But things change. Developments in the psychological sciences, I will argue, mean that our image of what we are as persons can no longer be the same as their image of what they were as persons. And thus the God they made can no longer serve as ours. We must let it go, but tenderly, and celebrate its going.

A goodly part of this book will consist in an attempt to substantiate my claim that we do in fact need such a fundamental re-think of what we mean by 'person', and thus by 'I'. And for reasons to do with what I take to be the nature of science (see Chapter 3), my task will be not to accumulate the evidence necessary to prove my claim as true, but to accumulate the evidence necessary to support my claim as valid. In other words, I will be rejecting both the non-scientific approach which is far too generous (believe all those things which cannot be proved as wrong—ultimately anything), and the non-scientific approach which is far too parsimonious (believe only those things which can be proved as right—ultimately nothing), in favour of the scientific approach which is just about right (believe just those things which can be 'supported as valid'). And if this is an attempt which succeeds, if others also come to think that mine is a valid claim, then many questions of quite huge proportions will inevitably emerge. In the course of what I write, I will try both to delineate just what I take such questions to be, and have a first go at trying to answer anyway some of them.

One among such questions will be this one: to the extent to which we are developing a new image of what a person is, a new image of what 'I' is, a new understanding of 'ipseity'—as I will call it—does it follow that we will therefore need a *new* image of a personal God, or *no* image of a personal God? Within the scientific community—and I want to try to place myself nowhere more firmly than within the scientific community —by far the most common answer is undoubtedly the second. But it is, nonetheless, an answer which raises some not insignificant problems.

Among these problems is one notoriously well rehearsed. Kant rehearses it when, in the *Critique of Practical Reason* (1778), he argues that 'the possibility of a highest derived good (the best world) is at the same time the postulate of the reality of the highest original good, namely the existence of God' and concludes that 'therefore it is morally necessary to assume the existence of God'. Dostoevsky rehearses it when, in *The Brothers Karamazov* (1880), he puts into the mouth of Ivan the now famous comment that 'without God everything is permitted'. I refer of course to the problem which all the various varieties of relativism pose for our understanding of the good. For amongst other things, the God idea has served as that which traditionally has conferred immunity to that relativising critique of the nature of the good which is now rampant. And many more than Kant and Dostoevsky have alerted us to the possibility that it is perhaps the shedding of the God idea which has necessitated the shedding also of any possibility of universal moral criteria.

If this is how it is, it is a very heavy price to pay for Godlessness. It leads some in the scientific community, for example, particularly some working at the biological end of the various psychological sciences right now, to question the very need for a concept of 'I' or 'person' or 'self' at all. It leads them to answer in the negative not only the question 'do we need an image of a personal God?' but also the question 'do we need an image of ipseity in any shape or form?': a seriously drastic philosophical move—or so I will suggest—which only compounds the difficulties, already considerable, of creating the universal moral criteria which would need to be fundamental to any possibility of moral science. But suppose it could be shown that this 'heavy price' is not one which it's necessary to pay. What then? Once it is understood that those thinkers, prophets and teachers of the Axial Period *made* their God-idea in their own *made* image for their own *made* purposes, it becomes easier to see a way forward. It becomes easier to see that what we can do is construct a *new-made* God-idea, in our own *new-made* image, for our own *new made* purposes—foremost among which is the *new-making*, but this time scientifically, of the ground on which can stand universal moral value.

It was in Daniel Dennett's *Darwin's Dangerous Idea* (1995) that I first came across the notion of a 'skyhook'. Whether it is original to him or not I don't know: I think I've also seen it credited to Richard Rorty, perhaps in his *Objectivity, Relativism and Truth* (1991). But anyway in my case borrowed from Dennett, who wanted to argue that no such things could exist, I think of a skyhook as a device which could lift us out of our particular belief systems to see the world as in some sense it 'really' is. Using such imagery, this will be my endeavour: that where Dennett wanted to advocate the use of 'scientific cranes' rather than 'religious skyhooks', cranes which bit by painful bit could lift us to greater understandings but always with their feet on the ground, I will want to argue for the use of 'scientific skyhooks'. And it will be as I develop my argument that I will try to 'unbaffle' my readers (the lovely word is from Dennett himself when he is attacking that mysterianism which I also want to counter) concerning such devices. Suffice it only here to say that of course they could not be devices we could go out and discover: they would have to be devices we would go in and invent. It would have to be because they needed such 'scientific skyhooks' that persons would have to make them. As spectacularly did those persons of the Axial Period, those persons of the monotheistic religions, when in God they began the making of the biggest, strongest skyhook the world has ever known. They made it, and they saw that it was good. And I will argue that the way in which, at the millennium, we should have been celebrating what was just one among what were many key moments in that making project, the moment chosen as the birth day of just one among what were many key architects of that spectacular creation, was with a making of our own.

For things that are made wear out. Their skyhook 'God', graceful and fine as once it was, is old now and shabbily worn. Every time they are undertaken, repair jobs are lasting for a shorter and shorter period. Fractured and fatigued, what's gone now, and beyond all repairing, is the underlying structure: what I will call the I AM structure. The 'God-shaped hole' gapes. And we have to work out, urgently, creatively, 'what is to be done'.

Much has of course been done already. Freud identified, but by no means uniquely, three functions as those which, across more than two millennia, have been fulfilled by the skyhook 'God'. It explained the origins and workings of the universe; it explained the origins and workings of life; and it gave us reasons to be good. Of the three, the first and second are now quite well secured elsewhere. In the making of substantial paradigms, first for the physical sciences, then for the biological sciences, we have made pretty serviceable hooks on which will hang our ideas about the origins and workings of 'stuff', both inorganic and organic. But the paradigm hook on which we might hang the psychological sciences, on which we might hang our theories about persons and selves, and with them our theories of what it means to be good, is as wobbly and ill-fitting almost as ever. It is a huge task which faces us. It is the task of 'making good' the hole in which that skyhook now fits so ill. And it is urgent. Borrowing the words from Carlyle, 'the doom of the old has long been pronounced, but alas the new appears not in its stead'. We need that new.

And that being so, it seemed to me that by far the best way I could try to honour the memory of one of the most significant among those who some 2000 or so years ago helped to fashion the skyhook 'God', would be to try some celebratory emulation. I would not go out hunting, with the postmodernists, to try to discover a substitute device, and then, because it's just not the sort of thing which can be discovered, flop back into relativism. Instead I would stay home thinking and try to invent one. We are the makers. And in the belief that by far the most important making task ahead of us at the moment—by far the most important making task we have ever faced—is the making of that neutral ground on which to stand in refutation of moral relativism and moral pluralism, I decided I would try to make a candidate skyhook. This book is the result of my labours.

It's in no way a job of resurrection. Faced with the collapse of an old paradigm, nostalgia for the past gets us nowhere. And certainly doesn't celebrate the achievements of that past. The apparent viciousness of Jesus' call that we 'leave the dead to bury their dead' is just that: apparent. For to take leave of that which is spent is to make progress. And I am going to argue that progress cannot now be made by resurrection: can only now be made by recreation. There is a sense in which the creation and recreation, the making and remaking, of God has been going on for as long as persons have existed. And what I will suggest is that where it has been partial, the task is to make it complete; where it has been an art, the task is to make it a science; where it has been interested, the task is to make it disinterested; where it has been ideological, the task is to make it scientific. It was Kierkegaard who suggested that 'deepest down in the heart of piety lurks the mad caprice which knows that it has itself produced its God'. In writing my book what I hope to show is that that is a caprice which, far from being mad, is quite spectacularly sane.

Although I had hoped it might not, by far the greatest difficulty I have had in writing this book is maybe already becoming clear. I have found it hard fully to assimilate linguistically what I think is a new and radical idea: to think with it, talk to myself with it, and make it thus my own. And I suspect that this will be a difficulty which others will have, but tenfold, when they read. It has been a struggle to domesticate my idea: to try to get it's language to function for me if not as one of what Mauss called those 'determining habits of consciousness, themselves unconscious', then at least smoothly. This has been so much a daily sweat and strain that when I came across Wittgenstein's comment that 'philosophy is a battle against the bewitchment of our intelligence by means of language' I laughed aloud with relieved recognition: until, that is, I realised that the fact that it is a battle for everyone is no guarantee that when you think the battle has been won you are not just—bewitched! I hope I am not bewitched. I think I am not bewitched. Nonetheless, while there is certainly an extent to which I think have won the battle (hence the book), I am also only too aware of an extent to which I have not (hence its no doubt considerable infelicities). If Francis Bacon was right to claim that truth emerges more readily from error than from confusion, let me hope at least for error. That anyway is a start. And I have developed a deep respect for a couplet from Goethe which reads: 'Whatever you can do, or dream you can, begin it / Boldness has genius, power and magic in it.'

A second difficulty has been to write clearly. It was of course the inimitable Oscar who said that 'if we write intelligibly we run the risk of being found out'. I nonetheless hope that I have written intelligibly. And if that means that found out I must be, then found out I will be. For it is central to my case that the number of philosophers in the world is roughly equal to the number of persons in it. And amongst them, there will be many who are pretty good humbug-detectors. If however there are also amongst them those for whom a condition of intelligibility is brevity, then for them I will not have succeeded. On this I can only take comfort again from an aphorism from amongst the professionals. When I found Hilary Putnam's comment that 'any philosophy that can be put in a nutshell belongs in one', I was again relieved: until, that is, I realised that it in no way follows that a philosophy that cannot be put in a nutshell would not in fact be better if it could! But I decided, if nervously, that that was a possibility with which I'd just have to learn to live.

A third difficulty has been for me to answer the question: 'for whom am I writing?' In the full spirit of one of my main arguments, I want to claim that I am trying to write *as* 'one' (before any such 'one' can really exist) and *for* 'all' (before any such 'all' can really exist either). But in anyway trying to write that way, I hope I can make some small contribution to bringing such a 'one-and-all' world into existence.

And a final difficulty has been with quotation. There is a very important sense in which all language is quotation. We think or speak or write, and thereby take part in an ongoing human conversation. With each thought or utterance or paragraph, we are each to some extent a mouthpiece for all those others who, to greater or lesser extent, have thought or spoken or written that which we now think or speak or write. The current buzz word for this understanding is 'inter-textuality'. But maybe Goethe already understood it when he said something to the effect that every idea has already been thought, but the task is to think it again. I have written not a scholarly book, but I hope a thoughtful and serious one. I have lived seriously, read seriously, thought seriously. And I try now to write seriously, but in a personal, discursive and I hope readable way. It is I know a book veritably stuffed with reference, often informal, to the work of others: declared when I recognise the source; not so declared (necessarily) when I don't. If you should find your own language, or that of another, unaccredited here, I would ask you to take it as the very great compliment it undoubtedly is. For all it means, as George Eliot put it, is that you have joined 'the choir invisible of those immortal dead who live again in minds made better by their presence'—if somewhat ahead of your time. Those who do recognise undeclared references to the work of others, should they be present, can smile at their own erudition. But both they and those who don't need blame only me for those things which are written without due care and attention. Or are just daft.

Chapter Two

GOD

What We Have Lost

In the previous chapter I argued that our recent celebrations of 2000 years of the Common Era should have been marked by some re-assessment. I suggested that in remembering the God-making project undertaken by the people of the Axial Period, we needed to acknowledge that theirs was a creation which has now come apart in our hands, and that that was a disintegration which faced us with serious philosophical problems. But in order further to consider the nature of those problems, it is first necessary to register a little more fully the nature, and particularly the extent, of that disintegration. It is such registration which is the aim of this chapter. What I will be arguing here is that what Nietzsche famously called 'the death of God' is a somewhat patchy phenomenon: that he is, if you like, dead 'only in parts'. Secularisation undoubtedly advances apace, if unevenly. But what I want to consider here are various interpretations of just what it is that has gone. I will write entirely from within the Christian tradition, for that is the one that I know. But I confidently believe that, at different stages of development and sophistication, there are going to be secularisation movements, with recognisable family resemblances to this one, within all the major religions. And that those more knowledgeable than I would have little difficulty in detailing them.

In the order of what I take to be the completeness of their loss, most to least, I want to consider what secularisation has done to Christ, to Jesus, to Christianity, to religion and to God. And I want to insist that although within our theologically unsophisticated post-Christian culture these five are commonly thought of as 'all-of-a-piece', it is better by far that we think of them as conceptually distinct. Mainly, of course, because they are. But also because since they have been lost to different extents, the loss of each needs individual assessment. It is therefore to stress such differential extent and assessment that I will consider the five under their five distinct sub-heads.

1. Christ

Called so after the book of that name edited by John Hick, I think we have lost most completely, to both incredulity and theological sophistication, 'the myth of God incarnate': that is, the concept of the divinity of Christ as Lord. In 325 CE the Council of Nicea, source of 'the Nicene Creed', defended the idea of 'the Trinity': the idea that God the Father, Jesus the Christ and the Holy Spirit are together 'one God'. An alternative idea, put forward by followers of Arius at the same council and denying the divinity of Jesus, thereby became 'the Arian heresy'. This was followed by more than a century of debate and argument on the issue, after which it was the Council of Chalcedon, meeting in 451 CE, which pronounced definitively that Jesus had indeed been God in human form. Thereafter, the construction of a tradition to that effect, involving of course the determined undermining of all heretical alternatives, occupied the new Christians for much of the next millennium. For one small example of just how long that construction took, consider the Feast of the Transfiguration. This is a feast commemorating the day on Mount Tabor when Peter, James and John recognised Jesus-the-man as Christ-the-Lord. But establishing August 6 as the feast day took until 1000 CE in the Eastern Church, and 1457 CE in the Catholic west.

But Chalcedon notwithstanding, heresies of course lived on. And modern versions of the Arian variety are now so widespread as to be almost mainstream. Arguing his case in two now celebrated papers which appeared in *Working Papers in Doctrine* (1976), the one called 'Does Christology rest on a mistake?', the other 'In defence of Arius', Maurice Wiles succinctly expressed the modern variant. He began by reminding his readers that belief in incarnation, in God becoming man in the person of Jesus, arose not in a vacuum but from within a more comprehensive system of beliefs: a system which, as well as a story of redemption via incarnation, carried also stories of creation and of fall. He then pointed out that although almost all Christians today take 'the creation' and 'the fall' not literally but symbolically, see them not as specific events in history but as truth-bearing myths, nonetheless the *doctrine* of creation has survived the death of the seven days story, and the *doctrine* of sin has survived the death of the Adam and Eve story. And he concluded by suggesting that, analogously, the *doctrine* of redemption could survive, albeit with perhaps an even greater sense of loss than in the two previous cases, the death of the Christ story.

For believers, this was strong stuff. And it was an argument which was developed, in the following year, by two books on the subject. In *God as Spirit* (1977), G. W. H. Lampe suggested that the model of in*carn*ation should be replaced by a model of in*spir*ation: that we should see Jesus as God-inspired rather than God-incarnate. And then in *The Myth of God*

Incarnate (1977), its six contributors and its editor John Hick produced the book on these themes which most captured the public imagination. In a phrase which echoed one previously used about the seven contributors to an earlier volume—to *Essays and Reviews* (1860) which had caused a considerable rumpus in the Church of England by suggesting that the books of the bible should be read in just the same way as any other books, that is critically—its co-authors were dubbed 'the seven against Christ'. And the new rumpus was at least as considerable as the old. What Hick and his contributors were arguing was that it was nothing short of a major development in theology which was called for: a development which, arising from a greatly increased knowledge of the origins of Christianity, entailed an acceptance of Jesus not as God incarnate, or the second person of a Holy Trinity, but instead as 'a man approved of God' (Acts 2.21). And they insisted that this was a theological development that was needed for two reasons: first, in the interest of truth; but second, in the interest of ecumenism. For a Jesus who, because he in some sense *is* God, has to be presented not as *a* way to redemption but as *the* way to redemption; a Jesus about whom it has to be said that nobody comes to the Father except through him: such a Jesus, they suggested, acted constantly to undermine 'our relationship to the peoples of the other great world religions'.

At much the same time, others were pointing out that, important as they were, such 'practical implications' of the old theology were by no means the only ones to consider. Feminist theologians were also on the case. In *Beyond God The Father* (1973), Mary Daly had already argued that what she memorably called 'Christolatry' must end, but this time in order to accord with the liberation of women. Indeed she went so far as to defend the notion of an Anti-Christ as a concept which, she argued, need have no necessarily negative connotations. And in *Woman Guides: Readings Towards a Feminist Theology* (1983), Rosemary Reuther added her voice to the growing clamour by suggesting that an exclusively male redeemer would necessarily alienate women from claiming their full humanity as women. Her emphasis was on the face. No matter how much Jesus is presented as representing 'generic humanity', no matter how successfully he is portrayed as a thoroughly anti-patriarchal male, if his remains the only *face* of the redeeming God, the only *face* of authentic humanity, then women will remain forever other, forever outside.

The consequence of developments such as these is that both within and without the Christian tradition there are now huge literatures, both specialist and generalist, on the question of the divinity of Christ, most of it of a thoroughly 'Arian' nature. And anyway in the more popular texts, two of the arguments most often mounted by those who would question any such divinity are those which focus on his character and those which focus on his claim.

First, the argument from character. Many commentators, among them the Jewish scholar Montefiore in *The Synoptic Gospels* (1927), have pointed to what they take to be weaknesses in the character of Jesus as man as argument against his divinity as Christ. Philip Toynbee, for example, has made reference to his excessive emotionalism, particularly his expressions of anger. In *Why I am not a Christian* (1957), Bertrand Russell suggests that Jesus exhibits 'a vindictive fury against those people who would not listen to his preaching', and argues that although such a stance is quite common among preachers, it does nonetheless detract from what he calls 'superlative excellence'. It is not, for example, to be found in Socrates who, says Russell, adopted a stance both bland and urbane in relation to those who refused to listen to him: a stance which, in Russell's opinion, is one more appropriate than that of furious indignation. And in *An Atheist's Values* (1964), Richard Robinson suggests that the character of Jesus is one from which the ideal of truth is lamentably lacking: that he poured scorn on professors of knowledge, going even so far as to suggest that the kingdom of heaven is hidden from the wise.

Be that as it may, there is, second, the argument from claim—or rather lack of it. New Testament scholars are now more or less unanimous that Jesus did not claim to be divine: that he never used the word 'Christ', or its Jewish equivalent 'Messiah', of himself. The first direct evidence of the use of the term in connection with him is from St. Paul. In *Jesus: the Unanswered Questions* (1988), John Bowden argues that it was in the writings of Paul, the man who more than any other might be called the founder of the Christian *church*, that the Jesus who *had a* message became instead the Christ who *was the* message.

In summary, what I am here suggesting is that Jesus as divine, as 'the Christ', is now an idea very far gone indeed. Incredulity will let fewer and fewer of today's more scientifically educated people accept the 'conjuring trick with bones' that it requires. And in addition, modern theological doctrine has been very active in suggesting that it is anyway either not necessary to, or even destructive of, the development of the religion which, perhaps increasingly anachronistically, bears its name. Thus does the tradition initiated at Nicea come to share the fate of all traditions. For paradoxically, the defence of something as 'traditional' only works when it does not need to be made. As soon as the traditional has to be argued for, rather than just taken as read, then the case for it, whatever it is, is well on the way to being lost.

2. Jesus

Almost as completely as we have lost Christ to incredulity and theological sophistication, we have lost Jesus to historical scholarship. In conducting 'the quest for the historical Jesus', a search so labelled after an influential book on the theme by Albert Schweitzer under that title, we have steadily lost more and more completely the story of Jesus the man. Following in the tracks of Strauss (1835), Feuerbach (1841) and Renan (1863), all of whom wrote on similar lines, Schweitzer published his book in 1906. His main conclusion was that the historical evidence concerning the life of Jesus was so thin that each person who wrote a biography just 'created him in accordance with his own character'. And in *Jesus and the Word* (1926), with his conclusion that 'we can know almost nothing concerning the life and personality of Jesus, since the early Christian sources show no interest in either', Rudolph Bultmann was widely considered to have brought 'the quest' to an end.

More recently, however, that which in some places has come to be called 'the new quest' has produced some renewed interest among Jewish scholars. The first publication, under the title *Jesus the Jew* (1973), was from Geza Vermes. It was followed by *Jesus and Judaism* (1985) and *The Historical Figure of Jesus* (1994), both from E. P. Sanders. It would be wrong to say that nothing of any historical interest has emerged from such texts. They raise, for example, the very interesting question as to why Jesus was crucified. Both authors are convinced that the traditional Christian explanation—because of a breach of Jewish religious law—has to be wrong. Sanders sees the most likely reason as his ejection of the money-lenders from the temple. Vermes suggests it was more likely to have been just the general commotion that his teachings caused 'in the most dangerous politico-religious location at the most dangerous politico-religious moment of the year'. But whichever, these disagreements and others like them do nothing to stop the image of the historical Jesus getting fainter by the decade. In an earlier book, Sanders himself had suggested a list of only six facts about the life of Jesus which are (only almost) historically indisputable: that he was baptised by John; that he was a Galilean who preached and healed; that he had twelve disciples; that he was active only among Israelites; that he was involved in controversy concerning the temple; and that he was crucified. It is a list which contains no stable or star or Bethlehem; no virgin birth or miracle or resurrection. And it is a list of that which is only *almost* indisputable.

3. Christianity

It might be thought that the extent of the loss of both Christ and Jesus would automatically be reflected in a commensurate loss of the religion to which they have been central. But I don't think that's necessarily so. Christianity has of course lost significance, but less so I think than either Christ or Jesus. So let me first offer just some snapshots, taken mainly over Britain, of the extent of this our third loss. And then try to suggest why I take it to be one somewhat less extensive than either of the previous two.

According to the 1851 census, 40% of the people of Britain attended church at least once a week: in a longitudinal study of religion in York, Rounder and Lavers placed such attendance at 36% in 1901, 18% in 1935 and 13% in 1948. In 1861 there was in Britain one Anglican clergyman for every 960 people: in 1961 there was one Anglican clergyman for every 4000 people. Between 1975 and 1979, this time in Europe as a whole, Protestant churches lost some 30,000 members a year and Catholic churches some 27,000. In 1995, fewer than 4% of the adult population of Britain were active members of the Church of England; 88% of its clergy were more than 40 years old; only 8% of the adult population attended any church. Not surprisingly, therefore, it is no longer possible for teachers to assume any knowledge of the texts of the Christian tradition in their pupils, and almost certainly no longer appropriate always to give 'Desert Island Discs' castaways a bible on their island. Whatever else such evidences as these will do, they will certainly justify the very modest claim that, although it is still a disputed issue, we have lost Christianity, at least to some extent, as an organised religious influence.

But only to some extent. What I want to suggest is that because the religion 'Christianity' is more important, in a greater variety of ways and to a greater variety of people, than either the concept 'Christ' or the person 'Jesus', its loss is less complete. Let me illustrate what I think is the greater complexity of this loss by considering, if a bit schematically, the way in which it is seen by four different groups of people.

Firstly, there are Christians whose position is that, to the extent to which Christianity has gone, that is good. In *Taking Leave of God* (1980), Don Cupitt, then lecturer in the philosophy of religion at the University of Cambridge and Dean of Emmanuel College, insisted, that 'for both the Christian and the atheist religion is a temporary phenomenon'. The Christian position, he elaborated, is that in paradise there was no temple and in the New Jerusalem there will be no temple. This is a position not so very far away from that of the atheist who maintains that just as our very early ancestors had no God, so the humanists of the future, liberated from religion, will have no God either. It's just the in-between time that's the problem. For the atheist, that is a time when, by helping to develop human subjectivity, God has importance as some sort of stop-gap: before such time as humans could 'do it for themselves', he is important as law-giver and as source of morality.

Also in this camp are many of the feminist theologians. In *Theology and Feminism* (1990), Daphne Hampson, then lecturer in divinity at the University of St Andrews, argued not only that Christianity is 'a myth which cannot possibly be true', but also that it is 'highly sexist'. She therefore would see its demise as only to be welcomed: both in itself and as prelude to a more honest spirituality based on 'an awareness of that dimension of reality which we name God, which is part of us but which also moves between us'. These and others like them are the sorts of texts which, over recent decades, have encouraged even believers to see that the loss of Christianity the religion can not only be borne but even welcomed. And with a growing number of Christian writers, among them many ordained ministers, backing fast away from virgin births, miracles, the resurrection, all once considered central tenets of Christianity, there is certainly a case for saying that some sort of loss of Christianity is, at least for some believers, not just real but also good.

But secondly, there are Christians who say that, to the extent to which Christianity has gone, that is bad. These, for the most part, are the orthodox and traditional believers. And their fundamental lament is, and has to be, that it is that which is true which is being lost to more and more people. Never mind social and political consequences, dire though many of these might seem to be to many of this persuasion: it is the consequence for truth which for them is the problem. If the religious belief based on the teachings of one who was not *a* way but *the* way, not *a* truth but *the* truth, not *a* light but *the* light, is retreating; if the religious belief claiming that none can come to God except through Jesus is in decline; if these things are the case, then by such folk they can only be lamented.

Third, and perhaps most vociferous, there are non-Christians who say that, to the extent to which Christianity has gone, that is good: maybe even very good. Some draw attention to a history of vicious crusades, torturing Inquisitions, the bigoted suppression of science, and ask us, with some reason, to rejoice. In *The Conversion of Europe: from Paganism to Christianity, 371–1386 AD* (1997), for example, Richard Fletcher gives an account of the often far from saintly conversion process by which people were sometimes persuaded, more often compelled, to embrace the new religion. It is not a pretty story. It was a conversion process most often undertaken, so he argues, by 'members of the most powerful families, disseminated through their networks, sustained by their influence and supported by endowments from their lands'. And it had so little to do with disinterested goodness that most of his readers could probably only be relieved that no such horrors are likely to be repeated: that Charlemagne's brutal destruction of the religion of the Saxons is something the likes of which we will probably never see again; that the landowner Pammachius' forcible dragooning of his peasants in Numidia into conformity, a dragooning liberally praised by St Augustine, is firmly a thing of the past; that the frequently unspeakable cruelty of the crusades will not return. That such things will now be no more, say such folk, is surely just excellent reason to rejoice. Others point to the many wars in which at least one side, and often both, fought in the name of Christianity. Thank goodness (rather than God), they say, that given that belief in the divinity of Jesus seems to be passé even among the clergy of the Church of England, belief in the religion founded on his teachings must surely follow him down, and soon, to decent oblivion.

But fourth, there are non-Christians who say that, to the extent to which Christianity has gone, that is bad. This is a group which needs division into two sub-groups. Firstly, there are those of a 'law and order' persuasion who somewhat pragmatically feel that while they themselves might be grown up enough to live without illusion, it is doubtful whether the same can be said for 'them', whoever their particular 'them' might be. It is a not at all uncommon stance among non-believers to see Christianity as some sort of late-lamented, lost police force. But secondly, and much more importantly I think, there are those who recognise that because of differences between the under-pinning philosophy of Christianity and the under-pinning philosophies of some of the things that would seem to have replaced it, it's not at all obvious that non-believers also should not be lamenting its passing. When Ernest Lucas of the Institute for Contemporary Christianity suggests, for example, that the basic differences between Christian beliefs and those of a variety of New Age persuasions are that those of the latter involve no room for salvation, no special status to human beings, a rejection of rationality, a replacement of either/or thought by both/and thought, and no time for an individual God, then even an atheist like me can rustle up some serious lament for its passing.

What I am trying to get at by outlining these four positions is that because of the much more complex relationship to Christianity that has existed, and in many places still does exist, amongst a wider assortment of a much larger number of people, particularly perhaps because of the possibility of *non*-Christian lament for the passing of Christianity, it lingers rather more extensively than does either Christ or Jesus. The loss of Christianity leads to bigger problems for more people. And that means that in deciding whether to let it go or, yet again, shore it up with a compromise or two, far more people come down firmly on the fence—and so residually, it lingers.

4. Religion

It is undoubtedly true that secularisation proceeds apace almost everywhere. The process whereby religious thinking, practice and institutions lose social significance is seemingly relentless. It goes on in tandem with, indeed as part of, that development of societies which brings greater education to more people at an increasing rate. For it is particularly in more developed societies, and among more highly educated elites in all societies, that religion is losing its grip most completely. It was this phenomenon which was revealed, for example, when in 1986 B. P. Beckwith published a review article in the journal *Free Enquiry* (vol. 6, p. 46). Searching through all the relevant US studies of the relationship between religious belief and level of education—there were forty-three of them—Beckwith found that in four of the studies no relationship had been found, but in the remaining thirty-nine it was an inverse relationship which prevailed: the higher the level of education, the lower the level of religious belief. Now if what replaces religious belief is higher standards of education, then it is not at all obvious that such retreat should be understood, as famously it was by Matthew Arnold in his poem *On Dover Beach*, as 'melancholy': as that 'long, withdrawing roar' of 'the sea of faith' which will reveal nothing but 'vast edges drear' and 'naked shingles'. Perhaps we should think instead with Feuerbach as he argued that religion, far from fulfilling a human life, actually thwarts its development: by setting up all that is actually the best in people as if it was really the best in God, it alienates humans from what he called their 'species being'. Or with Marx, as in *Towards a Critique of the Hegelian Philosophy of Right* (1844) he argued that 'the abolition of religion as the illusory happiness of the people is a demand for their true happiness', and that the 'call to abandon illusions about their condition is the call to abandon a condition which requires illusions'. A mainstream position within anthropology, after all, would give credence to the idea that religion can be seen as a form of surveillance: as an invention to keep in check the otherwise unobserved behaviour of people living in now bigger than face-to-face groups. If these things are so, where is the need to lament the decline of religion? I think these things are so, and there is no such need. But I also think that that is not the end of it. If religious decline is to be fully understood, there are two other things to consider: the first relatively unimportant; the second absolutely vital.

First it should be remembered, if no way dwelt upon, that the idea that we have lost religion in the sense in which Durkheim defined it, that is as a 'unified system of beliefs and practices relative to sacred things', could at first flush be thought absurd. It is absurd, for example, if one considers that between 1900 and 1982 the percentage of the world population giving adherence to Hinduism and Islam, but granted only those two, actually increased: the former by 1 per cent, the latter by 5 per cent. It is absurd if one remembers that there is still a theocratic state in the world, Israel, which until very recently went so far as to have a *Ministry* of Religion. It is absurd if one heeds a 1991 Gallup Poll in the US which revealed that 46 per cent of Americans still give credence to a religious belief suggesting that some time in the last 10,000 years God created us more or less as we are now. And it is absurd if one thinks of the growth of house churches and the evangelical movement in Britain. It would be possible to go on listing examples such as these quite lengthily. But this, I think, is a relatively unimportant issue. In the sense in which it was introduced first by Marx but accepted now much more generally, 'uneven development' is a concept more than capable of explaining such things within the general context of religious decline.

But second, and much more importantly, it should be thoroughly accommodated by any theory of decline, that there is a difference between the singular and the plural: between 'religion' and 'religions'. The results of a 1996 MORI poll commissioned by the British Humanists Association found that although the majority of Britons do not believe in God – only 43 per cent of those surveyed ticked the box 'I believe that God exists' – nonetheless 67 per cent of respondents believed themselves to be religious – and if you want further complications, 79 per cent believed in an afterlife! In the sense identified by Jung, who was wont to maintain that he had never seen a single patient whose problem was not at bottom a lack of religious belief; in the sense identified by A. N. Whitehead, who thought of religion as what the individual does with his solitariness; but above all in the sense identified by John Macmurray, who saw religion as growing out of our relations as persons: in *these* senses, while the decline of 'religions' is manifest, the decline of 'religion' is much less so. Indeed, there are ways of thinking about what might be called 'a religious attitude', which, I will argue, are clearly – and happily – not in decline at all. Contrary, say, to the traditional Marxist approach, which too often presented religion as some sort of 'bolt on' phenomenon administered to otherwise naturally God-less souls by rulers struggling to control them, there are ways in which it can be more usefully thought of, as in Macmurray's approach, as actually characterising what he called 'the personal' and I will be calling 'the personic'. In the eighteenth century, that which was called 'natural religion' was that which, differently from the religion of revelation grounded in mystical and essentially first-person experience, was deemed to be amenable to support from the natural sciences. By analogy, and using a terminology which later I will more formally propose and defend, it is possible to make the case that in the twenty-first century, that which might be called 'moral religion' will be that which, also differently from the religion of revelation grounded in mystical and essentially first-person experience, actually will be amenable to support from what by then will be the moral sciences. All of which, clearly, is premised on the idea that while 'religions' are for the most part in decline, 'religion', anyway as by some it is conceptualised, possibly isn't.

5. God

Nietzsche declared God dead in 1885. Not to be outdone, God declared Nietzsche dead in 1900. This, at least, is how a cartoon joker in *Philosophy Now* (issue 8, 1993/4) saw it. But is the joker right? Is it perhaps not rather that, as happened in the case of Mark Twain, reports of God's death have been somewhat exaggerated?

Well, even in 1885 such reports were hardly new. Many find them in Hegel, or in early Heine, for examples. For many, modernity is actually to be characterised by his death: for some—Gillian Rose, for example—post-modernity by the death of the Godhead also. And not only that. Laplace had 'no need of that hypothesis'. Stendhal argued that 'the only excuse for God is that he does not exist'. Schopenhauer believed that making creatures who you knew would sin—and you did because you knew everything—and then punishing them for that sinning was 'revolting'. And in much the same vein, Empson saw God's creation of creatures 'certain of damnation, and always corrupt, just to punish Adam' as the action of one who could only be thought of as 'laboriously vile'. If such a God is indeed now dead, what's the problem? I would suggest that there are two sorts of problem: one immediate and pragmatic; the other more long term and philosophical. Though as is so often the case, of course, what appear to be two separate problems can also and readily be seen as thoroughly interdependent.

Firstly, the pragmatic. A belief which maintains that there is one God—rather than, for example, the more than three thousand recognised in ancient Mesopotamia—and that all humans are made in the image of that God, is exceedingly good for social cohesion. Nature can be singularly red also in tooth and hand. And by calling into question many actions which come merely 'naturally' to us, but which can be fundamentally damaging to others of our kind, a belief in God can be spectacularly useful. In encouraging us to see the diminishing of a fellow human as the diminishing of her or his creator, and thus sacrilegious, it can perform a very valuable social function. For Christians it is the idea of the soul, which links humans, and only humans, irrevocably to God. And in the development of the idea, held for us now in such lovely images as that of 'all souls', the huge contribution of the Axial-period God-makers was that they began to make and articulate the idea that what was special about humans was their capacity for loving rather than just living. Adding to such an idea the notion that God himself is in some way also 'person' serves only to strengthen the message. In *A History of God* (1993), for example, Karen Armstrong, noting that a God totally removed from any humanity, a God such as Aristotle's Unmoved Mover, would be incapable of inspiring the spiritual in us, concludes that therefore all religion begins with one or another form of anthropomorphism. She sees this as a projection, and believes that, as long as it does not become an end in itself, it is thoroughly beneficial. She thinks, for example, that it is because the three major monotheisms share an imaginative portrayal of God in human and personal terms that they all also share an essentially egalitarian ethic. And contrasts this with the comparative absence of any marked social concern in, for example, the non-anthropomorphic Hinduism. It is this, she thinks, which explains the fact that, in the ancient world, it was the Jewish people who were the first to try to establish some sort of welfare system for all: an idea completely alien to most of the other civilisations of the time. These then are the sorts of considerations which mean that, at the level of the immediate and the pragmatic, the death of God could be seen, and of course has been seen, as in no way conducive of human decency.

But secondly, the philosophical. Here the problem is one of grounding. A belief in an outside-the-system God, it has been argued, provides a 'view-from-nowhere' position on which to ground absolute values and truths. It was what Kant was arguing when, in the *Critique of Judgement* (1790), he insisted that 'in order to set ourselves a final purpose in conformity with the moral law – it is also necessary that we assume – that there is a God'. For with no outside-the-system God, the absolute values propounded by such an entity become the merely relative values propounded by each situated individual: which situatedness results in the collapse of all moral universals.

When considering this problem, the philosophical problem, it is important I think to try to be as clear as possible as to just what this 'God', who is maybe 'dead', was supposed to be. In *The Concept of God* (1974), Keith Ward suggested that there are five basic views of God current among believers. There is 'personalism', where the image of God is of a person (a male person, of course), but on an altogether grander scale than human persons: Ward thinks this view is as misplaced as that which suggests that a human is just an animal on an altogether grander scale, and that in both cases the difference is qualitative rather than quantitative. Then there is Thomism, summed up in Aquinas's phrase 'esse suum subsistens', which is the view that God is what is, is 'beingness', is 'being' itself: but Ward finds this one intellectually incomprehensible, especially as God is said also to be unchanging. Thirdly, there is the God of what is called 'Process Theology', whose chief British spokesperson has been David Palin of Manchester University theology department: it suggests that God changes, and is both creative and responsive. Fourthly, there is the functionalist approach of such as Don Cupitt and D. Z. Philips: in Ward's vocabulary, they are 'anti-realists' because they think that God is a human construct. And finally, there is the type of idealism put forward by Hegel, according to which God is something like 'the infinite always'.

Adopting these categories at what would seem to be their face value, I think that it is the demise of the first God that would matter. If God is just 'what is' or 'being', then clearly both are still with us. If God changes, he is no more useful as a source of universal values than any of us are, so that his demise, had it taken place, would not have mattered very much. If God is a human construct – whether 'the infinite always' or some other – he is an idea, and thus never having lived, he simply cannot die. But if God is some sort of all knowing, all powerful, all good *person*, then his death would more than just matter. If he was a person who made the world and all that is in it; a person who has ideas about how those other persons, the ones he has made, should behave; a person who can, and does, intervene in their lives: if he was these things, then his death would be truly calamitous.

Cutting what I hope will be a clean and clear path through these various possibilities, I will try, across the rest of this book, to make the case for the defence of absolute value. It will be my argument that God-as-supernatural-being is of no philosophical consequence: such a God is not dead, not because he lives but because he has never lived. The question therefore becomes not 'is there a problem if God is dead?' but 'is there a problem if the God *idea* is no longer thought to be useful?' And what I will suggest is that if the God *idea* is fundamentally the one which grounds universal values, if it is the *idea* which makes such values immune to a relativising critique, then that God *idea* is one which, not by resurrection but by recreation, has to be defended. In that sense, *and in that sense only*, I will as an atheist make a case for the re-creation of God: not an outside-the-system God who made us, but an inside-the-system God which we make—and must keep on making. For in a very important sense, I will suggest, such making just *is* the human enterprise.

In later chapters, I am going to put forward and defend an hypothesis as to what a person is which, as far as I am aware, is a new hypothesis as to what a person is. It is an hypothesis which carries the implication that the old God, the one we have substantially lost, was based on an inadequate psychology. His creators made him in their own image all right. Montesquieu was right to quip that if triangles had a God he would be three-sided. In his poem *The Indian upon God*, Yeats was right to suggest that the God of the moorfowl 'holds the world between his bill', the God of the lotus 'hangeth on a stalk', and the God of the roebuck is 'the stamper of the skies'. But I will claim that when the Axial period God-makers used their image of what they were themselves as persons to make their God, they were using what is now being shown to be a quite massively faulty image. It is as though the triangle thought it was a square, the moorfowl an ostrich, the lotus a daffodil and the roebuck a Billy-goat, *and made their therefore faulty Gods out of such errors*. And I will suggest that if we are to escape the implications of a relativism which denies us access to universal value, then what we need to do is (1) to recognise the nature of the error in their image of themselves which caused the Axial makers to make a faulty God; (2) to adopt a new and more accurate image of ourselves than the one they had of themselves; and (3) to make out of that new and more accurate image of ourselves a new and non-faulty God in which we will be able to ground such value.

This work is urgent. We are living at a time suffused by the presence of the absence of God: an absence captured, albeit inadequately, by the now much quoted concept of a 'God-shaped hole'. In *The Structure of Scientific Revolutions* (1962), Kuhn suggests that, however inadequate it has become, an old paradigm is only ever completely sloughed off when there is a new one, or at least the bare outlines of a new one, to take its place. The old God paradigm is now in such a state: thoroughly undermined, but with as yet no coherent alternative to replace it. And that hurts. Concerning what I have outlined in this chapter as a hierarchy of loss, our coping mechanisms have for the most part coped with the lesser (and therefore more complete) losses of Christ, of Jesus and of Christianity. But not surprisingly, they have been far less successful at coping with the greater (and therefore less complete) losses of religion, and particularly of God. These are losses which have left many persons and societies fractured and bereft: a situation sufficiently painful and bewildering to enough people to make it urgent that we try to make that 'coherent alternative'. This book is my contribution to doing so. And I will begin, in the next chapter, by considering what it is that we have lost our losses to. It will be my case that this, of course, is science. But that far from defending us in our current vulnerabilities, anything other than a total acceptance of its hugely liberating potential will only ensure that we have no hope of making that grounding for universal value of which we are in such very dire need.

Chapter Three

SCIENCE

Observation, Hypothesis, Experiment, Theory

In 'On the Historical and the Dogmatic Method in Theology' (1913), Ernst Troeltsch suggested that what had led to the break up of the whole Christian outlook was not any particular historical event, but the historical world view as such. It had been cracked apart by an application of the historical method, as opposed to the dogmatic method, to the problem of understanding the past. The dogmatic method was one basing itself in certainties, fixed points and authority. By contrast, the three most characteristic features of the historical method, he suggested, were its reliance on probability, analogy, and correlation. And it was this historical method which had fundamentally undermined the Christian perspective.

If it is the historical method which has thus challenged the whole Christian outlook, it is the scientific method, of which anyway the historical is simply one form, which has undermined the very fundamentals of all religious thought. As just one small but telling example, in *Taking Leave of God* (1980) Don Cupitt asks whether St. Paul would have thought of the resurrection of Jesus as a fact, and suggests that because he would not even have had the concept 'fact' to use in his thinking, he could not have done so. If a fact is something which is independent of local cultural belief and human wishes, religiously and morally neutral, whose truth is testable, then such a concept, Cupitt suggests, would have been literally unthinkable for St Paul. Indeed, the inability to think that which is 'religiously and morally neutral' is a phenomenon still to be found way beyond the times of the saint. Some fifteen centuries later, in 1609, we find one of the greatest of scientists, Johannes Kepler, thinking within the same limitation. Having made what to him was the staggering discovery that the path along which the planets moved was elliptical rather than circular, he justified his discovery by arguing that since the circle represented the spiritual and the straight line the material, the combination of circle and straight line which is the elliptical indicated that the planets were trying to arrive at a divine ideal. And four centuries after that, such immanent thinking was shown still to be alive and well and living in Manchester, when the then chief constable of the city, in what became a notorious speech, referred to HIV infection in homosexual men, at that time the main sufferers, as the God-given punishment of those 'slopping around in a cess pit of their own making'. Even among those who would perhaps think of themselves as being members of 'the educated classes', such thinking was mainstream until very recently. And is commonplace still. But at least now it is in retreat. Anyway in advanced societies, the idea of a religiously and morally neutral 'fact' — such as that AIDS is a disease caused by a virus — is one which, while it is by no means yet characteristic of all human thinking, is nonetheless well in the ascendant. And that is not all. Slightly more complicated than the now simple idea of 'a religiously and morally neutral fact' is the idea that the factual content of a statement is the greater the more it rules out. But I suspect that even that one, while it might not yet be an idea which would occur unprompted, would nonetheless be one which, on explication, would be readily comprehensible to most people. This is human advance of a spectacular kind. And it is the phenomenon unhelpfully reified to 'science', more appropriately thought of as 'the scientific method', which primarily has brought it about.

In this chapter I will outline what I take to be the fundamentals of the scientific method. And will do so for two reasons: firstly to show just what it is to which we have, anyway to some extent, lost religious faith; secondly to indicate what will be the method I will use, in the second part of this book, when I undertake a scientific investigation of my own concerning the nature of the concept 'I'. But before I present that outline, let me first, and for emphasis, reiterate something. On the question of that religious faith, it is of course important that I do not ignore those facts, and they are several, which mean that our loss is still only 'to some extent', not yet 'entirely'. It just *is* the case that a survey published in *Nature* (April 1997, vol. 386, p. 435) revealed that 39.3 per cent of American scientists still believed in a personal God to whom they could pray. It just *is* the case that in 1916, the date of a previous survey which this one tried to replicate, the figure was hardly different at 41.8 per cent. It just *is* the case that in 1995, Paul Davies, the then professor of theoretical physics at the University of Adelaide, won one of the world's most prestigious *theological* prizes for his writings on the relationship between physics and religion. And it just *is* the case — and unlike these others, one to which later I will want to return — that in *Reason and Emotion* (1935), the moral philosopher John Macmurray, while agreeing completely that 'it is primarily the development of rationality in science that is revealing the childishness and irrationality of contemporary religion', nonetheless continued to defend the significance of the religious by arguing that contemporary religion was not all that was possible: that religion itself could grow up. These things, and others like them, of course must be acknowledged. For while 'Ockham's razor' (the useful implement invented by William of Ockham suggesting that 'entities must not needlessly be multiplied') is one which certainly *will* feature in what I want to say about science, 'Ockham's broom' (a devilish device identified by Sydney Brenner, one of the world's leading molecular biologists, as the tool we use to sweep unwanted facts under the carpet) is one which certainly must not. That said, and in the title words of one of the best known introductions to the subject: 'what is this thing called science'?

Before post-modernism poured its icy water on such projects, many 'grand narratives' about the history of ideas flourished. One such was that there were stages of development to be found in human history, such that humankind moved from a religious phase to a metaphysical phase to a scientific phase (Comte); or from a mythological phase to a religious phase to a scientific phase (Freud). Without here considering the validity or otherwise of these or other narratives—that is a task for later in the book—it seems to me reasonably unproblematic to say that religious belief precedes scientific thought. And that when the latter emerges, it creates at the very least some problems for the former.

The monumental writings of Joseph Needham on Chinese science notwithstanding, it is widely agreed that scientific thought emerged in Ancient Greece. And in the ultimately fruitless search for the first scientist, the honour is not infrequently given to Thales of Miletos. Around 600 BCE, he it was who, wondering what the world might be made of, came up with the thoroughly counter-intuitive suggestion that it might be made of water. To the history of what would much later (in the fourteenth century) come to be called 'science', what matters about this speculation is both the question and the answer. The question exemplifies a curiosity about reality not directly linked to immediate human needs. The answer is an attempt to find a unifying principle at work. And both of these, the search for how things are and the search for unity, are characteristics of the scientific enterprise to this day. What's more, if science has anything to do with getting a low assumption-to-explanation ratio, with explaining as wide a variety of phenomena as possible using as few principles as possible, then Thales' idea was certainly a winner. It is a full two hundred years later that Aristotle is to be found claiming that 'all men by nature desire knowledge'. But given that he does not seem to have had an anyway explicit theory as to how such knowledge is most reliably to be acquired, then despite the fact that the attempts he made to systematise such knowledge as had by that time *been* acquired also gives credence to the idea that science had a single origin in Ancient Greece, he perhaps cannot be identified as any sort of founder of what would come to be called scientific method. And maybe he wasn't. But then maybe the search for any such 'firsts' is not particularly informative. It is maybe also worth remembering, for example, that in the Christian and different tradition, many are now the scholars who would defend the translation of 'logos' in the first verse of St. John's gospel not as 'word' but as 'reason'. And the very idea of 'one God', of monotheism, has by some been seen as in some ways a spectacular precursor of all 'theories of everything', from Thales to Stephen Hawking.

But if there is a case for saying that 'science, the activity' began with the ancient Greeks, there is also a case for saying that 'scientists, the people' began with the Victorians. In 1840, the same year in which the British queen married her Albert, William Whewell wrote his book *The Philosophy of the Inductive Sciences*. In it he wrote that 'we need very much a name to describe a cultivator of science in general': and added 'I should incline to call him a Scientist'. And notwithstanding the addition of a 'her' or two along the way, we have all felt inclined to do so since. During all of the thus two thousand or so years that it took for 'scientists' sufficiently to have emerged to need a name, the growing sophistication of the scientific method, its growing use in more and more fields of human endeavour, the growing size and coherence of what has come to be called 'the scientific community', has put relentless pressure on faith to justify itself.

But such pressure, while it has had the consequence of considerably undermining faith, has not had the consequence of totally destroying faith. Faith lingers. And it lingers, or so anyway I will argue, because of an uneven development within science: because although the relatively *more* successful development of physical and biological science has undermined it, the relatively *less* successful development of psychological and moral science has left it room if not to flourish then at least to hang on. It is this state of affairs, inherently unstable, which I will claim is about to change. It will be the main argument of my book that, particularly as a response to the now rapidly accumulating scientific understanding of the workings of the human brain and the human nervous system, we are going to need a fundamental rethink of the nature of ipseity, of the meaning of the concept 'I'. And that when that rethink is in place, it will bring in its wake two things: first a flourishing also of psychological and moral science; second, and as a consequence of that first, a final demise of faith, anyway as currently understood. But before I can talk of 'this thing called science' at all, that is, of a phenomenon put so firmly into the singular, I must first provide some basis for the unity thus being claimed.

There are presently available perhaps three candidates for a hook on which to hang such unity. In descending order of the degree of reductionism involved, they are (1) that all sciences can be logically deduced from physics, and are therefore one; (2) that all sciences use the same method, *the* scientific method, and are therefore one; and (3) that all sciences use the same language, *the* language of science, and are therefore one. By far the most widely accepted of these, where any is accepted, is the second. And I will be accepting it too. It is a unity candidate which I think uses two very general and underlying principles: first, the very general 'generate-test-regenerate' heuristic, that which Richard Dawkins, somewhat imperially, calls 'universal Darwinism'; and second, the very general premise that what we are looking for are causes. Before getting into more detailed matters, let me first comment on these two very general claims.

On the first: I shall be respecting a very general way of thinking about the hypotheses of science, sourced in a very useful way of conceptualising an organism. An organism can be seen as an hypothesis, generated by its genes, as to the nature of its environment. The further along a continuum (from just about adequate to wonderfully accurate) is the hypothesis, the further along a continuum (from just about surviving to wonderfully flourishing) is the organism. Of what is generated, that which passes the environmental test of selection is regenerated. That which does not is annihilated. The comparison with our attempts to think scientifically, a comparison made to my knowledge by both B. F. Skinner and Karl Popper, but no doubt by many others too, is helpful. Persons generate hypotheses as to the nature of reality, test them, and then either junk them or continue to work with them depending on the result. Concerning such a process, Popper rather nicely commented that it was one which, by comparison with the somewhat more severe version to be found in nature, enabled our hypotheses to die in our places.

And on the second: I shall be respecting what has often been argued to be a crucial difference between 'how' questions and 'why' questions. In *Consciousness Reconsidered* (1992), for example, Owen Flanagan opens with this real stunner of a paragraph:

> Three of the greatest perplexities are these. Why is there something rather than nothing? How did some of the stuff there is come to be alive? How did some of the living stuff come to be conscious?

He is asking one 'why' question and two 'how' questions. And the critical difference between them is that while 'how' questions are seeking 'by-what-means' *causes* as their answers, 'why' questions are seeking 'to-what-end' *reasons* as their answers. But to ask for reasons is to presuppose purpose. To enquire as to the purpose of life, for example, is to assume that there is a purpose to enquire into. And that is the mistake. There isn't. Which is not of course to say that there is not a purpose of anything: just that there is not a purpose of life. Purposes are made. And makers come onto the scene long, long after the emergence of life. Reasons can only be required of persons—if even and always of them! To ask—of whom?—things like 'why are we here?', or 'why is there something rather than nothing?' is to offend against this reality: about the origins of the universe, 'why' questions are just illegitimate; it is just a mistake to ask them; it is a falling into the animistic thinking of the small child to ask them. Let me spell that out a little more fully.

I am suggesting that it is because 'why are we here?' and 'why is there something rather than nothing?' do not have the status of being questions that they cannot have the status of requiring answers. A question is not a question just because it has the appropriate syntactical construction: the point towards which Noam Chomsky was drawing our attention with his famous 'do green ideas sleep furiously?' variant. As well as being syntactically coherent, a question also has to be semantically coherent: it has to make sense. And demanding reasons for the existence of entities which emerged before the emergence of reason does not make sense. It makes nonsense. Pre-person entities have not *reasons for* their existence but *causes of* their existence. And this being the case, while the second and the third of Flanagan's 'three great perplexities' are *questions*, and thus can have *answers*, the first is not and can not.

But we need also to be clear about the significance of this difference. When science refuses to answer 'why' questions as asked of pre-reason entities, that is a measure not of its weakness but of its strength: it recognises that because they aren't questions they can't have answers. And when religion does answer 'why' questions, as asked of pre-reason entities, that is a measure not of its strength but of its weakness: it does not recognise that because they aren't questions they can't have answers. That the scientist's art is the art of the soluble is that which enhances rather than diminishes its seriousness: a reality which can be nicely illustrated by using a Piagetian example and some Piagetian vocabulary.

The great developmental psychologist tells the story in terms of the baby. The younger baby, operating with the grasping reflex, tries to grasp the sunbeam: that is, tries to *assimilate* the ill-understood nature of sunbeams to the immature demand of its grasping schema. The older baby, schemata differentiated, grasps the rattle and smiles at the sunbeam: that is, *accommodates* a more mature demand to the real nature of a world now better understood. Sunbeams can't be grasped: rattles can. But this is how it is with us all. The less mature adult response is the rain dance: a magical and *assimilatory* demand, made of an ill understood world, that it meet our need. The more mature adult response is the digging of irrigation channels and the saving of some of the waters of the rainy season for the dry one: a demand scientific because *accommodatory* of the real nature of the real world now more thoroughly understood. What this means is that the successful theories and laws of the scientific approach are those which leave the world exactly as it was before they were formulated. They are pure accommodation. And it will be my case that it is because the theory of 'I' which I will propose is one which, unlike the one currently in use, does just that—that is, shifts us from the magic of an *assimilatory* 'I' to the science of an *accommodatory* one—that it will be one which will give us at last the possibility of genuine psychological and moral science.

Having then considered, if indeed only briefly, what I take to be two very general principles underlying the scientific method, I want now to turn to some rather more detailed specifics. For both of the reasons with which I began this chapter—because I want to describe what it is to which we have lost religious faith, and because I want to outline the method I will try to use myself in the investigation which constitutes the second part of this book—I want now to present what I take to be the essentials of the scientific method, of thinking scientifically. And to make things clearer, I will do so as a series of numbered points.

1. The scientific method begins with observation

This insistence on the primacy of observation was a contribution mainly of the sixteenth century scientific revolution. It was propounded as an antidote to the armchair speculation methods of many (though by no means all) of the then 'natural philosophers'. The main problem with this formulation now, and it is a serious one, concerns beginnings. It has become increasingly clear that it is impossible to argue that observation can be theory-free. And that means that it has also become increasingly clear that, anyway on some of the readings of the concept, there is a problem concerning its objectivity. In *Theory and Evidence: The Development of Scientific Reasoning* (1996), Barbara Koslowski points out that most philosophers of science have now largely abandoned the view of the logical positivists that scientific method could be independent of theory. In making her case she quotes the work of such as the social constructivists Hanson and Kuhn, and the realists Putnam, Kitcher and Boyd. And claims that the empiricist tradition as a whole now widely acknowledges the problems that this development throws up. For if observations are always and only made *out from* some theory, *out from* some theoretical perspective, in what sense can the scientific method be said to begin (originate) with observation? If the observations with which it begins are 'always already' imbued with previous theory, can they really be thought of as providing any 'beginning' at all? In their making of this same point Patricia and Paul Churchland—in *The Churchlands and their Critics* (1996)—explain that animal understanding, including human animal understanding, is, even at its earliest and simplest, thoroughly theoretical from the start: by which they mean, they say, that it is 'speculative, systematic and corrigible'. This is a problem which will recur many times, and in many guises, across all the rest of this book. And by its end, or so I believe, will be found to have been resolved.

2. The observation is through the senses

This stress on 'through the senses' is going to be important for three reasons. First, because it challenges the credentials of claims to knowledge apparently arrived at by other and essentially mysterious means: by telepathy, or magic, or 'I-just-know' routes. Second, because it puts emphasis on understanding just how those senses work, and particularly on how they can deceive: on understanding, for example, that when a psychiatric patient says that Napoleon is sitting on the end of her or his bed, while he or she is not (necessarily) lying, neither is Napoleon on the bed. Third, because it plays a crucial role in that remould of the concept 'I' which is central to my concerns.

3. The observation is of material things and happenings

Why these two? Because together they cover all that there is. And that means that in restricting the observation stage of the scientific method to the observation of only these two, I am restricting it to the observation of anything and everything. There are of course huge arguments as to whether or not anything or everything can be studied scientifically. What of love, of morality, of aesthetic judgements, and so on? On such issues, I will be arguing that unless a phenomenon can be conceived of as either a material thing or a happening, it cannot be studied scientifically. But that since *all* phenomena can be conceived of as consisting in material things and/or happenings, it therefore follows that *all* phenomena can be studied scientifically. Of course I will have to try carefully to delineate what I mean by all these terms. And that will be my endeavour. In particular, I will try to show that it is not the physicality of the things studied which decides whether or not their study can be scientific: this is a misconception that leads to the ubiquitous elevation of 'proper' sciences such as physics over 'pseudo' sciences such as psychology. I will suggest instead that it is *methodological* rather than *ontological* considerations that matter. And that once that is accepted, the making of what I will call 'moral science', if still a very difficult undertaking, is anyway firmly on the agenda.

4. Every attempt is made to observe as objectively as possible

One emphasis here has to be on the 'as objectively as possible' phrase. It needs to be noticed that it is not the scientist, unless the bad scientist, who claims to deal in certainties: it was Einstein, after all, who said something to the effect that if a theory is valid it cannot be certain, and if it is certain it cannot be valid. Scientists are actually used to functioning always in uncertain worlds, always with knowledge which is flawed. It is that which sometimes makes them impatient of people who cannot cope with risk judgements as readily as they can themselves. Indeed, it is surely the critic of science, the one forever insisting that 'we can never be certain, you know', who is stating—a certainty.

But a second emphasis, and one to some extent contrary, must be put on something else. In his autobiography *No Discouragement* (1996), A.H. Halsey writes of science as 'a set of procedures which, over a range of activities and practitioners, has been shown to have been effective in diminishing subjectivity'. He goes on to observe that 'it is extremely difficult to diminish subjectivity to zero anywhere' and that 'it is extremely difficult to get it below a very high level in the study of social affairs'. Agreed. But such 'difficulties' in no way justify what is a thoroughly irrational response: a response which, in effect, says that because science cannot diminish subjectivity to zero, we should use *as well as* science (the moderately irrational position), or *instead of* science (the totally irrational position) those procedures such as 'common sense', or 'intuition', or 'just knowing', which cannot diminish subjectivity at all! As observed by Anthony O'Hear in his *Beyond Evolution* (1997), 'the fact that one does not have perfect knowledge does not make it more rational to act as if one has no knowledge at all'. Except insofar as it might be nice to study the processes of their generation so as to learn how better to generate better ones, it does not matter much where ideas come from. But it matters enormously which are to be the procedures by which we decide whether or not to believe them once they've arrived. Intuition might generate an idea: fine. Common sense might generate another: fine also. But to decide whether or not these are ideas to which we should give any credence, we need the scientific method. Why? Because it has shown itself to be by far and away the most reliable method by which we can most fully escape subjectivity, partiality and interest. That common sense which Descartes called 'the most widely shared commodity in the world, for every man is convinced that he is well supplied with it' is actually only about as common as common nonsense. It is the extent to which *scientific* sense is *uncommon* sense which is the problem.

5. It is assumed that valid inferences may be drawn from sensory experiences, and that material things and happenings may be explained in terms of rational principles and generalisations

One of the things it is important to register about science is that if it is to be done at all, if the scientific method is to be pursued, there are assumptions to be made. We have probably all, from time to time, become tangled in argument with a certain sort of 'no-assumptions' nay-sayer. He or she is one who says 'yes' water has only ever boiled at 100 degrees centigrade at sea level up to now, but 'no' that does not mean that we can predict that it will next time. Or 'yes' consciousness has only ever emerged from organic things up to now, but 'no' that does not mean that we can predict that it will next time. It is thus that the boiling of ice and the consciousness of computers can be defended. The problem is that any such science-fiction defence is completely out of kilter with the science-fact nature of the situation. For the science-fact nature of science (or the scientific method) is that it's not actually about making predictions. It's about making explanations. And while, as Sam Goldwyn is said to have put it, 'making predictions, particularly about the future, is very difficult', making explanations, particularly about the complex, is harder still. Making *them* involves, among other things, the prior making of a very particular assumption. But make it, and all is well. Assume only that there is a caused 'way that things are', and all is well. An explanation then consists in coming to understand the nature of that way. And what's more, even if there is a problem about how we know when we have arrived – there is, and it will be addressed later – nonetheless the very existence of that way also means that even an only steadily accumulating knowledge of its negation, of the way things aren't, will necessarily bring such explanation closer.

There remains, of course, the problem of explanation itself. It is notoriously difficult to explain explanation: so much so that some philosophical positions, logical positivism for example, have a tendency to shun it altogether as mere metaphysics. But anyway at this early stage of the process which is scientific thinking, at the stage of assuming that reality may be explained, suffice it to say that by explanation I am thinking of something like this: that the more complex may be understood in terms of the less complex; that what did occur in these circumstances tells us something about what would have occurred in those; that what is actually so might give us clues as to what must be so. These are all things which will be further elaborated as my outline continues.

6. Before any inference is drawn, a significant number of observations must be made of the material thing or happening under study

One of the major things that differentiates scientific from non-scientific thinking is the huge reliance of the latter on personal experience (the experience of one) and anecdote (the sample of one). The idea that such direct 'evidence' is incontrovertible, or anyway more significant than anything to be found in any number of mere books, is widely cited. It is of course a nonsense. If the comparison is between my personal experience and their personal experience, where 'they' are the persons who have written the books, it is a *somewhat* partial position. If the comparison is between my personal experience and validities arrived at by a scientific community using the scientific method, it is a *spectacularly* partial position. And the defence of such positions is just illegitimate: *somewhat* illegitimate in the first case, *spectacularly* illegitimate in the second.

But this is not the only reason why issuing the standard warning against arguing anything from a single observation, from a sample of one, is so important. It is I think also important because it draws attention to what is a rather more fundamental reality about the nature of scientific as opposed to non-scientific thinking. Confucius, modernised, is supposed to have said—and if he didn't, it doesn't really matter—that the common person is he or she who is interested in that which is uncommon, while the uncommon person is he or she who is interested in that which is common. It is a statement of course marred by a delicately nuanced elitism: can you not just hear the self-congratulatory tones in which it invites its listeners to identify with the 'uncommon' person? But for all that, I think it points to something important. For the common person—as at earlier stages of our acquisition of both general intellectual capacity and specific intellectual understandings we are all common persons—I think it is the case that the uncommon is often taken to be more significant. As it is also the case that for the uncommon person, for those only *as yet* few who acquire levels of both general intellectual capacity and specific intellectual understandings as to become uncommon, the common does indeed become more interesting. And I fully expect that this line of argument will prove differentially exceptionable depending on the generality of its application: a point which because the issues which it raises will recur many times in what I write is worth further illustration here.

If I suggest that tabloid journalism privileges the story of the individual who has committed the violent crime (uncommon) over the story of child-rearing styles which foster violence in those so brought up (common), I expect the (uncommonly) well educated, as they reach for the broadsheet, to concur: it is development, progress, hooray stuff, that moves us from *The Mirror* to *The Guardian*. But if I suggest that it will also be development, progress, hooray-stuff when that very broadsheet privileges knowledge of the (common) findings of science over knowledge of the (uncommon) masterpieces of art, I know I will get a more bumpy ride. So with just a tiny example of what I mean, let me defend myself in advance.

In his book *Poetries and Sciences* (1970), I. A. Richards castigates the poetry of those who are merely scholars. He claims that such poetry, arising as it does from merely the knowledge of technique and the desire to write, will be betrayed by its rhythm. For rhythm, he suggests, is no mere series of syllabic tricks. Instead it is that which directly reflects the personality of the writer. To be itself genuine, rhythm must arise from impulses genuinely aroused and experiences supremely ordered: a fact which makes it the most subtle of indices to the real interests of the poet. I consider this argument, and it sounds sort of right to me. But hold. Here is a (supremely confident!) statement purporting to say what *is* the case:

rhythm *is* what betrays the lack of impulses genuinely aroused and experiences supremely ordered. That is no small claim. An equivalent statement of what *is* the case in science would require validation. Its author would have to demonstrate that he or she had used a method of arriving at the statement which could be defended as valid. And notwithstanding the problems presented to scientific thought by the problem of induction—of which, more later—in order to *be* so defended there would have to have been 'a significant number of observations' of the phenomenon under study. In this case, the phenomenon concerned is a significant number of correlations between 'rhythm' and 'impulses genuinely aroused and experiences supremely ordered'. And the validity of the claim that it is the one which is responsible for the other would at the very least depend upon such things as a recognition that correlations are not causes, and an indication as to what each of the two is to be taken to mean. My point here is nothing to do with whether or not such recognitions and indications are possible in this particular case, though I think that they are. My point is that it is to those areas of our understanding where such clarifications *have* been made, where such definitions *have* been attempted, and where both *have* formed the basis of the subsequent hypothesis-experiment-theory stages of scientific method, that the 'uncommon' person might be expected to pay more attention. Such a position involves me in suggesting, for example, that contrary at least to conservative literary opinion, more *knowledge* about persons can now be gleaned from those parts of psychology which now *are* underpinned, at least to some extent, by the scientific method than can be found, for example, in one of Shakespeare's uncommon (even though uncommonly good) plays. It is a position which of course means that I must entertain, as across my book in various places I will, some highly contentious ideas about the relationship between art and science. When the scholar poet A. E. Housman suggested that science was 'an aristocratic affair, not communicable to all men, nor to most men', I will want to associate myself not at all with the elitism (or the sexism!) of the claim, but entirely with its implication that, anyway as far as acquiring knowledge and making truth is concerned, scientific thinking is a development of, an advance upon, thinking in other modes.

7. Careful records are kept of all observations, not excluding those that are unexpected

Scientific thinking not infrequently involves treating a material thing or a happening as a variable, and comparing its presence with its absence. And because memory is both motivated and fallible, knowing just when a phenomenon is present or not present means not just the careful observation of the phenomenon, but the careful recording of that observation for future study and use. Science is about evidence: justification is required not for *doubt about* that for which there is no evidence, but for *belief in* that for which there is no evidence. And unlike other varieties on offer, scientific thinking requires good evidence. It requires, for example, that the bigger the claim being made, the better the evidence there is to support it. But observation is not evidence. Evidence is made from observation: bad evidence from bad observation; good evidence from good observation. And in the case of the latter, at least some of that goodness consists in its being observation well-recorded.

I shall have occasion in a later chapter to consider the concept 'ideology', and to make some contribution to the vexatious task of trying to separate it unambiguously from the concept 'science'. It is perhaps worth registering here, therefore, that I think this 'careful records' point can be quite helpful in doing so. We can, I think, separate ideology and science rather better by 'method used' than by either 'type of subject matter studied' or 'type of finding arrived at'. The question 'what is science?' can of course be answered by saying that it is a method used to try to understand things. But then the question 'what is ideology?' can also be answered by saying that it is a method used to try to understand things. It's (just!) that they're different methods. And one of the ways in which they're different is at the level of record. The records of ideology are often kept, not easily accessible for scrutiny, in tradition, authority, myth, intuition. The records of science by contrast are kept, accessible to all, in the laboratory notebooks, data banks, published papers of scientific communities. And if, as I shall argue later, ideologies are systems of ideas whose function is to maintain partiality (interest), and sciences are systems of ideas whose function is to develop impartiality (disinterest), then one of the important ways in which each of these functions is carried out is via the keeping of record. The interested record is partial in both senses: it is not impartial; it is not complete. The disinterested record is impartial in both senses: it is not partial; it is not incomplete. And an important part of any training in the use of the scientific method has to aim to inculcate the care necessary to the production of such record.

8. Patterns discerned in the (significant number of) observations can then be used to make a general inference, known in this its untested form as an hypothesis

It is at this stage of the procedure, at the stage of the generation of hypotheses, that the Humean sources of the scientific method can most clearly be seen. What is involved is comparison: the comparison of a conceptual model of how things might be (an hypothesis) with a contrived experience of how they are (an experiment). As the result of such comparisons, the conceptual models – the hypotheses – are validated, or not validated. And scientific understanding grows – in either case.

Among non-scientists, and in somewhat romantic 'great man' histories of science, there has been a tendency to present the generation of hypotheses as in some way mysterious: the work perhaps of unfathomable genius. Some of those with arts-biased ideas as to where creativity is to be found, some who think that it is likely to be found somewhat nearer to them than to mere scientists, suggest that hypothesis-generation is quintessentially an artistic activity. It must be: it's creative, for heaven's sake! Thus frequently repeated tales of 'eureka moments', of Archimedes in his bath or Kekule at his fireside, fuel the notion that here at the level of the generation of hypotheses lies the invisible hand of the mysterious: of intuition, of inspiration from on high, of 'I just know' stuff. The model for the whole approach could be summed up in Pope's famous line: 'God said, "let Newton be", and all was light'.

The scientific method, by contrast, is one which thoroughly counters such self-serving romanticism. It requires us to remember that eureka moments, though still wonderful, have always been preceded by much pre-eureka slog; that such slog has been undertaken as much by countless unnamed others as by the person of the moment; that as Pascal put it 'chance favours the prepared mind'; and that at any particular historical conjuncture, socially structured, many minds are going to be prepared. Of course it is often only one such which gets into (Whig) history books. But that is perhaps more of a comment on the partiality of their authors than on the reality of the functioning of scientific communities. Referring back to a previous point, such authors are maybe those who, anyway as yet, are only capable of being more interested in the singular and uncommon than in the collective and common. In the history of ideas about natural selection, for example, many now know that we must add the name of Alfred Wallace to that of Charles Darwin. But there were, of course, also others. While Darwin was still aboard the Beagle collecting his specimens, the Scottish arboriculturalist Patrick Matthew published a coherent and meticulous description of natural selection in his book *Naval Timber and Arboriculture* (1831). Though being a scientist rather than a maker of great-man, Whig-history myths, when much later he realised that he had been preceded in this fashion, Darwin himself acknowledged the priority without reservation. As is the case with every considerable hypothesis, so with this one: the hypothesis concerning the nature of evolution, even at its outset, was a many-authored phenomenon.

It is widely believed that good hypotheses need four basic characteristics: they need to be useful; they need to be capable of refutation; they need to provide the basis for prediction; and they need to be simple. To try to establish the extent to which I think such beliefs are valid, I want briefly to consider each of these four.

First, usefulness. In the most general of senses, of course they need to be useful. Their testing is often costly, sometimes very costly, and it is difficult to justify the necessary expenditure of resources, human and financial, on study of the not-useful. But we need to ask careful questions of this 'useful' criterion. In particular, we need carefully to ask 'useful to whom?' and 'useful as what?' To be useful to the working scientist, it is not at all necessary that, from subsequent experimentation, a good hypothesis should be one which emerges as having been validated, only that it should be one which emerges as having been generative: of new data; of new ways of thinking about problems; of new experiments which might be tried. Many an hypothesis which never made it to the level of theory generated hugely useful knowledge on its way to extinction. And to be useful to the human enterprise of extending its understanding of reality, it is not at all necessary that, to be good, hypotheses need to be beneficial to some interest group wanting to make money. With these constraining and enabling understandings in place, of course hypotheses need to be useful.

Second, refutation. The subject of much debate among epistemologists, consideration of the refutation point in general is beyond the scope of what I want to deal with here. But two issues which frequently arise from it are not.

It seems to me that the Popper-derived insistence that we learn not from positive instances but from negative ones, not 'by trial and success' but 'by trial and error', and thus that scientific hypotheses are those for which there are practicable ways of attempting their refutation—making all historicist hypotheses, of course, thus thoroughly unscientific—is an argument which can misrepresent the level at which hypotheses operate. While it is common to talk about 'the refutation of hypotheses', the issue is much more usually about the refutation not of hypotheses themselves but of operational statements derived from hypotheses: things which would be true if the hypothesis were true. These are the things, not the hypothesis itself, which in reality are tested. These are the things, not the hypothesis itself, which it is perhaps useful to insist need to be capable of refutation.

It also seems to me that the also Popper-derived insistence that we try to falsify (or disconfirm) hypotheses, as opposed to trying to validate (or confirm) them, is less than helpful. In the edited collection of papers *The Philosophy of Karl Popper* (1974), Hilary Putnam argues, in his contribution on the corroboration of theories, that confirmation and disconfirmation are not, as Popper suggested, two distinct strategies. Instead they are two parts of the same strategy, each needing to be tested by the same sorts of tests. This is a stance itself 'corroborated' by work in psychology on the development of scientific reasoning. In *Theory and Evidence: the Development of Scientific Reasoning* (1996)—again—Barbara Koslowski, well known for her experimental work on cognitive development, shows that it is work which suggests that the confirmation-disconfirmation dichotomy is less salient to such reasoning than the 'conjecture-refutation' model would seem to allow. She points out that in trying to confirm an hypothesis, a well-trained scientist regularly takes into account possible alternatives anyway: after all, to make her or his favoured hypothesis that bit more convincing, it is positively in her or his interests to do so. And the same applies when instead the scientist tries to disconfirm a hypothesis: again, he or she will of course take account of possible alternatives precisely in order to show that it is perhaps one of them which makes better sense of the facts. What Koslowski is arguing is that confirmation and disconfirmation processes are much more similar, indeed much more one and the same process, than Popper's ideas allow. And it seems to me that hers is an approach to this question considerably better than anyway the orthodox approach to it. Apart from anything else, it does not rely upon ignoring that capacity for inter-personal co-operation and division of labour which is perhaps one of the most important characteristics of persons. I can try as hard as I can to validate my hypothesis (which involves me in trying to falsify yours): you can try as hard as you can to validate your hypothesis (which involves you in trying to falsify mine). This gives us attempted refutation all right, but in the course of attempted validation. It just does not seem to me to be at all necessary—which is just as well because I suspect it hardly ever happens!—that refutation of my hypothesis needs to be by me. Why won't 'by you' do just as well?

Third, prediction. The hypothesis which predicted that banging drums, shouting, and screaming as loudly as possible would cause the monster who was eating the sun to go away was right every time. Whenever terrified people in ancient cultures greeted total eclipses in this way, their prediction was always born out. The sun always reappeared. Or again, the hypothesis which predicts that people jumping from the fourteenth floor of a skyscraper will splatter dead on the ground, and the alternative hypothesis which predicts that while people jumping from the fourteenth floor of a skyscraper *up until now* have splattered dead on the ground, but *after now* will float—my variation on Lakatos's 'floater' story (see below)—each have exactly the same amount of validating evidence going for them. In the first case, is there anybody who thinks that a theory of sun-eating monsters could be derived from an hypothesis providing such a well-validated prediction? In the second, is there anybody who would not urgently advise any would-be 'floater' to think again?

The vital point here, yet again, is that science is not essentially about prediction. Because an hypothesis predicts well tells us nothing essential about its validity. There are sciences which cannot make a single prediction: evolutionary biology perhaps is one. And because science is essentially about explanation rather than prediction, it does not matter in the slightest. To stand a chance of being validated to the level of theory, hypotheses have to offer explanation not prediction. Prediction is epiphenomenonal to the worth of an hypothesis: often useful, never essential. It is explanation that is essential. And an explanation, to add a further point to those I made before on the subject, is always something historically specified: that is, it is proposed by and designed for an historically specified audience, for whoever are the currently relevant 'cognoscenti' if you like. It is also seeking to answer historically specified questions, using historically specific modes of enquiry. Indeed, this is an historical specificity which is true not only of what the professionals call the 'explanans' (the explanation given) but also of course of the 'explanandum' (the material thing or happening to be explained): to illustrate, we no longer look for explanations concerning the nature of the ether.

Fourth, simplicity. 'It is vain to do with more what can be done with fewer': one version, in translation from its original Latin, of Ockham's razor. Such already-mentioned bias towards simplicity in hypothesis construction, earlier found in Aristotle but much more commonly associated with the fourteenth century English Franciscan William of Ockham, is very widely accepted still as guide. Just what 'simplicity' means in any particular case—number of different kinds of entity? number of prior axioms? number of prior hypotheses subsumed?—is for debate. But the general principle seems to hold. So when, in due course, I offer my own hypothesis as to the nature of a person, I will be trying to respect it. Though I will also bear in mind a correction from Einstein which warned that one should certainly use as few entities as possible as will offer explanation—but not fewer.

9. In formulating hypotheses, we make use of inductive thinking

This makes reference to a problem of scientific method well known, if not notorious. The 'problem of induction', as it is called, is the problem of how one moves from particular examples to general statements. While, with deduction, it is logically the case that if all As are Bs and all Bs are Cs, then all As are Cs, with induction it is not logically the case that if all observed As are Bs, then all As are Bs. There might be some unobserved As which are not Bs. And we would not know it—because they were unobserved. While almost everybody recognises the problem to be a problem, not many conclude, with Nabokov, that 'general ideas are short cuts from one area of ignorance to another'. But then Nabokov was neither scientist nor philosopher. For most scientists and philosophers, the problem is just that: a problem. It is not a reason to throw up one's hands in despair and go home. The way in which I want to approach the problem is to question an assumption.

It seems to me that contained in the idea of induction as problematic there is a piece of invalid reductionism. There is a problem of induction only if reason is reduced to logic. Deductive thinking is capable of being understood using logic: inductive thinking is not. But that is only a problem if one has already made the prior assumption that the scientific method is based in logic: that is, that it is deductive thinking which is paradigmatically scientific thinking. But suppose one has not. Suppose one takes the position that all that Hume did was to point out that induction is not deduction—for which, much thanks. It then becomes possible to argue that there are some (relatively simple) processes of thought where the more limited tools of deductive logic will suffice. And there are other (rather more complex) processes of thought, where one reaches for the more powerful tools of inductive reason to do the job. The scientific method would then consist in the appropriate use, at the appropriate stage of the process, of one, or the other, or both tools. And the task for philosophers would then become the rather more tractable one of developing 'principles of induction' to accompany the already existent and widely accepted 'principles of deduction'. In *The Logic of Scientific Discovery* (1959), of course, Popper has claimed that there is not, and could not be, a valid principle of induction. But mine is an approach which responds by saying that such is the case only if 'valid' is taken to mean logically valid, rather than rationally valid. It is surely just an error based in the reduction of fundamental reason to mere logic to think that the only alternative to logical thinking is illogical thinking. In *Reason and Emotion* (1935), for example, John Macmurray convincingly defends the proposition that 'thought that is merely logical is never rational', that 'rationality is not to be defended in logical terms'. And adds, for good measure, the thought that 'lunatics are often superbly logical'. And in *Science in History* (1954), for another, J. D. Bernal, criticising what he calls Aristotle's 'verbal dodge' contained in the syllogism 'all men are mortal, Socrates is a man, therefore Socrates is mortal' adds that this is a nonsense 'still taught as logic today, as if we could ever know the general before knowing the particular'. It seems to me that these are the sorts of things we need to consider if we are to get mere logic into perspective vis-à-vis reason. And to emphasise the point, let me return to Lakatos's already-mentioned 'floater'.

The philosopher of science Imre Lakatos created 'the floater' as a man who puts forward the hypothesis that falling bodies fall with uniform acceleration up to time T, but 'float' thereafter. He takes time T to be now, and proposes to jump, forthwith, from the Eiffel Tower. Since all the predictions of the new hypothesis are so far the same as those of the old, it is as well corroborated (i.e. as little refuted). According to those who think there is a problem of induction, we all believe he is irrational. We all try to persuade him not to jump, and in the last analysis might very well restrain him against his will for his own sake. But we none of us can logically refute the claim of his hypothesis. Its predictions to date have been excellent. And however many past examples there have been of instant death in such circumstances, 'the problem of induction' ensures that we cannot claim such death as a logical necessity.

I have already suggested that a first thing to say about this example is that what sorts good hypotheses from bad hypotheses is not prediction but explanation. But a second is that we need to look at 'the problem of induction' in this light also. All problems are generated by hypotheses: that is, they are thrown up as a consequence of working within this, that or the other hypothesis. But since hypotheses can be good or bad, so too can the problems they generate be good or bad. From which it seems to me to follow that there is no (good) problem of induction, only a (bad) problem of induction: that Hume, or whoever, simply invented a bad problem by virtue of operating from within a bad hypothesis as to the nature of reality. Using the Piagetian language introduced earlier, the hypothesis was a bad one because it was assimilatory. It demanded of reality that which was not available to reality: that induction should have the same logical structure as deduction. An accommodatory approach would be one which did not require that there be a symmetry between induction and deduction where no such symmetry can exist. An accommodatory approach, therefore, would not generate the (bad) problem of induction.

Another way of expressing this is to examine the language we use. Remembering the point made in the previous chapter that it does not follow that simply because a question is good grammatically it is therefore also good semantically, we can say that while 'what is the logical under-pinning to induction?' is good grammatically, it's just bad semantically. Our language will let us ask it, but reality won't give us an answer: it's a 'question' rather than a question. And maybe it is the case that the problems thrown up by hypotheses generated from within the philosophical community are more often of this ilk than are those thrown up by hypotheses generated from within the scientific community. Whichever, hypotheses with a tendency towards the bad seem often to be those which forget that, as Pavlov puts it, language is the 'second signalling system' not the first. While by contrast, good hypotheses are those which more readily take into account something fundamental about explaining: which is that to explain something, your explanation has to be made to fit that something, not that something made to fit your explanation.

10. Once formulated, hypotheses must be able to be tested: and the way to test an hypothesis is to deduce from it statements which would be the case if it were valid, and then try to discover whether they are in fact the case by means of experiments or contrived experiences

To the person in the street, as perhaps also to many a working scientist, it is the idea of 'experiment' which, more than any other, characterises science. Indeed, the phrase 'the experimental method', referring to those procedures whereby science requires that we put our prejudices at risk, is quite frequently used as if it were synonymous with 'the scientific method': as if experiment were all. In the popular imagination, if experiment is not done, or cannot be done, then science is not done, or cannot be done. As example, take psychoanalysis. As some sort of story, akin perhaps to stories of 'the Fall' or 'the Resurrection', all of them able to provide structuring mythologies capable of bringing pattern, order, meaning to our lives, it might be interesting. But as science? Its critics are vocal. Where is the objective evidence for testable hypotheses? Where are the predictions? But, above all, where are the experiments? And this last would seem to be a question not lost on the psychoanalysts themselves: the relevant literature contains plenty of examples of people trying, in vain I think, to establish an experimental basis for its ideas. And that's fine. While it is never appropriate to make experiment everything, a stress on it as important is correct, is fine. But only as long as it is appropriately characterised. The characterisation I would wish to defend is that which thinks of experiment as 'contrived experience'. But because they are particularly relevant to my own experiment, as detailed in the second part of this book, I want to deal here with what I think are two problems with that characterisation.

First, I want to draw attention to the variety of experiments. Although it is a formulation of the relationship between 'us' and 'our experiences', contrived or otherwise, which will come to be fundamentally questioned by my own hypothesis as to the nature of persons, for now it suffices to think of contrived experiences, in a common-sense way, as those which don't just happen to us but which, in order to ask questions about the nature of reality, we deliberately give ourselves. And as such, their variety is considerable. They can be *natural* experiments: the world can throw up a pattern—two hospitals, perhaps, with consistently different procedures—which enables a researcher to come along and test an hypothesis as to the consequences of that variation upon outcome. They can be *laboratory* experiments: those in which, by making every effort to isolate, eliminate or control other variables which might confound results, experimental design aims to ensure that the variable under study is tightly controlled by the researcher, perhaps making use of a control experiment. They can be *thought* experiments: who has not heard of Galileo's cannonball, Einstein's elevator, Maxwell's demon and Schrödinger's cat? According to Francis

Bacon, if not to Karl Popper or Imre Lakatos, they can be *crucial* experiments: able to decide decisively between two competing hypotheses, as some theoretical physicists thought was the case with the Aspect double-slit experiments done in Paris in 1981 and 1982. But what seems to me vital to recognise about this variety of experimental types is not that we have to try to decide which is the legitimate one for all circumstances, but which is the appropriate one for each circumstance. The former is a practice almost always indulged in by a certain sort of physical scientist who aims to privilege the laboratory experiment over all comers. The latter is what we need.

Second, I want to draw attention to the level of experiments: or, perhaps better, the level of the hypotheses of which they are the tests. I want to reiterate my earlier point that experiments are testing not our *actual* hypotheses but operational statements derived *from* those hypotheses. It's as though just as big fleas have little fleas upon their backs to bite them, so big guesses (hypotheses) have little guesses (operational statements) derived from them to test them. And it is a use of experiment, both frequent and illegitimate, to act as if what has been refuted—where anything has—is not just a specific operational statement derived from an hypothesis, but the very hypothesis itself. My point will be perhaps better made by means of an illustration.

As a teacher of psychology, and derived from I know not where, I would often use the following borrowed example. Suppose your hypothesis was that carrot eating improved the eyesight, especially at night. Derive from that a statement which would be true if the hypothesis were true: perhaps that when driving at night, rabbits would have fewer car accidents than moles. Go out and test not your hypothesis about carotene: that cannot be done directly. Go out and test your operational statement about the accident records of rabbits and moles: as long as the animal police keep their records efficiently, that's easy. The results, as gleaned from the relevant statistics, would then be one piece of evidence amongst what would need to be many more, that your hypothesis was or was not worth continued investigation. From that hypothesis you would need to be able to derive not one testable operational statement but many testable operational statements. And your many testable operational statements would need to be tested on a number of separate occasions (to control for fluke), in a number of different ways (to control for method bias), by various and independent workers (to control for experimenter bias). Recognised techniques for achieving these controls, double blind trials for example, would need to be respected. And until at a chosen level of probability it can be shown not to hold, you would in each case have to accept the null version of your hypothesis. This is the one which states that the variables under study have a chance rather than a systematic relationship one to another: this being the technique, incidentally, which by requiring that one rejects the null 'in case' renders the so-called 'precautionary principle' unscientific. Above all, and throughout such a (very mini!) research programme, an appreciation of its levels is crucial. And one of the crucial things that that means is that an illegitimate use of experiment, the one to which I was trying to alert my students with my rabbits and moles example, must be avoided: it must be recognised that while it *is* OK to use the results gleaned from working at the level of an operational statement to *validate* an hypothesis, it is *not* OK to use the results gleaned from working at the level of an operational statement to *invalidate* an hypothesis. If moles were shown to be crash prone at night and rabbits to be excellent night drivers, that *could* be used to *validate* the carotene hypothesis. But if moles were shown to be excellent night drivers, and rabbits to be crash prone at night, that could *not* be used to *invalidate* the carotene hypothesis. Unexceptionable stuff—or so you might think.

But then consider some of the biggest hypotheses of the lot, especially some which Popper would label 'historicist. If your hypothesis is that capitalism contains within it the seeds of its own destruction, and one of your operational statements derived from that is that the chain of capitalist economies will break at their weakest link—I deliberately use a very celebrated example—then interesting things follow. If an anti-capitalist revolution occurs in Russia, that *can* be used to *validate* the hypothesis concerning capitalism containing the seeds of its own destruction. But if capitalism is later re-established in Russia, that can *not* be used to *invalidate* the hypothesis concerning capitalism containing the seeds of its own destruction. Though from reading much current 'end of history' literature, you might be forgiven for thinking that it can.

11. Hypotheses well validated in these ways come to be accepted as theories: open always to further validation, to modification and to rejection

Kurt Lewin somewhere maintains that there is nothing so practical as a good theory. John Searle somewhere maintains that knowledge advances less from the making of an assortment of factual discoveries, more from the making of a comprehensive theory. Both they, and the no doubt many others who have made similar claims, are I think right. But if they are, that requires us to consider both what it is that makes a theory good or comprehensive, and what it is that makes a good or comprehensive theory so very practical.

A theory can come to be appreciated as a good theory for some or all of several reasons. It can be very well validated (or corroborated, if you prefer: I do not mean to attach any particular significance to my own preference for one word over another here). It can be capable of making rather more sense instead of rather less sense of the relevant data. It can generate more and more useful predictions, rather than fewer and less useful predictions. It can be more rather than less simple and elegant, perhaps in terms of the number of assumptions it rests on vis-à-vis the quantity of data it explains. And no doubt there are other criteria too. But anyway in my opinion, none of this is critical. What is critical is, first, that our good theories are our best *explanations* (again!). And second, that our good theories are those claiming to be *valid* rather than those claiming to be *true*.

On explanations (again!): the already-much-mentioned approach that it is by its predictive rather than its explanatory capacity that a theory should be judged is, despite being in error, still widely believed. Perhaps a residue of the one-time hugely influential logical positivism; perhaps encouraged by the thorough-going separation of science and philosophy of science in most higher education; perhaps made worse, anyway in Anglo-American places, by traditional empiricism and pragmatism; perhaps encouraged by the notorious problems in coming up with explanations of, as opposed to predictions from, the theories of quantum mechanics; perhaps because, as already suggested, explanation itself is far from being an unproblematic concept: for whatever reason or combination of reasons, the seemingly over-determined 'don't-worry-me-with-explanation-just-give-me-prediction' position is still very much in evidence, especially among working scientists. On this, once again, scientific method requires us to understand that, while they offer prediction as an optional extra, good theories offer explanation as a sine qua non. As it also requires us to understand—for example when dealing with the related argument that because a single set of facts can be explained by an infinite number of theories, therefore theory is much over-rated—to get hold of the idea that far from speaking for themselves, facts don't speak 'til spoken to: by theories.

And on truth and validity: here we get to the vexed and vexing question of proof. How often must a triumphant 'ah, but can you *prove* it?!' have been thought finally to deliver the coup de grace which wins the argument for he or she who proclaims it? It doesn't, of course. Rather less because there is no 'proof', rather more because there is no 'finally', nothing is ever 'finally proved'. A good theory is always one well validated, never one finally proved.

Consider, to use an everyday example, the oft-heard proposition that because we live, move and have our being in the organic, that limits the range and nature of the understanding we can have of ourselves, and of the world of which we are a part. No others being available, this is a thoroughly theoretical statement: that is, it is constructed from within some particular and interested purchase on the world which is the theory concerning such things embraced, knowingly or otherwise, by its propounder. When trying to decide whether or not it is a good theory, what matters is the extent to which it is well validated vis-à-vis its competitors. For it does have competitors. There are other interests and other purchases which would generate other and different statements. 'Limiting' as compared with what, for example? Why not 'enabling'? Organicness is at least as readily construable as 'enabling' as it is construable as 'limiting'. Ask a rock.

So of course the 'we' of 'our theories' is always an historically situated we: what other we is there? And of course our theories are by and for us: who or what else could they be by and for? But would-be-theory-basher beware! There is no refuge to be found in refusing to accept a theory because it is not, and cannot be, finally proved. Such a stance does not involve staying, pure, in some neutral and theory-less place: there isn't one. It involves staying in some other and alternative theory which is not, and cannot be, finally proved either. But while coming to recognise this as the case is a perfect example of what development from an assimilative to an accommodative stance vis-à-vis reality is all about, it in no way follows from such recognition that we have to resort to the 'everything's-relative' and 'science-makes-no progress' positions so widespread today. The demand for an absolute and all-time proof of a theory is an immature demand: it is assimilative; it is trying to grasp sunbeams; it can't be met. But that does not mean that the mature and accommodative demand that a theory be well validated cannot be met either. It can. And the extent of that validation, perhaps particularly the extent to which it coheres with other well-validated theories, can and must be used as reason for preferring one theory over another.

The mistake has been to believe that if our theories are socially constructed (which they are) and thus historically contingent (which they also are) that must mean that there is a problem concerning their truth. It must mean that there is no way of deciding which of them are true and which of them are false: which of them are more valid, which of them are less. But it seems to me that there is not much of a problem here. To have solved it, wouldn't one simply have to demonstrate that there are some historically contingent social structures with a tendency to produce true theories, and other equally historically contingent social structures with a tendency to produce false theories? Surely it doesn't follow that because all theories emerge from the historically contingent, they all emerge from the *same* historically contingent with the *same* likelihood of offering up truth and falsehood? Only if we misread 'contingent' as 'arbitrary'—if we totally anachronistically see the fact that Aristotle did not elaborate the general theory of relativity as just happenstance, for example—are we forced into this murky corner. If instead we study the correlation between 'degree of success of theory' (as measured by 'does it explain?' and 'does it cohere with others which explain?' criteria) and 'type of social structure that produced it', we can both have our cake and eat it. And later I will be developing the case that there certainly are characteristic features of social structures which make them either more or less conducive of scientific thinking.

If you think that in the above I have fudged the issue concerning truth, that I have written of that which is true and that which is well validated as if they were one and the same, I have. Because I think they are. It is no use to argue that our statements have to be either logical truths or empirically verifiable. The statement 'our statements have to be either logical truths or empirically verifiable' is itself neither. And so are most of the others we utter. But 'truth' is a concept no less theory-dependent than any other. And the theory of that concept which makes most sense to me is the one which insists that 'truth' is of us: that if the whole human race dropped dead simultaneously two seconds from now, there would no longer be any truth. There would still be a caused way that things are. There would still be reality. But there would not be truth. For truth is not 'the caused way that things are', but an understanding of 'the caused way that things are'. And because there would no longer be any understanders, there would no longer be any truth. In other words, because the knowing is part of the being true, the theory of truth with which I will work, the one which seems to me most closely to cohere with the demands of the scientific method, is the one in accordance with which something cannot be true but not known to be true.

Turning now to a consideration of the practicality of a good theory, my own defence of such a claim—I don't know if it would have been Kurt Lewin's—would yet again home in on the nature of explanation, but this time somewhat differently. A good theory, by definition offering the best explanation to date, offers also that which maximally coheres with other theories. I take my illustration of this point from work of the physicist David Deutsch, who offers the thought that the theory of evolution and the theory that fossil dinosaurs were created *as fossils* and put into the rocks *as fossils* are both equally consistent with the facts. But it is not just the hugely greater explanatory power of the former theory which makes it the better bet: it also has the practical advantage of inter-theoretical coherence. What this means is that the more basic, the more numerous, the more well-validated are the theories which a discovery contradicts, the more well-secured does that discovery need to be if it is to be taken seriously—or at all—by the scientific community: so that homeopathy versus the second law of thermodynamics is just 'no contest'. It is above all this characteristic of a theory, what above I called 'the extent to which it coheres with other well-validated theories', which matters. It is the extent to which science is a fundamental unity, a mesh of inter-related and inter-dependent observations, hypotheses, experiments and (particularly) theories, that makes it possible to say, as perhaps again it was Einstein who said, that if the theory does not fit the facts, we should change the facts. He was serious. It is an approach so eminently practical.

For a good theory defends us against the barmy. And the barmy is almost always *practically* dangerous. Think just of the barminess of the most common, popular and televisual approach to the history of science, for example. Nourishing us on the gruel of 'great-man' history, as spiced with the sauce of 'cult-of-personality' journalism, it more often presents us with the outstanding individual scientist than with the whole scientific community. As an amusing but nonetheless serious example of such things, it was shortly after Einstein's death in 1955 that a cartoon in the Washington Post showed the earth among its surrounding planets, but distinguished from them by carrying the memorial plaque: 'Albert Einstein Lived Here'. This is barmy. Just as it does not really matter whether he said, or did not say, the words cautiously attributed to him in the previous paragraph, it does not really matter that Albert Einstein lived here. What matters, and matters enormously, is that the scientific method developed here; that its development required the development of the scientific community; and that the mesh of inter-related and inter-dependent observations, hypotheses, experiments and (particularly) theories developed by that community liberates us, and practically, from ignorance.

And a final observation under this 'theories' head. Later I will want to consider the significance of the widespread impatience with theory, especially with large-scale and general theory: particularly as it is found among conservative literary intellectuals; more generally as it is found among most of those thinking within the Anglo-American empirical and pragmatic traditions. I will be wanting to insist that the 'let-the-facts-speak-for-themselves' position is just that: a position; a theoretical position. I will be wanting to insist that the 'let's-use-common-sense-rather-than-all-this-theory' position is just that: a position; a theoretical position. I will be wanting to insist that the 'down-with-theory' position is just that: a position; a theoretical position. In short, I will be wanting to insist that these are ideologies. But it is perhaps here appropriate to preface such stuff by saying that there just might also be an element, if a lesser element, of unwillingness, or even incapacity, in such stances. It requires considerable effort and application to come to understand a systematic body of thought. And, quite apart from not having the time, not everybody has yet developed either the inclination or the capacity to undertake it.

12. A very well validated theory, or cluster of theories, within which almost all the relevant scientists conduct their work is called a paradigm

'Paradigm' is, famously, the language of Thomas Kuhn. He used the word to get hold of what might be thought of as a theory of theories: something not yet a theory of everything, but in its 'universe of discourse' – to mix my references – sufficiently large and all embracing to pull together and unify a huge area of knowledge. Since he wrote, as has frequently been pointed out, at least two new paradigms have come into being: in biology, molecular genetics; in earth sciences, plate tectonics. I want to say no more now about paradigms other than that it is frequently suggested that psychology, my own area of interest and concern, is still very much pre-paradigmatic. To the extent to which this is true, my case is that it is a less than adequate theory of the person which holds us back. And that what I will be trying to offer is something which might help to solve that problem.

* * *

In these twelve numbered points, I have tried to outline what I take to be basic to the scientific method. In summary, I would say that it is a method which involves a four-stage process of reasoning. First the observation of phenomena; then, using induction, the generation of hypotheses which might explain any regularities so observed; then, using deduction, the derivation from the hypotheses of testable statements which would be valid if the hypothesis concerned were valid; then the construction of experiments, or contrived experiences, which could validate (or fail to validate) the statement, and thus indirectly the hypothesis. A good theory, a theory such as the theory of evolution, is a well validated hypothesis.

So now, in concluding this chapter, I must return, if only very briefly, to where I began it. I must just glance again at what it was that was made by the great Axial thinkers of two thousand and more years ago. And do no more than gesture towards some of the ways in which the scientific method as here outlined has steadily unpicked their creation. In doing so, I should perhaps first make it clear that although 'the scientific method' has been summarised here as at one point in time – now – it has of course been, and is, an historically accumulating understanding.

Many are now the books in which a detailed application of the scientific method to our understanding of the five losses of the previous chapter – the losses of Christ, of Jesus, of Christianity, of religion and of God – make such losses intelligible, understandable, even inevitable. Much of the scholarship involves putting events into historical, and thus relative, perspective: for (more or less random and tiny) examples, *asking*, of the text 'in the beginning God created the heaven and the earth' whose is the speaking voice; *recognising*, when considering a virgin birth, that given the long tradition of unions between earthly maidens and heavenly gods of one sort or another, Mary and the Holy Spirit is a story not light years away from that of Leda and the swan; *identifying* the Eucharist with what, in *Totem and Taboo,* Freud called one of the higher motives for cannibalism, that is the belief that 'by incorporating parts of a person's body through the act of eating, one at the same time acquires the qualities possessed by him'. And so on, and so on, and so on. Perhaps now more or less complete, it has nonetheless been important work. In his essay 'What is Enlightenment?', Kant suggested that enlightenment was 'the emergence of man from his self-imposed minority', from 'his incapacity to make use of his own understanding without the guidance of another'. And together with very many other people who have thought about these things, though not perhaps with Kant himself, I think that it has been important work precisely because by far the most important 'Other' the shedding of whose guidance has marked the whole journey of modernity, is that 'Other' called God. All of which means that, as I write now, I think I can more or less take it for granted that most readers know these things, accept these things, have got used to living with these things.

And yet. I also think that a residual 'and yet' stance in relation to belief in God is both widespread and important. I suggested in the previous chapter that it is a stance which lingers for both practical and theoretical reasons. And to develop my case vis-à-vis the theoretical reasons a little more, let me offer two points of further elaboration: the one concerning an inadequate *understanding* (of the theories of the natural sciences), the other concerning an inadequate *development* (of the theories of the psychological and moral sciences).

On the first let me comment but briefly. Those great Axial thinkers made their God. They made him all powerful, all knowing, and all good: and in doing so, they made, they thought, a theory of everything. It was a spectacular creation. And in true Kuhnian style, it has lasted, albeit tattered now and well beyond its sell-by date, until there can be something somewhat better to replace it. As I have already suggested, in terms of the physical sciences, there is now that something. And that means that defences of the God of faith made on those grounds are now just errors. That they also have increasingly now the flavour of being 'just errors' I will illustrate by considering just two of the most commonly quoted of them.

One is the error of the 'just right' universe. When we consider the probably infinite number of possible universes, maybe all but one of them, this one, 'just wrong' for sustaining us, the quite staggeringly unlikely actualisation of this 'just right' one, so the argument goes, must require a miraculous explanation, a 'God' explanation. And so it must: until, that is, the historically accumulating understanding of scientific method gets us to the point of realising that the actualisation of any of the probably infinite number of alternative possible universes would have been equally staggeringly unlikely. And that it is only because we are of the one that made it off the drawing board, so to speak, that we think of it as somehow uniquely so. In simple statistical terms, it is in fact no more and no less unlikely than any of the 'just wrong' possibilities. The psychology of superstition relies heavily on ignorance of statistics: things such as belief in telepathy or the predictive capacity of dreams, for example, rely hugely on an underestimation of the likelihood of coincidence. As mathematicians, psychologists and statisticians are so frequently having to point out, our intuitions about probability are not good because our brains are designed to pick out things that do happen rather than things that don't. But as Galileo might have muttered it: 'they don't happen anyway!'

The other is the 'Ockham' error. In terms of parsimony, isn't it just spectacularly simple to explain everything with just one concept? The God concept, carefully construed, will do that for us. It will explain everything: until, that is, the historically accumulating understanding of scientific method gets us to the point of realising that an outside-the-system something which explains an inside-the-system everything actually explains nothing. Using the scientific method to give ourselves explanations requires us to manipulate variables: there have to be things with which we can compare other things. And in such a scheme, 'everything', if it does not include God, is not 'everything'.

On the second and much more important reason why a residual 'and yet' stance in relation to religious belief is so widespread, the one concerning our failure as yet to apply the scientific method to the study of persons, the whole of my book is my comment. I think of science as rational, reasonable and open: there is a scientific integrity; truth telling is at a premium. For these and other reasons, it seems to me that it has a claim to being our most morally demanding and acceptable activity. It is certainly our most successful. But its success has to some extent depended on the (relative) modesty of its aims. To date, scientists have not so often set out to answer the (relatively) difficult questions such as 'what is morality?', but rather the (relatively) easy ones such as 'what is the mass of the universe?'. Can we use the methods of this 'our most morally demanding and acceptable activity' to understand that very morality itself? As once we shifted from natural philosophy to natural science, can we now shift from moral philosophy to moral science? I will suggest that we can.

Returning then at its end to where I began this chapter, in *Taking Leave of God* (1980) Don Cupitt suggested that God is 'the religious concern reified'. And by way of introduction to all of the rest of my book—as also using, somewhat ahead of its careful definition in a later chapter, some of what will be my very particular terminology—I am going to suggest that it was because they reified themselves into I AM persons that the great Axial thinkers reified their religious concern into an I AM God; that by ridding ourselves of the first reification, we can rid ourselves also of the second; that this will be a move which will leave us free to develop a new and this time scientific understanding of what an I is; that armed with such a new and scientific understanding, we will be able fruitfully to reconstruct our concept 'person'; and that armed with such a new and scientific understanding, we will be able fruitfully to reconstruct our concept 'God'. I will argue that while the old God concept offered only an *ideological* formulation of that idea which is immune to a relativising critique, the new God concept can offer its *scientific* formulation. And I will suggest that, as a consequence of that *re*formulation, we will at last be able to ground a *science* of the good; at last be able to ground a *science* of morality.

In turning now to the second part of my book, I am going to undertake, across its six chapters, a scientific investigation. Following the outline I have here provided as to what I take the scientific method to be—that is, dealing in turn with observation, hypothesis, experiment and theory—my study will be of a phenomenon I will call 'philosophical conversation'. And when it is complete, I hope I will have convinced my readers that the new way of thinking about 'I' which by then I will have developed has enormous potential. It was Tolstoy who claimed that while it was his view that the question 'what shall we do, and how shall we live?' was 'the only question important to us', it was also his view that 'science is meaningless because it can give us no answer to our question'. The argument of part two of my book is that, sourced in a reformulation of our concept 'I', it most certainly can.

PART TWO

Chapter Four

OBSERVATION

So many 'ands' and so many 'us-es'

Being an on-going process rather than a thing, science does not really 'begin' anywhere: certainly like most other activities, perhaps like all other activities, it is irreducibly an 'always already' undertaking. And that presents me with a problem. If I am to make my study of philosophical conversation a scientific one, where do I begin? As summarised by Chalmers—in *What Is This Thing Called Science?* (1999)—it is now almost universally accepted by philosophers of science that positivism is irredeemably flawed because, as I pointed out in the previous chapter, 'observation statements must be made in the language of some theory': in other words, the 'observation' with which orthodox scientific method has traditionally claimed that it begins is observation necessarily from out some 'always already' (that is, prior) theoretical perspective. Or as Goethe put it: 'with every attentive look at the world we are theorising'. But equally irreducibly, if thoughts in a head are to result in words on a page, those words have to begin somewhere. And so given that, to the extent to which I find it possible, I want to engage in a study which respects scientific procedures, I am going to go along with what I take to be orthodox scientific method, and 'begin' with observation. The fact that, like all observation, mine also will be theory laden is an issue to which I will later return (see Chapter 9).

The object of my study, then, is philosophical conversation. In this chapter I will be *observing* philosophical conversation; in the next, offering an *hypothesis* concerning philosophical conversation; in the next three, conducting some *experiments* on philosophical conversation; and in the next, coming up with a *theory* about philosophical conversation. And although some readers might now expect me to begin with some sort of definition of just what I mean by philosophical conversation—particularly, perhaps, how I think it differs from just 'philosophy'—that is not my intention. It seems to me that definitions, and even then only provisional ones, come rather more easily towards the end of an initial airing of some of the issues involved in their formulation rather than at its outset. I will therefore begin with some such airings. And let the definition of the object of my study just emerge across the next few pages. It is difficult enough to say just exactly what philosophy itself is. Perhaps many of us might be happy to kick off with something like 'the subject that tackles all the *big* issues', or 'the subject that asks all the *big* questions'. And at its extremes, where anyway it would seem that general interest not infrequently settles, both the issues and the questions are certainly considerable. The former culminate in those of the cartoonist's 'life-the-universe-and-everything' sort: the latter in Leibniz's 'why-is-there-something-rather-than-nothing?' variety. Indeed, some wag has actually defined the philosopher as one who, had there actually been nothing rather than something, would still have wanted to know why: a question at one and the same time both barmy enough to raise a giggle, and acute enough to draw attention to the fact that it is surely not obvious that 'nothing' should be thought of as a less likely state of affairs, and thus less in need of explanation, than 'something'.

For professional philosophers, the big issues and big questions of their discipline have sometimes been grouped under four heads (Scruton, 1994): logic (the study of reasoning); epistemology (the theory of knowledge); metaphysics (the theory of being); and ethics and aesthetics (the theory of value). But as a non-professional, I have usually found it more revealing, anyway when thinking about philosophical conversation, to use a three-way rather than a four-way division: a division into the study of the good (ethics), the true (epistemology), and the beautiful (aesthetics). But wherever its boundary divisions and sub-divisions are to lie, if concern with the big issues and questions is what in some way constitutes philosophy, then perhaps the first thing to note is that the value of such concern, even the very legitimacy of such concern, is currently very much in question.

When that accolade was not being given to mathematics, philosophy used sometimes to be referred to as 'the queen of the sciences'. Leaving aside the somewhat odd use of language here—neither mathematics nor philosophy are today thought of as 'sciences' at all—such reference conveys something of the extent to which the subject, and its practitioners, were once revered. That is a reverence now more or less completely gone. The renowned Danish physicist Niels Bohr claims to have made 'the great discovery that all that philosophers have ever written is pure drivel'. The distinguished British geneticist Steve Jones, with echoes perhaps of Marx's comparison between philosophising and masturbating, suggests that 'philosophy is to science what pornography is to sex'. Even among its own practitioners, large chunks of its former content have been, or are being, rejected. Studying metaphysics, for example, once famously described as being like hunting for a black cat in a dark room where there is no cat, is regarded as of more than doubtful value by many of them. Indeed all those theories of everything (if you like the grand manner) or bits of joined-up thinking (if you don't) which have arisen much more often from amongst philosophers than from amongst most other groups, are currently under a dual, and some might say fatal, attack: specifically from scientists claiming to have taken over the elevated role of their construction for themselves; more generally from postmodernists, some of course themselves philosophers, propounding just sheer incredulity in the face of grand narratives of any variety. It is at least arguable that these are all developments with a tendency to undermine confident philosophical reach.

There is an 'endism' in the air. In his now notorious article 'The End of History' (1989), Francis Fukuyama pronounced that end, in anyway an Hegelian sense, as the end of the search for new and better forms of social structure. In so doing, he suggested that from now on it would be 'impossible to state a philosophical proposition that was both true and new'. And he has been followed by others: by John Horgan, for example, proclaiming *The End of Science* (1996), and by George Pattison announcing *The End of Theology* (1999). I don't know of any book called *The End of Philosophy*. But it would certainly not surprise me to find either that one already exists, or that one appears in the not too distant future. And while, if it did, it would no doubt be consciously using such an overblown title at least in part for effect (or sales), I think that by so chiming with an 'endist' zeitgeist, it would be pointing to something real about the current state of the discipline. There is plenty of evidence to support the view that philosophy currently knows something of crisis.

There is the long-running and seemingly irreconcilable split between 'analyticals' and 'continentals': a split seen by some (such as the reformed analytical Willard Quine) as one between those who see the right place for philosophy as being within science and those who see the right place for philosophy as being within metaphysical speculation; a split seen by others (such as the 'continental' Etienne Balibar) as one between those who have a model of science based in axiomatic theory and those who have a model of science based in proof. There is the major contraction of university philosophy departments: primarily in countries schooled in the Anglo-American tradition, but also elsewhere. There is the (belated?) recognition, Kuhn-led but perhaps Koyre-derived, that you can't do philosophy, or any other subject, as if its questions were timelessly a-historical: as if, as one Kuhn obituary put it, the lives of such as Aristotle, Kant, Hume or Leibniz had somehow been devoted to the puzzles generated by mid-twentieth century Oxford philosophers. There is the collapse, as the result of a Quine-led undermining of some of its major premises, of traditional analytical philosophy. There is what can sometimes seem to be the sinking of continental philosophy under a sea of incomprehensible jargon. In the light of massive scientific advance, there is even a very active doubting as to whether the discipline can be defined in such a way as to make it a legitimate enterprise at all. Philosophy, academic style, is not without its problems.

But this is not all. If it were merely that an academic discipline was in crisis, even that it was cracking apart, that might be no bad thing. Boundaries between disciplines do change. And when they do, it is at least as often a sign of intellectual progress as of intellectual retreat. If the still-deemed-legitimate concerns of academic philosophy were increasingly being addressed in places other than philosophy departments—in psychology departments, cognitive science departments, neurology departments, wherever—then to anybody other than would-be academic empire builders, that could well be a difference that makes no difference at all. And at least to some extent, maybe that is what is happening.

But if it is, something else is happening as well. The big issues, the big questions, are of enormous interest and concern to almost all of us. At some time or other, we all talk avidly about them. And it is this talking, this philosophical conversation, which I am going to study. In doing so, I will want to suggest that, whatever it is that is happening in the academy, philosophical conversation too shows signs of being in some way in crisis, in some way stuck. I don't think there's evidence of its being in any way in retreat. In pubs and clubs, on television and radio, in office and laboratory, 'travelling theories'—the phrase is that of Edward Said, and he is referring, somewhat dismissively I think, to the ideas which are passed around among those he sees as 'would-be' intellectuals—are still travelling. Indeed in Britain anyway, the last few years have seen something of a blossoming of 'philosophy cafes', 'philosophy for all' groups, glossy philosophy magazines, glossier television programmes. But it is certainly my experience that however extensive such philosophical conversation might still be, it is often to be found getting stuck in much the same sorts of grooves as those frequented by its academic fellow travellers. It is that 'stuckness' which I want to consider: initially to characterise; then to explore. But before I do so, I must first try to make one thing as abundantly clear as I can.

In case I could be misunderstood, it is vital that I clearly distance myself from something I regard as a serious disease process. I can't abide anti-intellectualism. Of all the fearful thoughts that keep us brutish, conservative and authoritarian, the fear of very thought itself is surely the greatest. How else would it be possible for 'intellectual' to have become a dirty word? If what is called 'the real world' is always posited as being somewhere other than in institutions of learning, we are in deep trouble. If puny common sense—that common sense which, in his poem *The Death of Alfred Hubbard*, Tony Connor so aptly designates as 'rules of thumb which were often wrong'—is allowed to masquerade as science or as theory, we are in deep trouble. If common nonsense is not understood to be every bit as common as common sense, we are in deep trouble. For if such attitudes prevail and go unchallenged, then a vital part of the human enterprise, the part that is about the steady accumulation of well validated knowledge to replace mere opinion, is under serious threat. I stress this in order to make it clear that in making a distinction between the philosophy which goes on under that name in the academy, and the philosophical conversation in which we all take part, I am in no way trying to indulge in that anti-intellectualism which I have heard designated, maybe with some justification, as 'the English disease'.

As anyway they pertain to philosophers, the symptoms of the disease, not infrequently expressed in the form of accusations, are several. They are accused of writing their books using a professional vocabulary sometimes difficult for the rest of us to grasp. But then where it is the case that such books are those 'whose difficulty has offended indolence' – to borrow my words from Yeats – we would perhaps do better to tackle the indolence than take the offence. They are accused – by Adam Smith in *The Wealth of Nations* (1776) – of being 'by nature – not in genius and disposition half so different from a street porter as a mastiff is from a greyhound'. But then Marx's huge enjoyment of the quip notwithstanding, he himself was just surely wrong to privilege changing the world over interpreting it. Some of them are accused – by Brenda Almond in *Philosophy Now* (issue 24, Summer 1999) – of advocating a return to 'ordinary language' as the arbiter of meaning, while nonetheless being 'inclined to treat with contempt – the ordinary person's understanding of what philosophy is'. But then not many philosophers would even claim, with A. J. Ayer, that they were 'ordinary language' types. And many of them are accused – as was the maybe apocryphal Oxford specimen who gave ten lectures on 'the meaning of life' and spent nine of them on the semantics of the word 'life' – of pedantry or worse. But then, as Ian Glynn points out in *An Anatomy of Thought* (1999), it doesn't necessarily follow that being able to deploy words appropriately implies any sophisticated understanding of the concepts they label, as neither does it necessarily follow that being able to share the conventions of intelligible conversation implies that those conventions are well-grounded: for as long as we are all confused in the same way, our conversation can proceed regardless. In summary, the point I am trying to make by using these few examples is that when academic and professional philosophers have something to say about the things we all talk about in the course of our philosophical conversation, then it probably does – at least sometimes – behove us to listen.

None of which however means that there are not also some important reasons for us to listen, but not necessarily to concur. For although the issues listed above as examples are not they, I do think that there are issues more serious to consider concerning the role of academic and professional philosophers in what I am calling philosophical conversation. And because they are issues which could mean that their contribution to such conversation might well turn out to be somewhat less than either they or others might imagine, they need addressing. So: before proceeding to an attempted definition of the 'philosophical conversation' I intend to study, let me give a brief airing to just four of them.

First, there is the unfortunate tendency in some academic places, but by no means only in philosophical academic places, to equate 'academic' and 'intellectual', and then to tie the 'would-be' label around the necks of those of the latter who are of the non-academic variety. This, where it occurs, is just nasty. It is pretension on behalf of those (many) academics who are not intellectuals: it is contempt for those (many) intellectuals who are not academics. And in studying philosophical conversation, I shall make it my business to have none of it.

Second, there is the question of the social base of academic philosophers. Going back only fifty years, Jonathon Ree has pointed out that the twenty leading Oxford philosophers of the 1950s could muster only one of their number who was not from what he called 'a high-bourgeois family'. In addition, only four of them were women; the sixteen men had all attended private and famous boy's schools; and all but one—who went to Cambridge—were undergraduates of Oxford. Ree's conclusion, which was that, even for Oxford, this was a spectacularly narrow social base, is surely not exaggerated. This, rather obviously, was not a 'view from nowhere'. And it could mean that the sociology of philosophy, perhaps a late developer among academic disciplines, will make interesting reading when it arrives. Of course this was fifty years ago. Of course it concerns just one academic institution. But fifty years is nowhere near long enough for the influence of such bias to have worked its way totally out of the discipline. And the one institution concerned was nothing if not hugely influential in shaping the form of that discipline. So that while things have clearly changed to some extent, if it could be shown, as I no more than suspect that it could, that the social base of today's philosophers, despite being wider, is not yet as wide as that in other disciplines, then that could well be something which would need to be registered as their contribution to philosophical conversation was weighed. Registered: but also well-considered. For C. S. Lewis's 'bulverism', the phenomenon whereby well-founded ideas are inappropriately rejected just because the people putting them forward have quite clear reasons for wanting them to be true, is also to be guarded against.

Third, there are their aims and objectives. As has often been pointed out, by the Oxford philosopher of history R. G. Collingwood among others, while new and naïve philosophy students might expect their chosen subject to provide them with ideals to live for, principles to live by, their professional teachers have almost always explicitly rejected any such aims. As I am positing it, and by critical contrast, philosophical conversation quite unambiguously *does* have such aims. As I am positing it, for example, philosophical conversation quite unambiguously aims not only to ask the famous 'how-should-I-live?' question of Socrates, but also to answer it. And to answer it, what's more, not only for the speaking 'I' but for all other 'Ies' as well! To the extent to which the academics reject such an approach, then to that extent *they* rule *themselves* out of any special role in philosophical conversation, as (or even *qua*) professional philosophers.

Fourth and last, there is the question of what is often called their 'grounding in the literature'. 'The literature' concerned consists in the papers and books of their academic peers. And if you want to get a quick insight into the narrowness of focus which that implies, look at some of the popular histories of philosophy: at the work of Bertrand Russell or Roger Scruton, for examples. Thoroughly internalist in stance, these are 'histories' which are not more than a chronological trudge through some selected writings of 'the big names'. Little or no attempt is made to situate what is offered in some wider economic, political, social or cultural context, with the consequence that, however thorough the 'grounding' in such 'literature' might be, it in no way necessarily implies an equally thorough grounding in the problems of the world which threw it up. This second grounding is one which has to be demonstrated to be the case rather than just assumed to be the case. And in making philosophical conversation rather than philosophy the object of my study, while I will use academic and professional contribution when it seems appropriate—which will be often—I will try never to make that totally illegitimate assumption. One consequence of such a stance, and a consequence which by now will probably be becoming only too apparent, is that I will be using, unashamedly and deliberately, what in the trade are known as 'secondary sources'. And 'popular' ones to boot. I will use them because that is often where philosophical conversation, as opposed to academic philosophy, is to be found. But I will also use them because that second and vital 'grounding' is certainly no less likely to be found, and occasionally rather more likely to be found, in such alternative places.

So now to an attempted definition of the object of my study: of philosophical conversation. I found the basis of the one I will use in a book by Charles Gore, the then Bishop of Oxford. In *The Philosophy of the Good Life* (1935), a record of his Gifford Lectures of 1929, he wrote:

> But if a 'philosopher'—that is, a lover of wisdom—means a man whose spirit can find no rest unless he can gain and keep some 'theory' or vision of the world of things and experiences, such as shall enable him to interpret its manifold phenomena as parts of one whole, and as expressive in some sort of one purpose, in which he himself is called to cooperate with

> will and intelligence—if that be the meaning of a philosopher, then, though defective knowledge and capacity may render me a poor specimen of the class, certainly I am a philosopher.

In defining philosophical conversation, I will base myself in what I anyway take to be an updated, modified and extended version of this comment from Gore. After him, I will use the word 'philosopher' (don't ask for the meaning, ask for the use) to mean a person seeking to gain and maintain a unified vision of the world of things and experiences, included in which is a vision of her or his self as seeker, which, if found, would enable her or him to answer the 'how-should-I-live?' question of Socrates for herself or himself, and for all other selves. And that will mean that, using now the jargon of scientific method, I will be making the term philosophical conversation *operational* as 'the conversation of persons seeking to gain and maintain a unified vision of the world of things and experiences, included in which is a vision of their selves as seekers, which if found would enable them to answer the "how-should-I-live?" question of Socrates for themselves, and for all other selves'. And with little fear of contradiction, let me suggest right here at the outset that such a definition means that (almost) all persons take part in philosophical conversation some of the time, and that some persons take part in philosophical conversation (almost) all of the time.

The significance of some of the aspects of my definition will emerge only as my study progresses. But I want here to draw immediate if brief attention to just four of its features. First, I want to stress 'seeking': my definition implies that, as Gore himself puts it, 'to acquiesce in the sceptical attitude which is content to find all views interesting, while abandoning the attempt to reach a conclusion or conviction of ones own, is to abandon the very aim of reason, which is the conviction of truth'. Second, I want to stress 'maintain': my definition implies that J. D. Bernal was all too right when, in *Science in History* (1954) he insisted that 'all that the neglect of philosophy will do is to hide a great deal of very bad, outworn and untenable philosophy'. Third, I want to stress 'unified': my definition implies that, when taking part in philosophical conversation, we reach not after being among Isaiah Berlin's foxes, who knew many small things, but among his hedgehogs, who knew one big thing. And fourth, I want to stress 'things and experiences': my definition implies here a duality which will take on some considerable significance to the case I want to make.

My final point before launching into some observations concerning my chosen object of study is one about its importance: about the importance of philosophical conversation. Why should it be worth studying? What do we get from it when it is good and miss in it when it is bad, or non-existent? When two or more are gathered together for more than minutes, why does some of it, however tenuously, almost always arise? I hope to be able to show that what we get from it is our world view, our basic understanding of how things are. And that the huge importance of acquiring a coherent world view, a coherent basic understanding, is readily recognised when we hear ourselves discussing without one. Such discussion has often a sort of mad eclecticism about it. Arguments are clutched at, higgledy-piggledy, from here and there, from everywhere and nowhere. Points are scored rather than made. The partialities of interest are often so blatantly and patently at work as to be embarrassing, not least to their 'owners', whose rising anger is as much with their own transparently partial selves as with those of others. These are the tensions that all of us surely know. And if we are to escape them, to grow straight in our understandings, then good philosophical conversation is what we need.

And perhaps particularly now. For we are at the moment veritably inundated with 'posts': we are post-this, post-that, post-the-other, post-the-lot. Together with the endism already mentioned, this is a 'postism' which means, for an example, that the term 'concept' is as likely to refer to the sales promotion of marketable gimmicks as it is to general, philosophical ideas; which means, for another, that with Stephen Dedalus 'the big words make us so unhappy' because we can no longer make them make any sense. Such a situation is one in which we are particularly in need of good philosophical conversation. Almost always more difficult than to find their solutions, we need good philosophical conversation to enable us to see the problems clearly. We need it because so much of received wisdom, because it is no longer thought to be wise, is no longer being received. We need it to extend our understanding that there is nothing necessarily sloppy about being amateurs. We need it because when Kant summarised the Enlightenment project with the words *sapere aude*, what we were to dare to do was not just to think, but each to think for ourselves. And given that we need it, it is important that we identify its nature, find out in what it consists, discover what hampers its development and fluency: in short, study it.

Let me then kick off my own contribution to that study by suggesting that, as a first approximation, one might be forgiven for thinking that philosophical conversation, even very philosophy itself, consists in nothing but 'ands'. 'Nature *and* nurture' 'free will *and* determinism', 'good *and* evil', 'freedom *and* equality', 'individual *and* society', 'right *and* wrong', 'appearance *and* reality', 'praise *and* blame', 'rights *and* duties', 'nature *and* culture', 'reason *and* emotion': is there any of the subject-matter of either philosophy or philosophical conversation which is *not* amenable to being thus additively summarised? 'Selfishness *and* altruism', 'ends *and* means', 'time *and* space', 'is *and* ought', 'identity *and* relationship': such 'and' pairs go – 'on *and* on'! Some of them, 'is *and* ought', for example, seem primarily the concern of the professionals: others of them, 'nature *and* nurture' perhaps, much more often the concern of those taking part in philosophical conversation. Not that there would not be a considerable overlap between the two: only some of the vocabulary of academic philosophy has been especially invented for the purpose; much more has begun its existence in philosophical conversation, and then perhaps been bifurcated such that it continues to flourish among all of us, but is also taken up, and maybe sophisticated, by the academics.

Now 'ands' are interesting things. 'The additive mode'—I borrow the phrase from Ong's *Orality and Literacy* (1982)—is characteristic, he there suggests, of pre-literate people. The transition from oral cultures to literate cultures, he argues, is a transition in language use (and thus in thinking) from the aggregative to the analytical; from the copious to the redundant; from the conservative to the creative; from the personal to the impersonal; from the agonistic to the co-operative; from the participatory to the distanced; from the present-oriented to the past/future aware; from the situational to the abstract: *and*—from the additive to the subordinative. The pre-literate persons who are small children, for example, will take their teddy *and* her hat *and* their breakfast *and* a spoon *and* a stool for teddy *and* a stool for me *and* some jam—breathless by now!—into the garden. They take thus additively because they are not yet capable of taking in a more sophisticated, subordinative fashion: of taking their teddy perhaps *wearing* her hat, their breakfast maybe *including* a spoon and some jam, and a stool *each*. The additive mode speaks of only a very basic classificatory system: there's this *and* there's that *and* there's the other thing. But how the relationships between them might be more fully structured, how one member of an 'and' pair might come to be preferred over the other, or subordinated to the other, for example, is not yet understood.

So is this also the case in philosophy and in philosophical conversation? Does their use of so many 'and' pairs suggest a certain primitiveness about them too? In the preface to the second edition of the *Critique of Pure Reason* (1787), Kant said that he aimed so to reform metaphysics that it would cease to be possible for metaphysicians to construct argument pairs—it was four such with which, in particular, he was concerned—such that although the reasoning of both was sound, their conclusions were mutually contradictory. Do such 'and' pairs as still litter philosophy and philosophical conversation bear a relationship to these his 'antinomies'? And has he, or anyone since, succeeded in bringing about the 'reform' to which he referred? Initially for almost no other reason than that it was possible to do so, I decided to begin my study of philosophical conversation with a very simple investigation of its 'ands'. According to at least one of its more distinguished practitioners, science is after all the art of the soluble. And starting with 'ands' solved my problem of where to begin.

I needed a list. And because it is both more restricted and better documented, I turned first to the territory of the academic philosophers, only later to the realm of philosophical conversation, to find one. Now lest I be accused of pedantry or worse, let me point out again that I would like this to be, as far as is possible, a *scientific* study of philosophical conversation. My list should therefore be thought of as the list of all my 'specimens'. To accumulate it, I went to *The Oxford Companion to Philosophy* (1995) and recorded all the 'this *and* that' entries it had to offer. Allowing myself to include four where it seemed to me reasonable to read a dash as in no way significantly different from an 'and', I finished up with a ninety-two item list. In its occasionally bizarre completeness—one entry, for example, reads 'dictionaries *and* encyclopaedias of philosophy'—it is to be found in Appendix A. It constitutes my raw data. And I have two reasons for starting with it.

First, it is both available and reasonably current. Of course *The Oxford Companion to Philosophy* is only one of several such texts. And while its dust-cover blurb claims that it is 'the most authoritative and engaging philosophical reference book in English', I am of course aware that that amounts to no more than being spoken of highly in its adverts. However, I did pay some careful attention to reviews less partial, and decided that it could probably stand as reasonably distinguished. Second, it is a huge time-saver. Simply by referring readers to their source in *The Oxford Companion*, I am saved from having to produce a (rather large!) supplement to what I write summarising the essence of the issues raised by each of these 'ands'. And this is doubly important. It is important because many of my readers will no doubt already be familiar with the main outlines of the philosophical issues lurking under most of them: and if they aren't, they can check them out. And it is important because the great bulk of these my specimens are about to be jettisoned, never to be encountered again in this book. It will be only when my list has been whittled down to those 'ands' which I will be maintaining, and maintaining because I believe them to have relevance to my study of philosophical conversation in general and to my hypothesis as to why it is so additive in particular, that I will offer my own and brief exposition of the debates raging under the heads of those the then selected 'ands'.

But more or less immediately, before I can even get started, I have to trip up at the very first hurdle over a problem already identified: over the dreaded 'problem of induction' as referred to in the previous chapter. Willard Quine called it 'the *scandal* of induction'. And as anyway applied to my list, to my specimens, I understand the problem (scandal) to be this. What can possibly be the status of these ninety-two items? If I propose to use my specimens, as I do, as the source of an hypothesis as to the nature of the human condition, no less, how can I possibly defend their particularity? They are, after all, just my *particular* specimens, selected from among the entries in a *particular* encyclopaedia, compiled by a *particular* editor. A different encyclopaedia, different editors, and especially given the fact that philosophy is widely acknowledged to be a thoroughly disputatious discipline, the list could be more or less guaranteed to come up different in at least some respects. In the face of such startlingly obvious difficulties, and consonant with my points concerning logic and reason in the previous chapter, this will be my strategy. When in due course I arrive at that list which it seems to me would allow much of the subject matter of philosophical conversation to be written up under its 'and' headings—and that will mean compiling later a second list of 'non-professional' pairs to add to the first—I will accept that it must be a list capable of passing the following test: that there could not be drawn up any other list which could be defended as allowing significantly more of the subject matter of philosophical conversation to be written up under its heads than could be written up under mine. To the extent to which it is possible, this would make the issue an empirical one: it would be open to others to construct and use such an alternative list. Of course I know this is in no way sufficient fully to get me off the 'problem-of-induction' hook. While nobody might challenge my list with another having such characteristics, so what? Nobody might want to be bothered to try. Or nobody might now, but somebody might then, and where 'then' could be read as 'any time'. These are all questions reflecting the more general 'problem of induction' towards which my stance, presented already in the previous chapter, is that it is a bad problem generated by a bad theory. So with the intention of returning to these issues again in Chapter 9, I can I think allow myself to proceed.

My over-riding impression of the philosophical conversation I engage in myself, of the philosophical conversation other people tell me of engaging in themselves, of the philosophical conversation I listen to or just over-hear, of the philosophical conversation I encounter in books, fiction and non-fiction, is that it is monumentally unsatisfactory. It is monumentally unsatisfactory because no matter from which specific issue conversation takes off, and often in only two or three 'moves', participants are back in the usual well-trodden 'and' grooves, rehearsing yet again the usual well-rehearsed 'and' positions, and with the usual lack of very much chance of achieving 'un-andedness'. When you're in them, the characteristic feel of these grooves is one of gridlock, or just helplessness. Here is what happens.

A good case can be made, and by somebody usually is, for the relative importance of one member of an 'and' pair: 'individualism' perhaps, when the conversation had begun by being about, say, taxation. As good a case can be made, and by somebody usually is, for the relative importance of the other member of the 'and' pair: in this case, 'anti-individualism', or maybe 'collectivism'. Philosophical battle commences. At some stage, someone who has so far said relatively little suggests that maybe there is something in both arguments, and that what we need to find is some middle ground, some compromise. This is a flabbiness almost always over-ridden—as often as not because it just *is* flabbiness—and battle continues. But then, neither side being able to muster a sufficiently convincing case to win the day, and nobody, not even the quiet soul who offered it up, really believing that the 'middle of the road' is any place other than one which ensures you'll get knocked down, the problem remains: as 'and-ish' and as vexing as before. According anyway to Etienne Balibar, there is a process which could be seen as equivalent to this one also in academic philosophy. Interviewed in *Radical Philosophy 97*, and commenting on the frequency with which philosophers have 'a first philosophy, and a second philosophy which destroys it', he suggests that this is because they choose one or other side of a debate—'individualism and holism' is the example he uses—and then 'push the real contradictions in the conjuncture to the extreme'. The result of this is that they 'usually succeed in destroying their original position'. Hence the necessity of their 'second philosophy'.

Perhaps I exaggerate slightly. I should not like to give the impression that, in the philosophical conversations I know about, no progress is ever made. That would not be true. But in my experience, in the pub, over the dining table *and* in the supposedly more rarefied atmosphere of the academic conference, some fundamental 'something' over and over again gums up the works; over and over again thwarts, subverts, distresses, frustrates; over and over again leaves the floor open to whatever is the currently fashionable 'ism' which wants us to rest in defeat before the complexity of things. It is on the track of that 'something' that I want now to begin the process of sorting my 'ands': of trying to isolate, for further consideration, the ones which might enable me to understand what it is that so frequently hinders the development of philosophical conversation. To that end, beginning from my raw-data Oxford list, I propose to cook it as follows.

First, I will remove from it all those 'ands', such as the already-mentioned 'dictionaries *and* encyclopaedias of philosophy', which offer no resource at all to my study of philosophical conversation. Second, I will remove from it all those 'ands', such as 'embraced *and* reluctant desires', which denote discussions specifically academic, with no obvious corollary in philosophical conversation. Third, I will add to it some 'ands', such as 'nature *and* nurture', which figure considerably in philosophical conversation but have no Oxford entry. And finally, I will whittle it down to a list of manageable length by maintaining only those 'ands' which figure in such conversation not occasionally but repeatedly. Each of these steps is based, of course, on only my own judgement. But my defence of the procedure is as outlined earlier: that it could then be an empirical question as to whether or not there might be drawn up any other such list which could be defended as allowing significantly more of the subject matter of philosophical conversation to be written up under its heads than could be written up under mine. Sorting my ninety-two-item list in this way produced the thirty-five-item one to be found in Appendix B. Twenty-two of them, asterisked, come from *The Oxford Companion* source. Thirteen of them, non-asterisked, have been selected by me.

This being now my operational list, each of its entries now *does* need an initial and brief explication. In each such, the total of which will make up the rest of this chapter, I will focus on two things: first, on what I am taking the problem presented by the pair to be; and second—where there is any difference—on what I am taking the problem presented by the pair to be *as it figures in philosophical conversation* rather than *as it figures in academic philosophy*. In what in the circumstances is the appropriate 'dictionary or encyclopaedia' form, I will use both alphabetical order and numbered points.

(1) acts and omissions (*)

If I lie and as a result you suffer, I am commonly taken to be blameworthy: if instead I don't lie, but rather fail to tell the truth, am I still so? If in my car I block the route of the ambulance to the road accident and the victim suffers, I am commonly taken to be at fault: if instead I don't hinder those who come to help, but neither do I help them, am I still so? If I kill someone, I am commonly taken to be morally culpable: if by my inaction I let someone die, am I still so? This is the way the 'acts *and* omissions' pair present themselves, to both academic philosophers and to those taking part in philosophical conversation. The argument goes back and forth on this one not because of a confusion over intended and unintended consequences. Instead, it persists because, even with intention present across the board, there seems to be a greater need to justify lying than to justify failing to tell the truth; a greater need to justify hindering than to justify failing to help; a greater need to justify killing than to justify letting die. And although it is difficult to say exactly why this seems to be the case, since it does so seem—and to very many people—discussion as to why does not easily abate.

(2) appearance and reality (*)

This is a pair established as such by Plato, but given its modern form primarily by Kant. It exemplifies the huge and very general problem of deciding what is true. If something appears to be the case to one observer but not to another, how are we to decide—*can* we even decide—which (if either) of our observers is seeing what, since Kant, are commonly called 'the things in themselves'? There is hardly a single philosophical conversation about anything at all which does not come up against this one. Amongst the various solutions on offer, there is 'coherence' theory (a statement is true if it coheres with other known-to-be-true statements); various 'intersubjective' theories (objective truth is the truth of agreeing subjectivities); 'phenomenological' approaches (that which is observed by any observer is as 'true' as that which is observed by any other observer): and various combinations of all three. I have never yet been present in a philosophical conversation where any of these 'solutions' was deemed sufficient to call a halt to a dispute.

(3) arts and sciences

As I have already suggested with the first presentation of what will be my own approach to this pair (see Chapter 3), it is one which raises many a hackle. Those involved in philosophical conversation invariably agree that art is older than science. Some pretty standard datings would take art, roughly speaking, to be as old as the decorated artefacts and cave paintings of the Cro-Magnon humans of the Upper Palaeolithic: that is, about 30,000 years old. And science, somewhat less roughly speaking, to be as old as the already referred to suggestion of Thales that the world might be made of water: that is, about 2,600 years old. But although there is widespread agreement on the facts, there is considerably less agreement on their significance. Are we to take them to mean that art is a less developed and more primitive way of knowing the world, even one which might die as scientific knowing supersedes it? It was, after all, no less a figure than Wagner who suggested that 'if we had life, we should not need art'. Is it the case, or is it not, that while what the major scientific theories have in common is that they have been validated *as* major scientific theories by general and disinterested scientific communities, the only thing that all the 'masterpieces' of art have in common is that they have been selected to *be* masterpieces by particular, interested and essentially political elites? Is Althusser right, or is he not, to contrast the pair by claiming that where art allows us to reach 'conclusions *without* premises', science requires us to understand 'conclusions *from* premises'. These are the sorts of issues which, anyway in my experience, inform and confuse most of the philosophical conversation concerning this 'and' pair.

(4) autonomy and heteronomy (*)

'Correlative terms, developed by Kant, of very wide applicability to moral theory', says our *Oxford Companion* source. And both terms, 'autonomy' as the rule of the self and 'heteronomy' as the rule of the other, are very widely used within developmental psychology in theories of moral development: theories such as the earlier and simpler one of Piaget; the later and more complex one of Kohlberg. In philosophical conversation, the questions they usually raise, often as people pore endlessly over such things as the now infamous Milgram experiments for example, are questions concerning human possibility. When for all of history there have only ever been various societies of the heteronomous, is it just wild utopianism to envisage it as being possible to create a community of the autonomous. For example, when Engels wrote of a future in which 'the government of people' would be replaced by 'the administration of things', was he serious? Could he really have believed that to be possible? It often seems to me that the reason why this one runs ands runs in philosophical conversation is something to do with the triumph of hope over experience. Most participants find it impossible to imagine such a state of affairs. But almost nobody wants to let the vision go as no more than a dream.

(5) chance and necessity

Although this pair labels not one debate but several inter-linked ding-dongs—for however they start out, 'ding-dongs' they almost always seem to become—the core problem, in effect, is that which concerns determinism. Where chance is present, how can there be cause? And where cause is not present, how can there be science—or indeed any other sort of reliable knowledge, if there *is* any other source of reliable knowledge? Things sometimes get both more sophisticated and more confusing when somebody points out that chance doesn't seem to sit comfortably within either a free-will or a determinist position. It contradicts determinism by requiring us to contemplate events not caused by other events. And it contradicts a free-will position by requiring us to contemplate events not caused by free agents either. And that is quite apart from the fact—think only of the millions spent on lottery tickets each week—that thinking about chance is notoriously difficult for human beings anyway.

(6) egoism and altruism (*)

This is one of those long-runners that every so often changes its form but not its substance—and there's another 'and' pair to think about! At the moment, in the context of quite huge developments in the biological sciences, its most common form is an argument about evolution: an argument given its perhaps unfortunate vocabulary by the 'selfish gene' title of Richard Dawkins' best-selling book. If a gene is to be understood as a necessarily selfish entity, aiming only to survive from generation unto generation, then how come altruism is ever possible? All biological entities, including human biological entities, become just vehicles for the propagation of their genes. And that means that the only way that altruism can enter the picture is via the notion of 'reciprocal altruism': a 'you-scratch-my-back-and-I'll-scratch-yours' sort of concept, developed by Trivers, which seems somehow unworthy of the name.

There would appear to be two sorts of strategies to mitigate these problems. One strategy is to challenge the current orthodoxy within evolutionary theory which says that natural selection works not at the level of the species, not at the level of the individual, but at the level of the gene. But quite apart from the small problem of the considerable evidence for the orthodoxy, I can't really see how such a challenge would help. If it is the possibility of altruism which is the problem, why should we be happier with a selfish species or a selfish individual than with a selfish gene? The other strategy is to claim that as 'uncaused first causes' or 'free agents', we are different: we can simply rise above our biology and *will* ourselves to altruism. But this is an escape route which more and more people would seem to find less and less satisfactory. I think that in 'egoism *and* altruism' discussions, the back-and-forth keeps going back-and-forth because, while most people want a way out, they don't find either of these strategies, or any other they can dream up, at all satisfactory.

(7) ends and means (*)

We are most frequently here in the territory of politics: in philosophical conversation about political change. In the academy, the argument is about whether or not 'means' can be judged bad or good based not on any characteristic intrinsic to them, but instead on their effectiveness in bringing about not bad but good 'ends'. In philosophical conversation about the political, the issues are ones such as the use of war (bad) to bring about peace (good), or the use of the restriction of liberty of the potentially violent mentally disturbed (bad) to bring about the safety of 'the public' (good). That the ends justify the means is frequently taken to be an illiberal (and therefore bad) doctrine by people who know quite well that in their next breath they will be supporting the preventive detention of terrorists. And it is because we all feel these tensions, in ourselves and in our arguments, that this once again 'ding-dong' issue is one which never seems to rest in any satisfactory resolution.

(8) fact and value (*)

One of the most basic and most troublesome grooves one can get into in philosophical conversation is that which requires clarity on the fact-value distinction, otherwise known, after Hume, as the is-ought question. The problem, at its most basic, is this: that even if what *is* the case can be successfully delineated, there does not seem to be any logically infallible way to get from that to what *ought* to be the case: 'ought says who?', or 'ought according to whom?', is perhaps the most frequent form in which the problem is encountered in everyday philosophical conversation. And if there is no satisfactory answer, then like it or not, the currently fashionable relativism seems to be the only way out. Until, that is, somebody points out that the statement 'everything's relative', being an absolute, is itself in contradiction with the relativism it propounds. And then—frustration, anger, angst!—there seems to be no way out of a considerable dilemma.

(9) freedom and determinism (*)

Arguably the biggest of all the big issues, this is the one which most frequently dominates discussion in both philosophical conversation and academic philosophy. If all is cause and effect, if every later event is produced as the consequence of former events, is determined by those former events, how is it that just one entity, the human person, escapes such a universal law? How can free will be made to fit into a determined world? And if it can't, if humans don't do what they do as free agents, then how can they be held personally responsible for those doings? Among most educated people, who in a growing number of parts of the world are a growing number of people, the old story that we, uniquely, were given free will by a divine creator just won't wash any more. So where *does* it come from? Or does it even 'come' at all?

(10) good and evil

Another troublesome duo! Simpler versions of the debate under this head are theological: if God is all powerful, all knowing and all good, then why does he permit evil? The most common modern way to get out of this one is not to be in it in the first place: that is, not to believe in God anyway. More complicated versions concern moral relativism. Many people, in the good company of such as Socrates with his 'he who knows what good is will do good', are now happy to acknowledge that 'evil' does not really exist: that is, that that which is called 'evil' is just the absence of good; that it is impossible for someone actively to choose evil over good; and that all apparently evil acts are thus the products of ignorance. But that doesn't help with the problem of moral relativism. The problem of finding the grounds on which decisions concerning the nature of that good remains.

For Plato, 'the Good' was, bar none, the highest of all his eternal forms. For many of us, the nature of the good is, bar none, the biggest of all our big issues. But it is my belief that there is, as yet, no philosophically respectable theory of the good at all. And that as a consequence, it is in the terrain of moral philosophy and ethics that both philosophical conversation and academic philosophy stumble about most thoroughly lost. At least among academic philosophers, there would seem to be a clearer idea than among those taking part in philosophical conversation as to what it isn't: as to what it can't be. Most of them, it would seem, are convinced that something like G. E. Moore's 'naturalistic fallacy' holds, and that that which is good can in no way be simply read off from that which is natural. But Moore concluded, in his *Principia Ethica* (1903), not just that the good had not been satisfactorily defined, but that it was in fact just indefinable. And in my experience, very few of those engaged in philosophical conversation, even those who don't still slop about in a 'good equals natural' morass, want to follow him that far. As neither do I. But just how we *are* to define it, or better to theorise it, causes perhaps more glitches in philosophical conversation than any other 'big issue'.

(11) I and thou (*)

In his famous book *Ich und Du* (1923), translated into English as *I and Thou* (1937), Martin Buber contrasted the 'I-thou' relationship with the 'I-it' relationship. For Buber, an 'I-thou' can subsist of course between a person and a person, but also between a person and a non-person: between, as he illustrates it, a person and a tree, for example. And this is because he puts 'relationality' first, both subject and object second. In my experience, this is not the way in which 'I *and* thou' issues are most frequently aired in philosophical conversation. Although frequently discussed under the same 'I *and* thou' head, the issues which there seem to matter are more usually one or other of two alternatives. Sometimes they are issues concerning first-person vis-à-vis third-person views of reality: in which case we are back with both 'appearance *and* reality' and 'fact *and* value'. And sometimes they are issues concerning whether 'thou' (with whom I share at the very most 50% of my genes, unless 'thou art' my identical twin) can ever be as important to me as 'I' (with whom I 'share' 100% of my genes): in which case we are back with 'egoism *and* altruism'. Either way, it is more heat than light which is frequently generated.

(12) identity and relationship

This pair figures when people are trying to make sense of change. If the basic units of reality are taken to be things (identities) embedded in structures (relationships), then change is most readily thought of as change in those things. If the basic units of reality are taken to be structures (relationships) embedding things (identities), then change is most readily thought of as change in those structures. The problem is one of capturing most appropriately what is actually constant flux. 'Identity' people tend to accentuate the stabilities, however temporary, which constitute 'things': 'relationship' people tend to accentuate the changes, however subtle, which constitute 'structures'. The fact is that it is just very difficult to think coherently about change. And that means that arguments on this one tend to go rather more with general (often political) bias, than with anything remotely like carefully considered theory: hence their frequent noisiness.

(13) ideology and science

We are dealing here with interest and disinterest. Ideological stances are taken to be those based in interest: those adopted by some power group, for example, which sees it as being in its interest that things should appear to be one way rather than another. Scientific stances, by contrast, are taken to be those based in genuine *dis*interest. There are, as always, particular versions of the ideology-science dichotomy. Perhaps the best known is the Marxist concept of 'false consciousness', whereby a class subordinated against its own real interests nevertheless accepts and internalises that subordination, without coercion, as the consequence of a 'false' understanding of reality. But whatever the version, with ideologies presenting as belief systems based on appearances (and therefore false) and sciences presenting as knowledge systems based on reality (and therefore true), arguments on this one go straight back to those concerning 'appearance *and* reality'. To get into this groove in philosophical conversation is to get into one already well-populated by those circling in vain around the 'real-according-to-whom?' questions of the 'appearance *and* reality' debate. Everybody wants to know where the view from nowhere can be found. But although there are some candidate locations on offer, none compel sufficiently to remove the nagging, frustrating and alarming doubt that the answer, not infrequently causing tempers to fray, might just be 'nowhere'.

(14) individual and society

This is a real chicken-and-egg debacle. In its crudest simplicity the question is this: do individuals make societies (through processes akin perhaps to social contract formation), or do societies make individuals (through processes akin maybe to socialisation)? One form of such discussion is to pit, say, a Hobbesian model against, say, a Marxist model. According to the former, we are to envisage 'always already' individuals in a state of nature, living lives nasty, brutish and short, and therefore agreeing to sacrifice some individual freedoms in order to come together under the rule of law. According to the latter, we are to envisage an 'always already' society, in a mode of production which determines its relations of production, such that the individuals made by it are 'the ensemble of their social relations', and can only gain their freedoms by changing the society that makes them. And there is nothing remotely esoteric about all this. At their extremes—and even quite near their middles!—these two positions seem fundamentally philosophically antagonistic, and thereby hugely politically divisive. If you think people make societies, you have to change people to change societies: if you think societies make people, you have to change societies to change people. Some of the obvious absurdities of both positions emerge quite clearly (and perhaps especially) from such a potted summary as this one. But at least in their more carefully presented and nuanced forms, neither is in any obvious way absurd enough to rule itself totally out of court in favour of the other: or even in favour of some nebulous 'third way'. So battle tends to continue.

(15) individualism and anti-individualism (*)

Among the professionals, this pair labels a very specific debate concerning the way in which a person is to be conceptualised. Individualism is that approach, sometimes labelled 'internalist', which holds that there is a psychological reality 'person' which is independent of its physical and social environment. Anti-individualism is that approach, sometimes labelled 'externalist', which holds that there is not. And the debate is largely conducted through a series of excruciatingly complex twin-earth thought experiments initiated by Hilary Putnam.

When all this figures instead in philosophical conversation, it is recognisably the same debate, but generalised. Thus an internalist approach to the history of science (great-man history, Whig history), would suggest, as I have already mentioned, that Newton made the scientific breakthroughs that he did because he was a genius ('God said let Newton be, and all was light'), and that if these internal-to-Newton factors had not been present, if Newton had not lived, then maybe we would never have understood the laws of gravity. An externalist approach, a Marxist one for example, would suggest, as did the Russian historian of science Hessen in his now notorious paper 'On the Social and Economic Roots of Newton's *Principia*', read at the International Conference for Historians of Science in London in 1936, that Newton made the scientific breakthroughs that he did because the seventeenth century needed its mines to be drained, its cannon-power to be improved, and the load-bearing capacities of its ships to be increased. And that while these external-to-Newton factors actually produced Newton, they would anyway have produce someone who would have given us the law of gravity. Necessity, after all, is widely acknowledged as the mother of invention. These are issues which, without an adequate theory of the person—and I will argue that we are indeed without an adequate theory of the person—are fundamentally beyond resolution. In the language of internalism and externalism—yet another pair of philosophical 'ands'—just pose Wittgenstein's 'external to what?' question, or its 'internal to what?' companion, to see just how complex an analysis of this 'and' would have to be to stop us running round in philosophical circles.

(16) is and ought (*)

The case here is the same as that previously presented under the head of the 'fact-value' distinction, and thus needs no further explication.

(17) justice and benevolence (*)

A former Archbishop of Recife said, now famously, that when he gave alms to the poor he was called a saint, but when he asked why there were any poor he was called a communist. This one anecdote can give us all the flavour of this telling pair. In the language of at least some among the professionals, both benevolence and justice are presented as 'other-regarding' virtues: Levinas, for example, talks of 'the relation with the Other that is justice'. But in philosophical conversation, the issue is frequently the very specific one concerning the giving of charity. Those at one end of the spectrum argue that because it depends on sentiment, even whim, and is ultimately optional, benevolence is no substitute for justice: that only justice can ensure that which is rationally and morally impartial. And those at the other argue that because that which is rationally and morally impartial is also that which is without human feeling, justice is no substitute for benevolence: that only benevolence can ensure the sentiment and optionality, if not perhaps whim, that are actually appropriate to the governing of moral relations between persons. The tendency to end discussions on this one in a 'both-are-important' consensus is particularly strong. But rarely strong enough, I believe, to bring all to some satisfied closure. It has been my experience that it is not at all uncommon for at least some of the participants in such a debate to be left with a strong sense of there being unfinished business in such a formulation. And I would be of their number.

(18) knowledge and science (*)

'All men desire to know.' But the particular form of knowing called 'science' is usually thought to be not as old as humankind, but instead as old as either the Thales hypothesis mentioned in the previous chapter, or as those events of the sixteenth and seventeenth centuries commonly grouped under the head of 'the scientific revolution'. In philosophical conversation, the debate that goes on under this head is in effect the same as the one already briefly considered under the 'arts *and* sciences' pair. Because it is the only reliable kind, does all knowledge ultimately have to be scientific knowledge? Or are there other knowledges—knowledges artistic, religious, philosophical or common-sensical—which can be retained? And if they can be retained, then as what can they be retained? As knowledge of the detail of what is, where science gets only at the abstraction generalised from that detail? As knowledge of realities just not amenable to scientific investigation in any shape or form? Yet again, this is a discussion which frequently generates considerable disagreement.

(19) liberty and equality (*)

In our *Oxford Companion* source, Richard Norman provides a neat and succinct one-paragraph summary of this one. On the one side there are those who maintain that these two values necessarily conflict: that because people differ in their abilities, some will always be more successful than others, and this means that equality can only be achieved by restricting their freedom to thereby become more prosperous. On the other side there are those who counter that it is only if everyone shares equally both wealth and power that the conditions necessary to freedom can be met: that far from being in conflict, therefore, these two values are in fact complementary. Indeed. But since neither side has as yet produced any argument sufficient to foreclose on this one, the clash of conflicting opinion remains quite frequently raucous.

(20) master and slave (*)

In this form, in these words, this is a debate, now famous, initiated by Hegel. In his *Phenomenology* (1807), the 'master *and* slave' allegory is the best-known part of the best-known section of the book. The master-slave relationship is depicted as resulting from a battle, which goes on in all human dyads, for the recognition that is bestowed by status. The slave loses, and thus fails to secure recognition. The master wins, but thus also fails to secure recognition—from anyone other than a slave, which, when what is sought is the recognition of a peer, will hardly do. Hegel goes on to argue that things anyway then get turned 'upside-down' because while the master's very dependence on the labour of the slave makes him the real slave, the slave's very labour sets him free to be the real master. Nietzsche's 'will-to-power', Adler's 'inferiority complex', and many other notions concerning face-to-face inter-human conflict are captured by Hegel's 'master *and* slave' story. And in philosophical conversation, it most commonly figures, thinly disguised, as an argument about equality. Is it possible for anybody to be free, slave or master, while their relationship is constituted out of their inequality? And if real freedom is dependent on equality, then given that in 'the real world' such equality is never possible, and on other grounds maybe not even desirable, is real freedom also a chimera? 'Discuss.' And we do!

(21) mind and body (*)

Closely allied, of course, to the 'freedom *and* determinism' pair, here is the site of another major battleground. We all know what bodies are. We all know that brains are parts of bodies. But just what it is that minds are, if anything, is a matter of huge debate and contention. Are minds what brains do? Can they be conceptualised in a way that would mean that computers could have them too, eventually? Do the higher non-human animals have them? Yes of course, if they are just 'what brains do'. But in that case, does it mean that they have as much free-will as we do, and it's OK to put a chimp on trial for killing another chimp? Or does it mean that we have as little free-will as they have, and it's not OK to put a human on trial for killing another human? In both academic philosophy and philosophical conversation, this is a debate currently very much alive. And what for all of us gives it its fascination, not to say dread, is the way in which it foregrounds the question as to whether or not we are to be held morally responsible for our actions. There have very recently been developed some massively improved technologies for studying brains. But perhaps because there is still no *formulation* of the mind/body problem which commands general agreement, there is still no *solution* to the mind/body problem which commands general agreement either. So that this one too is an 'and' which tends to leave most of us similarly puzzled.

(22) natural and cultural

Whether we take it in its tilling-of-the-soil meaning, in its making-of-artefacts meaning, in its making-of-art meaning, or in its making-of-a-whole-way-of life meaning, 'culture' is essentially of the human. Natural evolution is Darwinian: where there is no use of language, characteristics acquired by learning in one generation cannot be preserved to pass on to the next. But cultural development is Lamarckian: the acquisition of language, and more particularly of literacy, means that characteristics acquired by learning in one generation can now be preserved to pass on to the next. Natural evolution is slow: cultural development is fast. Albeit contentiously, natural evolution *can* be characterised—as Stephen J. Gould famously characterised it—as in no way progressive. Reasonably obviously, cultural development can be characterised in no other way than as progressive. So far, so clear: although I will later make the case for the use of a different language to describe these realities.

But concerning this 'and' pair, the disputes which occur in philosophical conversation are primarily about the way it figures *within* the human, rather than the way it figures *between* the human and the non-human. To what extent, if any, is a human to be thought of as an animal? To what extent, if any, are humans appropriately to be studied by biologists? What, if any, are the criteria by which a clear distinction can be made between the animal nature of a human and the human nature of a human? Is there indeed any such thing as 'human nature'?

This is an 'and' pair provided by me (to reflect philosophical conversation in general) rather than by my *Oxford Companion* source (as it reflects academic philosophy in particular). The actual issues involved in any discussion of the 'natural *and* cultural' pair are, I think, of as much concern in both places. But perhaps because of a greater awareness of the multi-layeredness of the issues involved, this is a pair which, in academic philosophy, is perhaps less often considered 'in toto': perhaps more often considered in bits and pieces under other and various 'and' heads. But whether as one big question or as many smaller but linked, the way to conceive of the relationship between the natural and the cultural is one of the 'all time (contested) greats' when it comes to furious and passionate discourse. And when somebody tries to require, as I will, that it must be a question also involving the *super*-natural, the super-natural in effect *as* the cultural, then neither the fury nor the passion is at all likely to diminish.

(23) natural and moral

Although also a duo provided by me rather than listed by the encyclopaedia, again the issues raised by this pair are, I think, important to both philosophical conversation and academic philosophy. And as I have already argued under my 'good *and* evil' entry, I think it is reasonably obvious that philosophical conversation lags somewhat spectacularly behind academic philosophy in making coherent sense of these issues. Among most of the professionals, it has, I think been long understood that there is no reductionism possible here. Among them—and following some now famous treatments of the issue by such as Hume and Mill, as well as by Moore—there would seem to be widespread agreement that what is moral can in no way be read off from what is natural. Where many of those engaged in philosophical conversation quite readily accept the 'natural' goodness of everything from breakfast cereals to non-genetically engineered bodies, from mother-reared infants to heterosexuality, from family values to retributive justice, academic philosophy, correctly it seems to me, will for the most part have no truck with such stuff. Its position, registering that the 'fact-*and*-value' and 'is-*and*-ought' problems exist as problems precisely because the moral cannot, at least in any obvious way, be 'read off' from the natural, is usefully and succinctly summarised in the second chapter of Janet Radcliffe Richards' book *The Sceptical Feminist* (1980). It is a summary which makes it clear that when Kant refers to 'our niggardly step-mother, nature', or in a letter to J. D. Hooker dated 13 July 1856 Darwin writes of 'the clumsy, wasteful, blundering, low, and horridly cruel works of nature', academic philosophers would, for the most part, have no problems with such less-than-positive images of the natural. For the most part, they would see no reason to assume, for example, that because some of our intuitions about how to behave derive from instincts that were evolutionarily sound for our ancestors, that therefore implies that they are likely to lead to actions morally sound for us today. But this, emphatically, is not the case in philosophical conversation. Exacerbated, anyway in European politics, by some of the developments within the 'green' movement, many philosophical conversations are those in which whether or not 'natural' equals 'good' is still, unlikely as it might seem, a live issue causing real dispute. It is for this reason, therefore, that it must be included among my 'specimens' here.

(24) natural and social

This one can take on almost as many forms as there are conversations in which it is discussed. The form to which I want to draw attention here is the one which concerns science. With relentless regularity, and not only among and between the different practitioners of the different disciplines, the debate as to whether or not the physical and biological sciences are *the* sciences, while the social sciences are only the *pseudo* sciences, waxes if not lyrical then at least long. Ultimately, and in the disputes generated by the issue 'ultimately' tends to arrive quite quickly, this is the 'freedom *and* determinism' debate again: if science is about the reliable, the repeatable, the law-like, then is the idea of a social (or human) science coherent? Only, surely, if the human is as determined as all the rest. And that can hardly be so: can it? Surely it's just self-evident that the weak determinism of experience, even the strong determinism of the genes, cannot allow us to predict in reliable, repeatable and law-like fashion, what the essentially *free* being, which is a *human* being, will do? Nonetheless, social (or human) sciences, in anyway an institutional sense, do exist. And that means that, in philosophical conversation, this is an argument which often hinges on whether or not that existence is legitimate. In short, it's an 'if not, why not? if so, how so?' sort of debate.

(25) needs and wants

In everyday parlance, needs are for things a person must have, wants for things he or she would like to have, but are not essential. This is a usage consonant with work on what is called 'personal growth' in a psychotherapeutic context. One of its seminal and still best-known expressions was in Abraham Maslow's concept of there being a 'hierarchy of needs', such that we only become free to want fulfilment, say, or self-actualisation, when more basic needs for food, or drink, or shelter, or sleep have been met. While these are sometimes the sorts of issues under discussion when the 'needs *and* wants' argument figures in philosophical conversation, it is more usually questions concerning issues more overtly political which take centre stage. This is a duo which is debated, for example, in discussion concerning the benefits system. When such a system is assessing needs, how are they to be differentiated from wants? *Can* they be differentiated from wants. *Should* they be differentiated from wants? Are needs and wants always relative, so that a government using a concept such as 'subsistence level' in its welfare policies needs constantly to be re-assessing its historically specific meaning? Or are there some absolute needs, based perhaps in something called 'human nature'? In the absence once again of a non-contested theory about such things, the politics of the 'needs *and* wants' pair is more often than not something of a battle-ground.

(26) objectivism and subjectivism (*)

At first flush, it looks as though we are back here with 'appearance *and* reality', and that that means there is little more to say. Such a conclusion I think would be wrong. For while it is at least reasonable to take 'reality' to mean 'that which is' and 'appearance' to mean 'that of that which is which is observed by observers', I suspect it is considerably less reasonable to take 'the objective' to mean 'that which is' and 'the subjective' to mean 'that of that which is as it is observed by observers'. It is probably less reasonable because I am trying to characterise not the nuances of academic philosophy but the guts of philosophical conversation. And in such conversation, while 'reality' is almost always something which is 'out there' somewhere, 'the objective' is almost always not. With the one purged of my partialities, special pleadings and misinformations, and the other riddled with my biases, prejudices and ignorances, in such conversation objective and subjective judgements are both 'in here'. And because this means that what we're really looking at here is yet another argument about the extent to which humans can find the view from nowhere, it also means that the territory we're really in here is that of the 'realism *and* anti-realism' pair, which will be outlined below.

(27) praise and blame

'Freedom *and* determinism' again, but with a twist. If people are free to choose their actions, then they can be *either* praised *or* blamed for them: if people are not free to choose their actions, then they can be *neither* praised *nor* blamed for them. This is the simple dichotomy. It is complicated by a third position, mentioned already in Chapter 3.

This is the stance whereby the two members of the 'good *and* evil' pair are conceptualised as being related asymmetrically: the good as the presence of something; the evil as the absence of that same something. On such a reading, the 'praise *and* blame' duo will need considerable recasting. Given that people can only do what they can do, and that learning to do that which is right is an essentially developmental process, 'praise *and* blame' will also need to be conceptualised asymmetrically. Because it is about celebrating the acquisition of something good, 'praise' will still be able to be seen as appropriate. But 'blame' will no longer be able to be seen as appropriate at all. Because it is about castigating the only *as-yet* non-acquisition of that same something good, then in Spinoza's word it will need to be seen as just 'vindictive'. Where the total inappropriateness of punishment rests in the total inappropriateness of blame, this is the stance which underlies all of Skinner's 'operant conditioning'. What he and those working within his tradition have been able to demonstrate is another and derivative asymmetry: that while being 'praised' (reinforced) for the way we behave *will* make us behave that way *more* frequently, being 'blamed' (punished) for the way we behave *won't* make us behave that way *less* frequently. But while for perhaps most of those with any psychological training, these things are now close to becoming almost self-evident, in most of the philosophical conversation of which I anyway have experience, they most surely are not. So again it is a mis-match between the professionals and the rest—this time, between the psychologists and the rest—which is the cause of most of the heated argument which circles around this particular 'and' pair.

(28) public and private (*)

This pair is most often met in political philosophy, where its locus of application has changed quite dramatically in the recent past. Traditionally, that is within a liberal conception of something most frequently called 'private life', the family has epitomised the private. The state, it has been thought, has no place in regulating relations between husbands and wives, children and parents. But particularly with the arrival of the 1970s round of feminism, with its 'personal-is-political' slogan, this is now substantially contested ground. Linked to the 'autonomy *and* heteronomy' pair, it is increasingly being argued that if a 'dependency culture' is bad when that dependency is on the state, then surely a 'dependency culture' is just as bad, if not worse precisely because it is private, when that dependency is of a woman on a man, or of a child on a parent. Thus the 'public *and* private' pair features in philosophical conversation of a mainly political kind, where of course it quite easily and raucously extends into the historically older argument about the public (societal) or private (individual) provision of such things as education or health care.

(29) quantity and quality (*)

In its more specific meaning, as given by academic philosophers specialising in logic, this is an 'and' which in my experience is almost never encountered in philosophical conversation. However, in its more general meaning, it is not only encountered there but encountered in ways which raise important and contested issues. They are issues which will be particularly important to the form of the hypothesis I want to offer, in my next chapter, as to why there are so many additive (rather than subordinative) and therefore primitive (rather than sophisticated) formulations in both academic philosophy and philosophical conversation. Briefly, discussion of this pair is frequently discussion of what I will there refer to as the 'emergence' problem: in essence, in systems which are characterised by gradual quantitative change, is it ever useful to see particular stages of that change, while continuing to have their step-by-step characteristics, as nonetheless having also, and momentarily, a qualitative suddenness? The celebrated argument about 'punctuated equilibrium' in evolutionary theory is just one among several well-rehearsed examples where these issues rage.

(30) realism and anti-realism (*)

A huge topic, already much encountered in the specifics of smaller ones. In what can only be inadequate outline, the issues are these. Is what is real—a major question, of course, in its own right—real independently of appearing so to situated and therefore 'somewhere' observers? And if it is, how can such situated and therefore 'somewhere' observers know that it is? How can they get to a 'view from nowhere' position which would be the only one capable of giving them such objective knowledge of what is really so? And if they can't, isn't 'anti-realism', always and necessarily some form of 'relativism', the only viable option? These are all so obviously issues raised many times before in my list that I will give them no extra and particular treatment here.

(31) reason and emotion

A common pairing of philosophical conversation, the set-piece battles here are over such things as whether women are more emotional than men; whether being emotional is a 'good thing' or a 'bad thing'; whether emotions hinder or help the finding of rational solutions to problems; and, less often, just what do we mean by 'emotional' and 'rational' anyway. In academic philosophy, there is also a debate concerning rational*ism*: an argument about the extent to which something called 'reason' can be viewed as intrinsic to humans, and therefore contributory to human understanding in ways independent of empirical experience. In the case of language, for example, such rational*ism*, taking its modern form primarily from the work of Chomsky, argues that it is the existence of 'linguistic universals', innate to humans, which make very reason itself possible. But this is an issue which figures much less frequently, anyway in my experience, in philosophical conversation: there, it is much more frequently the question of how reason and emotion are to be related which exercises folk. And as in the case of all my other 'specimens', the fact that they are a pair which are so often presented additively is I suspect an indication that it is our weak theorising of their relationship which underlies our confusions concerning their nature.

(32) reasons and causes (*)

A very active area of concern for academic philosophers, for example for Donald Davidson and for Derek Parfit, this is an 'and' which also figures, if somewhat elliptically, in philosophical conversation. If one asks, with the professionals, 'are reasons causes?', one is clearly once again in 'determinism *and* free-will' territory. If all is determined, the reasons I give for my actions are just descriptions (or mis-descriptions) of their 'real' causes. On the other hand, if I have free-will then those actions become events physically uncaused, and thus thoroughly problematic. It is in this seemingly ubiquitous 'freedom *and* determinism' guise that the 'reasons *and* causes' debate most often flourishes in philosophical conversation. As mentioned in the previous chapter, such debate is not infrequently cast in terms of a discussion of the differences, if any, between 'how' questions (asking for the *causes of* things) and 'why' questions (asking for the *reasons for* things). But whichever, when these are the issues under discussion in philosophical conversation, a closure acceptable to all is seldom (if ever) achieved. And next time around, the same old issues circle once again.

(33) right and wrong

Although they do not agree on just what the distinction is, this is a pair which, for the professionals, is distinct from the 'good *and* evil' one already considered. Amongst them, 'consequentialists' are those who think that 'the right' is instrumental in bringing about 'the good': 'deontologists' those who think that some things are 'right' just intrinsically. These are issues which need detain us here not at all. In philosophical conversation, where 'the right' is that which is 'good' and 'the wrong' is that which is 'evil', they are distinctions which, in my experience, are hardly ever drawn. And that means that given the 'good *and* evil' duo has already been introduced, this 'right *and* wrong' one needs no different attention for the moment. Suffice it to say that the fundamental problem of moral philosophy, the problem of how that which is 'good' is to be determined, alters not at all by calling that which is good that which is right. The issues remain stubbornly the same, and stubbornly just as intractable.

(34) stories and explanations (*)

In philosophical conversation it is now a commonplace to talk of myths not in the older sense of 'those things which are not true but false', but in the newer sense of 'those things which are not explanations but stories'. The number and variety of such stories is considerable. There are the creation myths found in almost every culture. There are the folk myths found in nursery rhymes, legends, traditional ballads and songs. Among ever-growing numbers of Christian believers, stories of 'the Fall', 'the Resurrection', 'the Virgin Birth' are now widely interpreted in this way. Among many psychologists, so too are the psychoanalytic narratives derived from such as Freud, Jung and Adler. The critical difference between the older and earlier 'stories' and the newer and later 'explanations' is in the nature of the patterns created in them. Stories create narrative patterns: patterns made out of intentions, motives, purposes, 'grand designs'. If we can tell a coherent story about events, we are satisfied that we have made some purchase upon them, gained some understanding of them. Explanations, by contrast, create non-narrative patterns: patterns devoid of intention, motive, purpose, design—'grand' or otherwise. Where stories are works of art, using local and specific methodologies, explanations are works of science, using a universal methodology. And perhaps the still most significant shift, historically, from reliance on stories to the search for explanations occurred with the Enlightenment: it was perhaps then that the search for practical policies, derived from general principles, resting in human universals, began in earnest.

It is now a search not at all without its many and vocal critics. In *Contingency, Irony and Solidarity* (1989), for example, Richard Rorty seems to reject the very idea of universalism. Still in the tradition—albeit in its later stages and forms—which saw the nineteenth century Romantic Movement reject eighteenth century Enlightenment thought, his is one among what are many contemporary philosophical voices questioning any such search. And the big issue here is, once again, the status of the human. Can we only ever tell *stories about* human behaviour—because humans are free agents who might have done otherwise? Or will we ever be able to give *explanations of* human behaviour—because humans are determined entities, concerning whom a science of their doings is a realistic possibility? [And it will not have escaped your notice that here, in 'Enlightenment *and* Romantic Movement', we have yet another of the 'ands' of philosophical conversation.]

(35) time and space

'Unless we compare like with like, my horse will always be faster than your pig is fat'. Thus, amusingly, did a letter in *New Scientist* (3.7.99) draw attention to the ubiquitous 'category mistakes' we so often make in trying to discuss philosophical issues. And it has often seemed to me that this 'time *and* space' pair is a bit like that horse-and-pig variety. It has been my experience that, in philosophical conversation, 'time *and* space' figures in such a variety of different guises, to such a variety of different extents, in such a variety of different ways, that confusion just has to be endemic. One sort of discussion, and we are in the realms of general relativity and space-time; another sort, and we are dealing with the 'change' dimension of time vis-à-vis the 'variation' dimension of space; yet another sort, and what emerges is a usually very rarefied discussion of the relationships between past, present and future. All three of these discussions—and perhaps a few more—will be relevant to different stages of my argument in this book. Suffice it here to say that, whichever it is which is under discussion, climbing to ever higher levels of confusion, whether as the result of category mistakes or not, is for most of us, most of the time, the order of the day.

* * *

These then are my thirty-five specimens: the thirty-five 'ands' which are perhaps the most frequently encountered in philosophical conversation. They constitute the observations concerning which, in my next chapter, I offer my hypothesis: an hypothesis which tries first to offer an explanation of the very existence of so much and-ness, and second to offer a way of escaping both it and the stuck-ness it induces. But before shifting to that next chapter, I want to conclude this one with two final explanatory points.

First I must make it clear that when I talk about academic philosophy, it is a very particular academy to which I refer. The considerable differences between what is taught as philosophy in the Anglo-American world and what is taught as philosophy 'on the continent', especially in France and Germany, are well known. But given the fact that the philosophical conversation in which most people in this country take part is conversation which takes place against the backdrop of the one tradition rather than the other, I have situated my argument in that context: my *Oxford Companion* source, for example, comes quintessentially from the Anglo-American tradition. But two caveats are immediately necessary. First, such situation will *not* mean that my critique of the arguments constituting such conversation will not draw, in part, upon the continental tradition as well: it will. And second, such situation will *not* mean that that same critique will not draw, in part, on sources outside both traditions: given that what the professionals (of either ilk) are up to is not my main concern, again, it will.

And second I must remind you—and also indeed myself—that as the sub-title of this chapter makes clear, the observations I have made concerning philosophical conversation reveal not one feature but two which are in need of explanation. As well as so many 'ands' there are also so many 'us-es': we also have just so many words for ourselves. We call ourselves, variously, 'humans', 'human beings', 'humankind', 'humanity', 'the human race', 'the human animal', 'people', 'persons', 'ourselves', 'we', 'us', '*Homo sapiens*', 'mankind', 'earthlings', even 'worldlings'. That is odd. And particularly odd is the fact that, concerning this rich abundance of vocabulary, just which 'we' we are, in any particular discourse, even in any particular sentence, seems to be much less a matter of picking exactly the right word for the context, much more a matter of the near haphazard. I will be suggesting that these are confusions absolutely central to the and-ness of philosophical conversation: in brief, that the problems represented by that and-ness can be either solved or dissolved by a more careful delineation of just who and what 'we' are, and thus who and what 'I' is. I will suggest that while, anyway so far, our usage of the several words we have to indicate ourselves has been sufficiently within a shared convention for us to be mutually intelligible, that is now steadily ceasing to be true. Consistent with the point made by Ian Glynn, and mentioned earlier in this chapter, I will argue that that shared convention has not in fact been based on an appropriate reading of our situation. And that what this is coming to mean is that, as more and more of us find it more and more difficult any longer to partake of what have been our mutual confusions, attention is beginning—not before time—to focus upon them.

Chapter Five

HYPOTHESIS

Eureka!

In this chapter I will present my hypothesis: that is, I will present the way in which I find pattern in, and thus a candidate explanation of, both the profusion of 'and' pairs and the proliferation of 'us' words in philosophical conversation. Before I do so, however, I must just point out that, in its formulation, 'Ies' is the best I can do as a plural for 'I'. 'We' is not such a plural. 'We' designates 'I plus you'. I need a word which designates 'I plus I'. So 'Ies' it has to be. And as my argument will show, the fact that, because it is not in the language, such a word has to be invented is itself comment on what I take to be the considerable problems with our current 'I' concept.

I will present the hypothesis in three stages: first, in one sentence, I will present it in its shortest possible form; then, in the bulk of the chapter, I will present it in rather more lengthy outline; finally, in (the same) one sentence, I will present it (again) in its shortest possible form, but now with also a name. My aim is that between its first and its second presentation, the reader will have gained two things: first, a general understanding of what it is I am proposing; second, an initial exposure to the language in which I am proposing it. The remaining chapters of the book will then try to take things further in the following way. The rest of Part Two will consist in an exploration of my hypothesis, first in experiment, then in theory: an exploration which will involve a more detailed elaboration of its meaning; a more lengthy defence of its appropriateness; and a more specific account of the way in which I think it either solves or dissolves those problems of philosophical conversation made manifest in the 'and' pairs and 'us' words to which I have drawn attention. And then in Part Three I will return to the issues raised in Part One, and present some ideas as to how, in the light of the study in Part Two, they could now be both better understood and further developed.

The Hypothesis

That with reality defined as 'all that there is', the coming into existence of its several parts can be sequenced into three levels and six stages, such as to reveal not one human condition but three human conditions, not one 'I' but two 'Ies': and that the removal of the confusions arising from thinking outside such a sequencing pattern solves (or dissolves) the problems of philosophical conversation outlined in the previous chapter, and makes possible the creation of the moral sciences.

Its Explication

In everyday conversation, 'the real world' is a phrase used almost always as a put-down of one thought not to be in it. The argument is simple. Wherever I am, the world is real: wherever you are, he is or she is, it might or might not be—depending on how near or far you, he or she is from me. And as with 'the real world' in everyday conversation, so with 'reality' in philosophical conversation. Only here, the archetypal 'I' is the physicist (and her or his apologists), who are not so much the only ones *in* reality, but the only ones whose subject matter *constitutes* 'reality'. My thinking begins from a totally other premise.

I take reality to mean 'all that there is': cod, figments-of-the-imagination, helium, verbs, tulips, my thumb-nail, the universe, God, fairies, soap, the square root of fifty-three—just 'all that there is'. Being simply defined as 'all that there is', such an all-inclusive concept of reality is a device which then pushes to another level such questions as 'real in which way?' or 'real as what?' There are I think two obvious problems concerning such a definition: the one terminological, the other conceptual. The terminological problem is that anyway since Einstein, before whose usage it was a word applied to anything from the Solar System to the Milky Way, it is the word 'universe' which has now been largely adopted to carry that 'all that there is' meaning: so why dissent from that convention? The conceptual problem is that, as many people have pointed out, since concepts will only function at all as a consequence of their differences, any concept which tried to mean 'everything in general' would end up meaning 'nothing in particular'.

My response to the first problem is that it is precisely because I want to challenge what I see as the aggrandizing claims of much contemporary physics—its claim, as expressed by Rutherford for example, that within the natural sciences everything is ultimately physics, and without the natural sciences everything is ultimately stamp-collecting—that I want to use an 'all that there is' word which precisely is *not* the physicist's 'universe' term, which term will be subsumed within it. For maybe paradoxically or maybe not, it is my contention that the derision contained in the 'stamp-collecting' quip is one which actually *feeds* that non-scientific approach, but this time to the human, which it presents as deploring. And my response to the second problem is that I agree that my term 'reality', because it covers everything, means nothing in particular: it's just that it means everything in general, which is what I need it to mean. This is a point not to be confused with one already raised, in Chapter 3, concerning the God concept: what was at issue there was the claim of such a concept not to identify but to explain an everything *in which it was not itself included.*

I have called such usage 'a device'. And to illustrate its role as such, let me develop the point by considering just the title of a book from one of those aggrandizing physicists. David Deutsch called his 1997 book *The Fabric of Reality*. What I am suggesting with my 'device' is that, as a physicist, he should instead (and more modestly) have called it *The Reality of Fabric*. He could then have left perhaps a biologist to tackle *The Reality of the Awareness of Fabric*: a book about sentient organisms. And perhaps a psychologist to have a go at *The Reality of the Awareness of the Awareness of Fabric*: a book about conscious persons. Of course neither the 'organic' reality which would need to be considered in the biologist's book, nor what I will be calling the 'personic' reality which would need to be considered in the psychologist's book, would involve either the creation of any new inorganic matter-energy stuff, or the destruction of any old inorganic matter-energy stuff. But both would involve the creation of new reality. Which means that when, by contrast with Deutsch's title, I let 'all that there is' be real, let 'all that there is' constitute reality, my first task must be to classify: that is, to ask 'real in which way?', or 'real as what?'

But not before registering that such an understanding of 'reality' also, of course, puts me at odds with some schools of academic philosophy. In 1994, Richard Rorty wrote an article entitled 'Reality As It Is In Itself'. The capital letters are a way of getting at the point, very widely accepted since Kant, that it is a hopeless project for me (or anyone else) to try to compare a non-conceptualised world ('Reality As It Is In Itself') with the concepts that I have about it, for the simple reason that a non-conceptualised world cannot exist. The world I experience is and always must be 'always already' conceptualised, by me. The position I take on this issue is radical. Let me make now the first of what will probably need to be several attempts to explicate it: other attempts, and in many guises, will occur across much of the rest of this book.

My position is that lurking in such a formulation, as in almost everything we ever say or think about, there is a residual, closet and illegitimate dualism. If we posit 'I' and 'my experience' as two different entities, related in what I will later be calling 'the having mode' (I *have* my experience), then of course 'Reality As It Is In Itself' is never available to the first of those entities ('I') other than via the second of them ('my experience'). But if instead we posit 'I' and 'my experience' as one and the same entity, if we say not 'I am that which *has* experience' but 'I am that which *is* experience', then the problem of direct access to 'things in themselves' just dissolves. In other words, thinking of me as on one side of an opaque glass wall, 'things-in-themselves' as on the other side of that opaque glass wall, and 'my sense perceptions', 'my experience', 'my concept system' as constituting the opaqueness of the glass and thus preventing my direct access to said 'things', then what I am suggesting is that the way to get that direct access is not to dismantle the wall but to dismantle 'me'. If 'I' am dismantled as an I which *has* my experience, re-mantled as an I which *is* my experience (and my concept systems, and my sense perceptions, and my beliefs, and my whatever-other-of-such-things need to delineated separately), then 'I', positioned now where once *was* the wall, posited now as identical *with* the wall, have got my direct access to said 'things'. There is no gap between 'me' and 'my experience', because 'me'—other than as equated with 'my experience'—does not exist. To repeat: I do not *have* my experience, I *am* my experience.

This is the position I will take. And the language I will use to take it in, language to be further explained in due course, will be as follows. When I am referring to an 'I' posited as that which *has* experience, I will call it an 'I AM': when I am referring to an 'I' posited as that which *is* experience, I will call it an 'I IS'. As I'm sure will already be dawning on some readers, such a dissolution of the 'direct-access-to-reality' problem comes at what, at first flush, will seem no small cost. Its price is indeed a thorough-going determinism. But a major part of my task will be to suggest that that is no cost at all.

I am proposing that 'reality', formulated as 'all that there is', came into existence across time in such a way that it is revealing to delineate its emergence across a series of levels and stages: specifically, across three levels and—because each level is sub-divided into two stages—six stages. The whole scheme is outlined in Table One.

The rest of this chapter will consist in an explication of this scheme. The level of detail of that explication will depend on three things: on the extent to which I believe the material will or will not be *familiar* to most readers; on the extent to which I believe the material will or will not be *contentious* for most readers; and on the extent to which I know the material will or will not be *important* to my later concerns. In general, that will mean that both each succeeding level and each succeeding stage will be presented in greater detail than the one previous to it. When I reach level three, stages five and six, where as far as I know I am offering something original to me, that will mean that *their* coverage will be the most detailed of all. In addition, two further introductory points are needed. One: this is a classification system offered as a model. But while the whole thing, in its three levels and six stages, is *my* model, individual claims within it, as for

Level	*Stage*	*Starting with*
LEVEL ONE REALITY The universe (*the inorganic*: to be studied by the physical sciences)	Stage One Stage Two	The big bang (15,000 million years ago) Formation of the solar system (4,600 million years ago)
LEVEL TWO REALITY The biosphere (*the organic*: to be studied by the biological sciences)	Stage Three Stage Four	Emergence of (irritable) life (3,500 million years ago) Emergence of (sentient) life (1,000 million years ago)
LEVEL THREE REALITY The world (*the personic*: to be studied by the moral sciences)	Stage Five Stage Six	The emergence of I AM consciousness (100,000 years ago) The emergence of I IS consciousness (currently under construction)

Table One: The levels and stages of reality

example the claim that the universe came into existence some fifteen thousand million years ago, are obviously adopted from already existing and other models specific to particular parts of this one. My strategy will be that I will only comment on such an adoption if it is of something not considered mainstream by the majority of those working in the relevant field: that is, if I adopt current orthodoxies, I will do so without comment; if I adopt current non-orthodoxies, I will try to explain myself. And two: since I am well aware that to think in stages of development at all is in itself a highly contentious idea, I will address that issue, in its own right, but later (see Chapter 9). In that same chapter I will also address, and at some length, what I take to be the related problem of 'emergence'.

Writing in the *New Scientist* (27.4.96), Paul Davies presented what is the first of my adopted orthodoxies this way: that the universe came into existence some 15 billion years ago without, anyway in the normal sense, any prior cause. This was not because it had some other than normal cause, perhaps some supernatural cause, but because there is simply no 'before' time or 'before' place in which any causative agency, whether supernatural or natural, could exist. This the so-called 'big bang' model outperforms, and thus has eclipsed, the alternative 'steady-state' hypothesis, put forward in 1948 by Fred Hoyle and Herman Bondi. According to what is now known as 'the standard model', in the beginning a single superforce separated out into the four forces we see today: *gravity* across astronomic distances; the *electromagnetic force* holding electrons around protons and neutrons; the *weak force* within the atomic nucleus; the *strong force* holding quarks in trios to form protons and neutrons. Subatomic particles came into existence in an orderly sequence—quarks, protons, neutrons, electrons—such that all things physical are now thought to be made up of six types of quark and six types of lepton. Given the four forces, the six types of quark and the six types of lepton, it is possible to work out which particles should exist, and then go out and find them. And on that basis, it is postulated that it was at about 300,000 years after 'the big bang', when temperatures were down to about 3000 degrees centigrade, that atoms emerged. And stars, galaxies, molecules and planets followed thereafter. That stars are made of atoms of the same kind as those we have here on earth, of the same kind as those we have here in our own bodies, was a discovery which Richard Feynman called 'the most remarkable in all of astronomy'.

There are about 10^{12} galaxies, with 'our' galaxy containing about 10^{11} suns. One of those 10^{11} suns is the sun at the centre of 'our' solar system. It is a sun which converts 596 million metric tons of hydrogen to 532 million metric tons of helium every second; releases the difference between the two as fusion-sourced energy; and acquires from the process a 6000 degrees centigrade surface temperature. This which is 'our' solar system came into existence some 4,600 million years ago. Its planets, forming over a period of about 10 million years, eventually came to consist in the inner, smaller, more dense, terrestrial ones (Mercury, Venus, Earth, Mars); the outer, larger, less dense, Jovian ones (Jupiter, Saturn, Uranus, Neptune). [Pluto, far out and very small, is since 2001 no longer regarded as a 'real' planet at all.]

This inorganic reality, which in energy terms consists in four forces, in matter terms consists in just over one hundred elements: one hundred and eleven now identified, with numbers 110 and 111 having been 'found' in 1995 in Darmstadt, Germany. The lighter elements came into existence first, the heavier ones later: indeed, the latter arise from a transformation of the former during the coming into existence of stars. And although it was initially thought that one of the elements, plutonium, existed only because in the course of the Manhattan Project of the early 1940s humans had made it, naturally occurring plutonium has now been identified. It is these forces and these elements which constitute inorganic reality as it exists everywhere: on the earth; on other planets; in other solar systems and galaxies; in stars; everywhere. The helium in the sun is the same stuff as is used to fill the balloon: the hydrogen in the sun is the same stuff as combines with oxygen to form water. This is the inorganic reality which physicists are calling 'everything' when they, somewhat immoderately, claim to be about to produce their 'theory of everything'—or TOE. When they succeed, what they will actually have done is produce a theory of all the forces: a theory which will unite the four as four different manifestations of some singular, underlying force. Which will be wonderful, but hardly a theory of 'everything'.

In dividing this Level One Reality into its two stages, I am of course choosing one division where others could perfectly well be defended as legitimate. In particular, it would seem that, basing myself in 'the standard model', a very good case could be made for a two-way division such that Stage One covered that period when the newly created universe was dominated by radiation, and Stage Two began at that point (when the temperature was down to about 4,000 degrees centigrade) when it began to be dominated instead by matter. But for my present purposes, as opposed to any other and different purposes it would have been possible for me to have, dividing at the point at which Earth came into being is more appropriate. And is anyway a division which at least echoes the other. For the process by which the radiation and matter of the universe decoupled, the matter began to dominate, and the radiation eventually became the weak background stuff we know today, is the process in the course of which the clumps of matter which became galaxies, suns and planets actually clumped. If the mass-energy which is the universe we know today was to emerge, this 'clumping' tendency needed to be not too little (because no stars would then have formed) and not too large (because fast-forming stars would have kicked the planets from their orbits). That it actually *was* so 'not-too-little-and-not-too-large' is what lies behind the frequently heard but nonetheless fallacious 'just-right-universe' argument already considered (see Chapter 3). For now, what matters is just that it was 4,600 million years ago, give or take some 10 million years or so, when one particular clumping clumped. And the earth we now call 'home' began to take shape.

All of the above is now in the realms of the well-established commonplace. Perhaps the only thing I am adding is that we mark the division between this the inorganic, and the organic to come, in more clearly distinct language. I am proposing to refer to the inorganic as 'of the universe', and the organic as 'of the biosphere'. The term 'biosphere' was coined in the nineteenth century by Eduard Seuss, an Austrian geologist, who used it to describe what he called 'the thin green smear' of livingness which coats the surface of planet Earth. And although it would have been nicely neat to be able to talk about the three entities corresponding to the three levels as 'the universe, the *earth* and the world' — as first I did — the confusions inherent in the use of such a middle term cannot I think justify such sloppy (even if appealing!) usage. Especially in the light of the singularly unhelpful Gaia hypothesis, the one that asks us to think of the earth itself as a living system, 'the universe, the *biosphere* and the world' it has to be.

The basis of the division between the inorganic and the organic is, of course, the emergence of that which lives: of 'life'. Of course that which lives is also subject to the four forces: there is no additional or 'vital force'. Of course that which lives is also made from some of the one hundred and eleven chemical elements: there is no additional 'vital fluid'—or gas or solid, come to that. But it will be my case that a theory of the organic, that which will explain the organic, while it cannot contradict theories of the inorganic cannot either be reduced to theories of the inorganic. As mentioned already at the beginning of the chapter, with the emergence of that which lives no new forces or elements came into existence: but new realities, and thus new things needing explanations and theories, did. And explanations and theories, I will argue, have to be derived from the specific study of the specific entity they seek to understand. Just as we derive theories concerning the inorganic reality of the universe from study *of that inorganic reality*, so we must derive theories concerning the organic reality of the biosphere from study *of that organic reality*. There is no short cut such that theories of organic reality can be read off from theories of inorganic reality, without going via that organic reality itself. So: what then of that Level Two organic reality?

In his *Matter and Consciousness* (1984), Paul Churchland defines life as 'any semi-closed system that exploits the order it already possesses, and the energy flux through it, in such a way as to maintain and/or increase its internal order.' In *The Life of the Cosmos* (1997), Lee Smolin defines life as 'a distinguishable collection of matter, with recognisable boundaries, which has a flow of energy, and possibly matter, passing through it, while maintaining, for time scales long compared to the dynamical time scales of its internal processes, a stable configuration far from thermodynamic equilibrium'. Such life, he adds, is both 'maintained by the action of cycles involving the transport of matter and energy within the system and between the system and its exterior', and 'stabilised against small perturbations by the existence of feedback loops which regulate the rates of flow of the cycles'. Such 'collections of matter', which according to Ilya Prigogine can be shown to be associated with non-equilibrium processes by the analysis of biochemical oscillators and cellular rhythms, are stable and open and dissipative. Defined in ways such as these, life on earth began some 3,500 million years ago: that is, after the period during which heat generated by the impact of huge asteroids bombarding the earth would almost certainly have made its emergence impossible.

About the living things thus now existent, orthodoxy both firmly accepts two things, and firmly rejects a third. It accepts, firstly, that they are all carbon-based. Perhaps because no other element can even begin to match it for the huge range of bonds it is able to make, and such a range is essential to the chemical complexity of organic systems, carbon, in particular a carbon which as the result of a changed ratio between carbon-12 and carbon-13 isotopes is lighter than the inorganic variety, forms the basis of all living things. It accepts, secondly, that they all share three defining characteristics: that they all reproduce themselves; that they do so autonomously (thus, for example, excluding viruses from their number); and that they behave in ways which promote their survival. But it rejects, thirdly, any idea of there being some sort of 'vital force' that distinguishes them from their inorganic precursors. In *An Anatomy of Thought* (1999), Ian Glynn summarises this last point thus: 'that all known forms of terrestrial life have a common origin, that the substances found in living organisms can, in principle at least, be synthesised from materials not derived from living organisms, and that the behaviour of such substances can be described by the same thermodynamic laws that describe the behaviour of substances not found in living organisms.' It is perhaps worth adding here, apropos of his maybe unnecessarily tentative 'in principle at least' phrase, that Friedrich Wohler synthesised urea in 1828; that Adolf Kolbe followed with acetic acid in 1845; that from about 1865 onwards, Pierre Berthelot synthesized organic compound after organic compound, and then went on to make a few that did not occur in nature at all; and that Stanley Miller produced several of the twenty amino acids which are the basis of proteins in the laboratory in 1954, and several more more recently.

In all such living systems, the cell is the basic unit: the smallest piece of anything that is alive. The first breakthrough for life was thus unicellular. And because living systems acquire energy rather than lose it, this was a breakthrough against all the thermodynamic odds. It was made by single-celled bacteria, each only thousandths of a millimetre long, but each with its own genetic instruction kit. In traditional classification, as established by Linnaeus, these pioneers did not figure in their own right at all. His division of living things into two main kingdoms, animalia and plantae, was just 'tidied up' by including fungi and bacteria as plants, protozoans as animals. But in part as the result of the realisation of the huge and distinct role of the bacteria, this all changed. In 1959, R. H. Whittacker of Cornell University proposed a new classificatory system with five kingdoms: monerans (bacteria); protoctists (protozoans and seaweeds); fungi; plants; and animals. It was a system nearer the mark than that of Linnaeus. But as the result of some contemporaneous and major developments in biological science, it was one which turned out to be very short lived indeed. The developments were threefold: the huge growth of molecular biology following the work of Crick and Watson, culminating in the discovery of the structure of DNA in 1953; the elaboration of the technique of cladistics (classification by common special features) by Willi Hennig in the 1960s; and the steady discovery of the existence of many more species than had been dreamed of even just decades earlier. As a result of these developments, biologists now work not with two or five 'kingdoms' but with three 'domains': Bacteria, Archaea and Eucaryota. The last-named category, that of creatures with cells which have nuclei, contains between ten and fifteen 'kingdoms', two of which are the plants and the animals. This is a system which, with its three major 'domains', certainly ensures that those founding bacteria have at last come fully into their own.

With neither a nucleus nor mitochondria, bacterial cells are the more primitive: with both a nucleus and mitochondria, eukaryotic cells the less primitive. The membrane-bound nucleus of the latter can carry thousands of chromosomal genes: their sausage-shaped, energy-producing mitochondria—since the work of Lynn Margulis, now generally accepted as having originally been separate bacteria which were at some stage incorporated by symbiosis into their host—just a few. And it is possible to argue, as does Lewis Wolpert for example, that second only to its very origination, it is the eukaryotic cell which is the major breakthrough in the evolution of life. Once such cells are in existence, multicellular evolution requires little else to 'take off'. And up to and including *Homo sapiens*, an adult individual of which species consists in about 100 trillion of them, they are the basic building blocks of all but the very simplest of living things.

Perhaps the next significant landmark in that evolution might be thought of as the one made some 2000 million years ago by the plants: the organic form which still accounts for 99% of the earth's biomass, and explains Seuss's already mentioned 'thin-*green*-smear' definition of his 'biosphere' word. In the fastest chemical reaction known, green plants trap sunlight by means of chlorophyll molecules. They then use the energy thus captured to extract electrons from water, thus making possible that synthesis of carbon compounds which feeds almost every living thing on the planet. The distinguishing feature of this 'vegetable' life-form is its synthesis of cellulose via photosynthesis: a process which involves the taking in of carbon dioxide and the pumping out of free oxygen. For nigh on 1000 million years, this was the photosynthetic process which, by so transforming the atmosphere, made possible the eventual emergence of a new 'animal' life-form: a form for which that plant-made oxygen was a basic condition of its very existence.

It is at this stage of the story that the first signs of a significant lack of consensus start to emerge. The oldest animal *fossil* so far identified is about 800 million years old. But as recently as 1999, researchers at Pennsylvania State University claimed to have traced the origins of the nematode worm to 1200 million years ago. One of the first organisms to have its whole genome sequenced, and known affectionately as just 'the worm' to the many scientists working on it, this is a claim which, if substantiated, would mean that animal origins would have been pushed back to a time several million year earlier than previously thought. And there have been other such claims for other creatures. For my purposes, however, it is sufficient simply to register the debate, and proceed. Beginning at about 1000 million years ago then, animals, creatures with directed movement, non-photosynthetic metabolisms and a blastula (hollow ball of cells) stage in embryonic development, began to emerge. First the invertebrates: creatures such as centipedes, coral, sponges, mites, scallops, segmented worms, spiders and, by far the largest category, insects. Still today almost 90% of animal species, about 800,000 of them, are invertebrates. And among them, 275,000 are species of beetle! Then came the vertebrates: animals which make up the other 10% of animal species, and which solve the problem of how to 'stay together' as one of Smolin's larger 'collections of matter' not by building a hard shell around their outsides but by constructing a hard scaffolding along the length of their insides.

The emergence of animal species would seem to have reached some sort of peak around 500 million or so years later, at about 500 million years ago, in what palaeontologists call, somewhat contentiously, the period of 'the Cambrian explosion'. The contention is there because of a current dispute about rates of evolution. In this dispute, the minority position is one suggesting variable rates over time: the majority position one suggesting pretty constant rates over time. Spectacularly reminiscent, for a social scientist, of the argument between those advocating revolutionary social change, and those stressing what somebody called 'the inevitability of gradualism', the 'gradualist' evolutionists, the majority, sometimes wince at the 'explosion' metaphor. I understand their concern. There is certainly a need to counter what in his book *Life: an Unauthorised Biography* (1997) Richard Fortey amusingly portrayed as a widespread image of geological history among non-geologists: an image according to which there was, in quick succession, a shift from primeval soup, via mysteriously extinguished dinosaurs, to humans wearing off-the-shoulder animal skins and wielding clubs. But I suspect it is better to go on struggling for a better public understanding of the time scales of geology and palaeontology than to give up on the 'explosion' metaphor. Coming to understand, via an appreciation of the role of bacteria for example, that the Cambrian was not an explosion which *created* life; coming to understand that it was an 'explosion' which exploded gradually (!) across some 35 million years: these are surely understandings well worth having. But be that as it may, all that I really need to register is that most of the animal forms familiar to us today appeared over probably a period of some 35 million years, and about 500 million years ago. And that one current suggestion as to how that might have come about, an hypothesis offered by the invertebrate biologist Andrew Parker of the Australian Museum in Sydney, concerns a development he believes to have taken place in the ancestors of the crustaceans which he studies. From a type of reflector known as a surface diffraction grating, many of them produce an iridescence which he believes might well have been related to that much contested 'explosion'. He sees what in palaeontological terms is the sudden introduction of vision into the lives of the animals of the period as crucial to this which was the first big expansion in biodiversity. The need to adapt to a biosphere in which there were creatures which could both see and be seen, he argues, might well have heralded myriad adaptations. And such adaptations could well have been predicated upon the emergence of those genes, known now as hox genes, which were capable of directing the cells of developing embryos to front or back, top or bottom, near or far, in more sophisticated fashions.

Just which of the previously mentioned defining characteristics of an animal—directed movement, non-photosynthetic metabolisms, the presence of a blastula stage in embryonic development—is taken to be defining depends on context. For example, in the current debate over whether the Ediacarans, living things that dominated the earth immediately before the Cambrian, are to be taken as animals or not, it is the 'blastula' criterion which is most often used. But things get easier later. By the time we get to the level at which sentience has emerged, nobody to my knowledge, not even Prince Charles, thinks that we are anywhere but among the animals. Life forms without nervous systems are referred to as 'irritable': that is, they are able to respond to their environments, as does a sunflower, for one example, when it turns its face to the sun, or an amoeba, for another, when it encloses its sustenance with its pseudopodia. But the 'directed movement' of which animal life becomes capable would have put a considerable premium upon the development of the considerably more powerful 'sentience'. To be able to feel the impact of a stimulus which can be neither approached nor avoided is perhaps gratuitous. But to be able to feel the beneficial impact of a stimulus which can be approached, the dis-beneficial impact of a stimulus which can be avoided, could be more than a little useful. Thus with some type of 'approach-avoidance' behaviour and/or 'fight or flight' mechanism present in otherwise widely different animal species, it is perhaps possible to see in sentience the evolved consequences of that capacity for directed movement. And as Charles Sherrington pointed out, once sense organs were in place, they could exponentially increase the evolutionary benefit they conferred if they could grow first a series of ganglion concentrations, then a single and fully-fledged brain, which could use the information they provided to predict the immediate future.

When now we approach the Level Three Reality of my chart, things begin to get complex, contested and confused. But not perhaps immediately. Immediately, when considering just the emergence of the human from the non-human, fairly widespread agreement can still probably be said to prevail. From about 20 million years ago until about 6 million years ago, the period of the Miocene, apes were the dominant animals on the planet. But during the period, a group we now label the hominids emerged and developed from amongst them. Its first development involved the divergence from the common stock of a sub-group which became the lesser apes (gibbons and siamangs); its second, the divergence of creatures which would become orang-utans. The third and last division of the remaining stock, at about 6 million years ago, produced two lines of development: one to the great apes (gorillas and chimpanzees) and one to hominids; to humans. This, at least, is current orthodoxy.

The first real step down their own and separate line of evolutionary development made by these newly-emergent hominids is now accepted as being bipedalism, or upright walking. But this is in fact a relatively recent change of perspective. Forced upon us by the slowly emerging detail of the fossil record, it is a change which nicely illustrates the crucial importance of that record to studies of early humans. It used to be thought that bipedalism, large brains and tool making emerged together: that the greater intelligence available to our larger brains made tool-making possible; and that bipedalism freed our 'arms' and 'hands' for the task. But when the fossil record showed that bipedalism preceded both, perhaps by as much as a million years, such an hypothesis had to be abandoned. It has been replaced by one which suggests that once our ancestors were out of the forests, bipedalism would have been a stance which, because it meant that a smaller surface area of the body would be presented to an overhead and tropical sun, would result in less over-heating: a development which would open up the possibility of that all-day hunting that would confer a comparative advantage over competitors. It would also, of course, explain why our ancestors retained protective hair on the top of their heads, as do most if not all of us to this day.

At different times during the evolution of the line leading to modern humans, their genus, *Homo*, has consisted in several different species. *Homo habilis*, the earliest user of stone tools, emerged some 2-3 million years ago, and was almost certainly the first of them: though some recent scholarship has advanced the possibility that *habilis* was the last of the australopithecines rather than the first of the hominids. The so-called Oldowan technology of *habilis* consisted in simple, sharp stone flakes and flaked cobbles. Tom Wynn, an archaeologist at the University of Colorado, and Bill McGrew, a primatologist at Miami University in Ohio, suggested that the making of such tools was well within the capacities of apes, whose cultural abilities had thus been underestimated. But Nick Toth, an archaeologist at Indiana University in Bloomington, has used Kanzi, a bonobo chimp we shall have occasion to meet again later, to show that the 'hard-hammer percussion' technique needed to make them, while it is completely beyond his capacities is not (after much effort!) beyond those of his human trainers. In this Oldowan technology, therefore, we have that which is a strong candidate as answer to what is arguably the central question of anthropology: the question of the origins of human culture.

After *Homo habilis* others followed. *Homo ergaster* and *Homo rudolphensis,* living at much the same time, are perhaps distinct species, perhaps not. *Homo erectus,* including what used to be called 'Peking man', more certainly is. Emerging at about 1.8 million years ago, *Homo erectus* continued to live in East Asia until about 300,000 years ago. *Homo heidelbergensis,* a descendent of *Homo erectus,* flourished at about 500,000 years ago. *Homo neanderthalensis,* emerging in Europe and the Middle East about 150,000 years ago, and surviving, in some parts of Europe, right up until about 30,000 years ago, was at one time thought to be our ancestor. But in 1997, the testing of DNA extracted from fossil bones showed reasonably unambiguously that this was not the case. They were instead our 'distant cousins', as the *New Scientist* headed its coverage of the story: which coverage also floated the possibility that we killed them off in what would have been, had it happened, 'one of the bloodiest episodes in human prehistory'. *Homo sapiens* came onto the scene some 150,000 or so years ago, and is both the most recent hominid species to emerge, and the only one now extant. Although it had perhaps not yet reached Europe, by about 50,000 years ago it was a species—sometimes now referred to as 'anatomically modern humans'—which had spread to cover most of the globe.

In *The Origin of Humankind* (1994) Richard Leakey summarises and systematises all this into an overall agreement among researchers that we can identify four distinct stages in the evolution of the human: first the initial emergence, perhaps seven or six million years ago, of a species still ape- like but with bipedal locomotion; next a period perhaps five million years long when, via the process of adaptive radiation, many such species proliferated, each adapted to its own distinct niche; third the arrival, perhaps three or two million years ago, of the new genus *Homo,* as exemplified by species with markedly larger brains than the others, and eventually by *Homo sapiens;* and finally the emergence, perhaps no more than 100,000 years ago, of language-using, 'anatomically modern' humans.

The major *disagreement* about early humans, though even that seems now to be receding, is not about the existence of the various species, but about the way in which they spread, and thus the way in which they are related to each other. There are two main models. According to both, *Homo erectus* evolved in Africa between one and two million years ago, and from there began to migrate to Asia and Europe. But then opinions divide. The 'single origin' model, sometimes called the 'out-of-Africa' or 'mitochondrial-Eve' or 'population-dispersal' model, is that adopted by a now growing majority of palaeontologists and anthropologists. It suggests that

Homo sapiens repeated the dispersal pattern of *Homo erectus*, but much later. It also emerged in Africa, but less than 200,000 years ago. And from there it migrated: eventually to all other parts of the world; eventually replacing all the species of earlier waves of migration, including *Homo erectus* and *Homo neanderthalensis*. The alternative 'multiregional evolution' model, adopted now by fewer and fewer palaeontologists and anthropologists, suggests that *Homo sapiens* arose more or less simultaneously throughout many parts of the world, and did so with *Homo erectus, Homo neanderthalensis* and the others as ancestors. However, the Neanderthal DNA discovery mentioned earlier is now considered, by many, decisively to rebut such an hypothesis.

From now on in, the previously mentioned complex, contested and confused state of the study of the specifically human must change the sort of summary I offer here. With a few nods in the direction of dissenters, my hypothesis has up to now simply accepted what is the current orthodoxy, based on the current state of scientific understanding. But from now on in, I will have to be presenting that which, increasingly, is just my own position. And to do that, I need to explain some of what is specifically my own terminology, beginning with a different-from-standard use of the term 'consciousness'.

Among psychologists, and their related ilk in psychiatry, cognitive science, philosophy and other such places, it is a commonplace to distinguish 'consciousness' from 'self-consciousness'. Roughly speaking, consciousness is taken to mean 'being aware': 'self-consciousness' to mean 'being-aware-of-being-aware'. Or as the seventeenth century poet John Donne put it: 'the beast does but know, but the man knows that he knows'. Thus dogs, it is argued, are conscious. They are aware of their world: they feel it impinge upon them. But they are not *self*-aware—hence their habit of looking behind the mirror for the dog which is barking at them. This is difficult territory, as in later chapters will to some extent emerge. But at the moment, I simply want to record that what orthodoxy calls 'consciousness' I will call sentience': and what orthodoxy calls 'self-consciousness', I will call 'consciousness'. I take my usage to be following the spirit if not the letter of that of William James, who I take to have been suggesting that where sentience is the experience of things, consciousness is the experience of experience. But more importantly, I adopt this usage for two reasons central to what will later be my case as to the nature of a person.

First: to say that a non-human animal is conscious, but not of its-self, can I think mislead us into thinking that it in some sense *has* a self to be *not conscious* of . But since I will be arguing that the relationship between selves and consciousness is essentially constitutive, that not 'having' a self but 'being' a self is constitutive of not 'having' consciousness but 'being' conscious, my position is better served by the sentience-consciousness terminology. It is a terminology which means that much of the current flurry of academic interest in consciousness, much but not all of it (if the *Journal of Consciousness Studies* is anything to go by) among what have been called 'consciousness physicists and ecclesiastical neuroscientists', becomes instead a flurry of interest in 'sentience'. Whether that flurry turns out to be one directed at a real biological problem or a pseudo biological problem remains to be seen. But whichever, the concept of consciousness which I will use will not refer to a *biological* problem of any kind, real or pseudo.

Second: my usage preserves etymologically that which will later be important to my case semantically. The 'knowing with' (con scio) roots of 'consciousness' are also the 'knowing with' (con scio) roots of 'conscience'. I will argue that the language out of which selves are generated emerged not for simple communication but for more complex co-operation. And that such complex co-operation, facilitated by the emergence of what is now often called a 'theory of mind' among its interlocutors, is unique not to human animals but to human persons. That being the position I will defend, the preservation of the lovely 'knowing-with' imagery contained in both the 'consciousness' word and the 'conscience' word for only the human personic is appropriate. My hypothesis therefore proposes that human persons are the first (and, anyway so far, the only) conscious entities. And that what that means is that they are the first (and, anyway so far, the only) linguistic entities.

It also therefore proposes—it just follows definitionally—that not all humans have been (or are) conscious. Though obviously thoroughly sentient, two categories of pre-lingual humans are thereby taken to be non-conscious: first, those members of earlier species within the genus *Homo* which were without language; second, infants of the species *Homo sapiens*. In other words, although some students of these things think a case can be made for Neanderthals having language (see below), the genus *Homo* became conscious (with the acquisition of language) probably only at the level of *Homo sapiens*. And individual members of the species *Homo sapiens* become conscious (with the acquisition of language) somewhere between one and two years of age.

The second proposition, the one concerning *Homo sapiens* infants, will be considered later (see chapter 9) when what some see as the problematic moral status of the 'without language' infant will be explored. Suffice it now to say that, in terms of a concept something like Habermas's 'speech community', an infant, though not yet able to *produce* speech, is already able to *respond to* speech, and is maybe therefore best thought of as a member of such a community *before* it is an interlocutor. This, after all, is what a concept such as 'interactional synchrony', as it figures in child-developmental psychology, is all about.

But the first proposition, the one concerning earlier species in the genus *Homo*, needs a little more fleshing out here and now. The general anthropological piecing together of evidence about how the various *Homo* species lived, together with the more specific (and incredibly interesting) detective work on the fossil remains of the relevant bones, particularly jaw-bones, is work which I anyway have come across mainly in reading Lieberman (1984) and Mithen (1996). It is work which would seem to suggest that what Mithen calls 'enhanced vocalisation' might have begun to emerge right back with *Homo habilis*. This would have been made possible because the upright position needed for that bi-pedalism already characteristic of this the earliest *Homo* seems to have led directly to the descent of the larynx. No such descent takes place in chimps: indeed it is precisely because they are therefore such poor vocalisers that American Sign Language (ASL) is so often used in experimental 'animal language' work with them. And neither does it take place in *Homo sapiens* infants until they are about three months of age. *Homo erectus* probably had a vocalising capacity much greater than that of modern apes, but there is no suggestion that it was a language user.

More recent evidence concerns *Homo ergaster* and the Neanderthals. Early in 1998 Ann MacLarnon of the Roehampton Institute in London reported on work which showed that fossil remains of *ergaster* have narrow canals in the thoracic vertebrae of the ribcage: canals which are similar to those of today's non-human primates. This is by contrast with early *Homo sapiens*—from about 100 000 years ago—where the canals are visibly wider. And MacLarnon suggested that the increased space might have been needed to carry the extra spinal cord necessary to the fine-tuned control of breathing, and thus to efficient speech. So perhaps we could narrow things down a bit and say with some confidence that *Homo ergaster*, living in Africa some 1.6 million years ago, almost certainly did not use language. Also in 1998, Richard Kay of Duke University in North Carolina reported on work measuring the width of the hypoglossal canal in modern chimpanzees, in the fossil skulls of Neanderthals, and in modern humans. It is at its widest (about the width of a pencil) in modern humans, who do speak: it is at its narrowest (about half that) in chimpanzees, which don't. Since the Neanderthal canals fell within the same size range as those of the modern humans, and the canals are used to carry the nerves that control the tongue through the base of the skull, it is Kay and his colleagues who have suggested that maybe they also could speak: a suggestion which, if it were true, would mean that complex vocal abilities might have evolved more than 300,000 years ago. This, however, is not the current orthodoxy in the field.

The brains of *Homo sapiens* had evolved to their modern size and shape by about 100,000 years ago. And it is this fact which has made so many people home in on that '100,000-years-ago' date as the one which probably saw the emergence of language: or anyway of the sort of language (discussed at length in Chapter 9) defined by criteria such as those of Hockett (1960) or Aitchison (1983), and involving grammatical rules and some considerable lexicon. Of course what we are talking about is a process not an event: the linguist Derek Bickerton suggests that the syntactical aspects of language may have emerged later than other and simpler ones; the linguist Leonard Rolf thinks it is possible to identify no fewer than eight stages in the evolution of grammar. It would nonetheless seem to be pretty safe to say that language use is a phenomenon at the very most only some 500,000 to 200,000 years old, rather more probably a phenomenon only some 100,000 to 50,000 years old: or as old as the youngest *Homo*, which is ourselves. But if such an hypothesis is the best we've got so far, it is not without its problems. And not least among such problems is the problem of 'the gap'.

It was between about 50,000 and 25,000 years ago that, throughout the Old World, our ancestors began to decorate their bodies, their tools and the walls of their caves. For my relatively simple purposes, a couple of the better known examples will suffice to give the flavour of what this means. There is, first, the example of a site dating from 28,000 years ago in Sungar in Russia, which revealed the bodies of two children and an adult who were decorated with thousands of ivory beads, estimated to have taken more than 10,000 hours to make. The little boy wore also a belt made from the teeth of 60 foxes. There is, second, the more celebrated and recent example of the cave paintings found in January 1995 by the French cave researcher Jean-Marie Chauvet. Made of ochre, charcoal and hematite, these were some 300 prehistoric rock paintings, and just about as many engravings, which he had come across in a cavern at Vallon-Pont-d'Arc in southern Ardeche. They depicted several species of animal, perhaps twelve or so, some of which had never been seen before in such rock art. And carbon dating revealed some of them anyway to be some 31,000 years old. Given that the previously oldest (reliably dated) examples were the 27,000-year-old handprints in the underwater Cosquer cave near Marseille, this made the Chauvet paintings the oldest so far found, and by some 4000 years. For a while, the main debate that ranged around them was whether or not, given their great age, they could have been the work not of *Homo sapiens* but of the Neanderthals: there is evidence that Neanderthals lived alongside modern humans until about 34,000 years ago in central France, and until about 27,000 years ago in Spain. But most researchers – Clive Gamble, for example, a pre-historian at the University of Southampton – are unconvinced, and continue to think that the first artists were descendants of the modern Cro-Magnon humans who arrived in Europe from Africa some 40,000 years ago.

It is these examples, and others like them, which highlight 'the gap' problem. For given this and other evidence that the artistic abilities of anatomically modern humans took some sort of 'great leap forward' maybe only 40,000 to 30,000 years ago, how can we make this tally with the idea that those same anatomically modern humans had been using language for so much longer than that? If it is language use which makes all the difference between 'us' and 'them', then what were our ancestors up to for the 60,000 or so years intervening between *its* origins and the origins of our artistic abilities? With the use of the symbols which are language in place, the use of the symbols which are decorated bodies, tools and caves would appear to have been unreasonably long-delayed. What's more, if we take seriously the ideas of such as Corbalis (see Chapter 9) that language might initially have been gestural, only eventually vocal, the problem becomes, if anything, more intractable. It would seem that this question, the question if you like as to why the Upper Palaeolithic did not begin 100,000 years ago rather than 40,000 years ago, is basically just that: a question. And one to which there's as yet no reliable answer. Before leaving it at that and moving on, however, it is perhaps important to register something about this problem: to wit, that it's one which, whatever its eventual solution, is singularly important to issues concerning the relative significance of the biological and the cultural in human affairs.

But anyway having now reached the Palaeolithic, the period running from about 3 million years ago to about 1200 BCE, it feels as though we are almost home and dry. After all, on the time scales with which we have just been operating, the 'Old Stone Age' of our schooldays is only yesterday, and my potted summary can perhaps become very potted indeed. During the Mesolithic period there came the spectacularly significant development of agriculture (at about 8500 BCE in what is now Turkey), and the highly significant discovery of metal smelting (at about 4500 BCE in Mesopotamia). During the Neolithic period there came the spectacularly significant development of writing (at about 3000 BCE in Sumer). And in Egypt, Babylon, the Indus valley and China, we can shift our attentions to the first human civilisations. So: language at about 100,000 thousand years ago is the jumping-off point which shifts me from my Level-Two-stage-four age of sentience to my Level-Three-stage-five age of consciousness. And wherever it more precisely emerged—an empirical question which will only be answered by further research—the thing in my position concerning language that really needs defence, which is thoroughly contentious, is that it is precisely its coming which *constitutes* the coming of consciousness. This is what I am suggesting. And to do so a little more systematically, I need to get in place some more of the terminology by means of which I will suggest it.

I am going to suggest that the idea that there is 'a' human condition is fundamentally mistaken. And that the mistake involved leads to almost all the messes that philosophical conversation gets itself into. I am going instead to argue that there is not 'a' human condition but 'three' human conditions. And that carefully distinguishing them, and holding the distinction clear in our heads, is vital to our understanding of our selves. That this is an understanding not clear in most of our thinking, most of the time, is given away by the second of the two observations of my last chapter: the observation that we have so many ill-differentiated words for ourselves. Let me therefore introduce, in one (I hope not fell) swoop, all the 'us' terminology I intend to use, with a first and brief indication of how I intend to use it. While some of it will almost certainly seem a little odd at first, I think the attempt at precision makes it well worth the effort to try to get hold of it.

Firstly, when considering the human I will work with four categories: one 'not-human' category and three 'human' categories. The one not-human category is to indicate not *everything* that the human is not, but just everything *organic and sentient* that the human is not. The three human categories are to indicate the three *different* things which, at succeeding stages of its development, the human is. To indicate the not-human category, I will use the phrase 'animal-animal'. To indicate the three different human categories, I will use the phrases 'human-animal', 'animal-human' and 'human-human'. I will use the phrase 'human-animal' about two categories: about whole species of the genus *Homo* which were without language; and about individual members of the species *Homo sapiens* which are without language. I will use the phrase 'animal-human' about members of the species *Homo sapiens* who, via their use of the I AM language which identifies a self as that which *has* experience, are posited as persons with I AM consciousness. And I will use the phrase 'human-human' about members of the species *Homo sapiens* who, via their use of the I IS language which identifies a self as that which *is* experience, are posited as persons with I IS consciousness. I will suggest that all three human categories are extant: infants, the post-lingual senile and the unconscious are in the 'human-animal' category; most of us most of the time are in the 'animal-human' category; some of us some of the time are just beginning to reach after the 'human-human' category.

Secondly, I will use the words 'person' and 'persons' in particularly defined ways. I will not use these words to refer to those in the 'animal-animal' or 'human-animal' categories. I will use these words to refer to those in the 'animal-human' and 'human-human' categories. Respecting what are perhaps the 'per sonare' (through sound) roots of the word, the restriction of the term 'person' to apply only to humans with language makes 'personness' a uniquely Level Three phenomenon. Such phenomena are those which I will call 'personic': a term coined for the purpose because the alternative 'personal', having been long in the language, has acquired meanings and associations—with 'the private', for example—which are inappropriate to my purposes.

Thirdly, I will use the words 'self' and 'selves' in particularly defined ways. I will limit their use to refer only to stages in the development of persons. Stage theories of the development of persons have been elaborated for the most part by psychologists operating from within the traditions of developmental psychology. They have been associated with the names of such as Freud, Erikson, Vygotsky, Piaget, Maslow and Kohlberg. And in adopting that usage whereby across a lifetime a person is thought of as existing always as a particular type of self, which type is understood always as a development upon the one previous, I am adopting not any one of their schemes in particular but the spirit of them all in general. I will work with a model consisting in a sequence of 'animal-human' selves, called 'normal' selves, each with a quantitatively more developed form of I AM consciousness, and a single 'human-human' self, called a 'moral' self, with a qualitatively higher form of I IS consciousness. And to complete this 'self' terminology, I will borrow a lovely word from a Gerard Manley Hopkins poem, and refer to the whole process of development across selves as 'selving'.

And finally, I will use the word 'experience' in a particularly defined way. I will use just 'experience' to mean the summed-to-date consequence of all the previous moment-to-moment states of the nervous system of a *sentient* being. And I will use 'linguistically-structured-experience' to mean the summed-to-date consequence of all the previous moment-to-moment states of the nervous system of a *conscious* being. Thus for 'animal-animal' and 'human-animal' sentient beings, it will be their 'experience' which constitutes their non-personness. And for 'animal-human' and 'human-human' conscious beings, it will be their 'linguistically-structured-experience' which constitutes their personness.

With these four explanatory paragraphs in place, my final and crucial working definitions can now emerge. A person becomes 'the linguistically structured experience of a human being': with a normal person being the linguistically structured experience of an I AM animal-human; and a moral person being the linguistically structured experience of an I IS human-human. All of which means that I can now return, 'word-armed', to an explication of the thoroughly contentious proposition that it is the coming of language-use which precisely *constitutes* the coming of consciousness.

I want to suggest that the 'knowing-with' (con-scio) phenomenon which is consciousness, a 'knowing with' both the past-and-present selves of other past-and-present persons in a shared speech community, and the past-only selves of our own present-only person retained in memory, is only available to brains linguistically-structured: that is, is only available to brains with the capacity to hold a representational system which refers contingently. In the history of reference, as non-contingent signs give way to contingent symbols, so those contingent symbols come to be able to refer not merely to their referents but also to each other. And I want to suggest that it is such inter-symbolic reference which actually *constitutes* that inner world which we call our consciousness: that consciousness, therefore, is not something 'expressed in' language but 'made out of' language. In other words, it's not that human-animal non-persons *have* non-linguistically-structured experience which, when language emerges, is then clothed in words. It's that animal-human persons and human-human persons *are* linguistically-structured experience which, before language emerges, just aren't. They don't exist.

This is an approach to an understanding of consciousness which while contentious—which approach isn't?!—is not one which I anyway take to be particularly peculiar to me. In my reading on these things, I have come across positions not a million miles from this one in several places. Indeed, in working my way towards my own formulation, I have used the ideas—as best I understand them—of a very long line of people: of Leibniz with his 'knowing-that-I-know' use of the concept 'apperception'; of Claparede, Janet and especially William James; of the battling Piaget and Bruner, the battling Skinner and Chomsky, and the newly influential Vygotsky; and among the 'moderns' I've come across, of such as Nick Humphrey, Ned Block, Keith Oatley, Dan Dennett, Paul and Patricia Churchland, and especially Benjamin Libet (see Chapter 6). In stringing out this list I in no way want to claim that my simple 'consciousness *is* language use' position is, anyway in its bald simplicity, also theirs. For one thing, being frequently to be found in noisy combat with one another, 'they' don't have a position. It's just that, before leaving these issues for later consideration (see Chapter 9), I want to suggest that, anyway as I understand it, mine would be a stance in no way fundamentally alien to some parts of several of theirs.

And then, by contrast, to add that that would certainly *not* be true, or so I believe, of my final and maybe most radical claim. This is a claim which, as far as I know, is mine alone. I am wanting to suggest that that language which, as it emerges, simply *constitutes* persons—constitutes them, that is, as 'linguistically structured experience'—emerges in two stages: first, as the language of what I am calling I AM persons; second, as the language of what I am calling I IS persons. I will argue that the I AM language, and thus the I AM consciousness, of I AM persons—that is, persons functioning at the lower stage of the personic—is characterised by dualism and voluntarism: that is, that they are persons for whom there is a double entity consisting in 'my non-determining experience' (1) which *is had by* a free-willing something called 'me' (2). And that the I IS language, and thus the I IS consciousness, of I IS persons—that is, persons functioning at the higher stage of the personic—is characterised by monism and determinism: that is, that they are persons for whom there is a single entity consisting in 'my determining experience' (1) which *is* a non-free-willing something called 'me' (1 again). And that the transition from the lower-level and incorrect dualist and voluntarist thinking to the higher-level and correct monist and determinist thinking is an on-going process which will involve huge and far-reaching consequences for our understanding of ourselves. Somewhat now spelled out then, this is the hypothesis which, in my next three 'experiment' chapters, I will try to put to the test. I hope I have succeeded in presenting some sort of first and general understanding of just what it is I'm proposing. And some sort of initial and specific exposure to the language I'll use in proposing it. Of the earlier propositions in its outline, those which are really no more than nods in the direction of current orthodoxies, I expect most of my readers now to be thinking: 'yes, of course'. Of the later ones, those which I think become more and more specifically just mine, perhaps something more like: 'you must be joking'. Of these second, it will be my task in the rest of what I write to persuade you instead to think 'oh I see!'; to think 'yes of course!'; with a bit of luck, to think—'eureka!'

So: the hypothesis again—

That with reality defined as 'all that there is', the coming into existence of its several parts can be sequenced into three levels and six stages, such as to reveal not one human condition but three human conditions, not one 'I' but two 'Ies': and that the removal of the confusions arising from thinking outside such a sequencing pattern solves (or dissolves) the problems of philosophical conversation outlined in the previous chapter, and makes possible the creation of the moral sciences.

—and for the sake of simplicity, its name

In the rest of this book I will refer to this hypothesis as 'the I IS hypothesis'.

Chapter Six

EXPERIMENT ONE

From 'acts-and-omissions' to 'I and thou'

So firmly is experimentation lodged in the public imagination as what science is all about that 'doing science' and 'doing experiments' are often thought of as if they were one and the same thing. About the nature of experiments, this is a confusion which goes along with quite a few others besides: that they are done in laboratories by people in white coats; that they prove things; that they can be done on all sorts of material things (like potassium, or water, or bridges, or elephants), but on no sorts of non-material things (like democracy, or figments of the imagination, or human nature, or God). What's more, these are all public perceptions which many in the scientific community are inclined to share: particularly those many who *do* work in laboratories, *do* wear white coats, and *do* think they prove things. And when to these which are general misconceptions are added two which, though particular, are just as widespread, misunderstanding can and does flourish.

The first such particular misconception concerns a tendency to over-rate their importance vis-à-vis other parts of the scientific method. This can result in an over-valuing of the ingenuity involved in the design and execution of an experiment, and an under-valuing of the imagination needed in the formulation of the hypothesis. For while experimental ingenuity can often be quite spectacular, the hypothetical imagination, without which there could be no experiment to design or execute anyway, can sometimes be just awesome. No less true for being the truism that it is, the appropriate delineation of a problem is almost always more difficult than the finding of its solution. Indeed it can be argued that once a problem is so articulated as to make sense, its solution is just a matter of time.

The second is to the contrary. It is a tendency to under-rate their importance as activities that enter into almost every aspect of our lives. Origins can help here. In Latin, 'experiens' is the present participle of the verb 'experiri', meaning 'to try out' or 'put to the test'. And in this sense, there is a way in which our every action is an experiment: that is, a 'trying out' of a particular way of doing things in order to give ourselves 'experience' (same root) of the consequences.

Experiments are controlled experiences: or as J. D. Bernal puts it in his *Science in History* (1954), 'an animal can learn by experience; a man, in using science, goes beyond this and experiences to learn'. But it is sometimes suggested that these 'controlled experiences', these 'experiments', work differently at the three different levels of reality: in particular, that experiments at level three are fundamentally different from those at levels one and two. There are differences. And some of them will later emerge. But right now, what I want to draw attention to is a fundamental sameness. Experimenting on the phenomena of level one (the universe) leaves the universe and the biosphere the same—but changes the world; experimenting on the phenomena of level two (the biosphere) leaves the universe and the biosphere the same—but changes the world; and experimenting on the phenomena of level three (the world) leaves the universe and the biosphere the same—but changes the world. Here is the fundamental sameness. And given how important it is to my argument that this be understood, let me consider a very basic example.

Physical reality is a level one phenomenon. If all persons simultaneously ceased to exist two seconds from now, it would still be there: and would consist now, in anyway some small part, of millions of *Homo sapiens* corpses! However, the *study* of physical reality, the doing of physics and the making of physical laws, is a level three phenomenon. If all persons simultaneously ceased to exist two seconds from now, it would *not* still be there. Where not even every living *Homo sapiens* hosts a person, dead ones never do. The physical laws are descriptions made by us about physical reality: they are not physical reality itself. It is important to register immediately here that this fact says absolutely nothing at all about whether they can be true descriptions: that is, it says absolutely nothing at all about the fundamental problems of epistemology. It just says that they are descriptions: that if 'truth' is available at all, it is available as a level three phenomenon, not as a level one phenomenon; that 'truth', if it is available at all, is made. And because only that which is made, rather than that which just is, *can* be true (as opposed to false), *can* be good (as opposed to bad), *can* be beautiful (as opposed to ugly), this is a characterisation of truth which, far from diminishing its significance, enhances it manyfold. In short, it means that where physical *things* are constitutive of all of the universe, physical *laws* are constitutive of some of the world.

With these few paragraphs by way of introduction, I can turn now to the guts of my experiment. I am going to undertake a 'trying out'. Specimen by specimen, I am going to try out the I IS hypothesis as a possible solution to, or dissolution of, the problems of philosophical conversation represented by my list of 'ands'. My task will be to read various parts of that hypothesis, but particularly the part concerning the difference between I AM and I IS conceptions of the self, into all of the 'ands' of my list. And my aim will be to see what if any difference might be made to the problems presented by those 'ands' by their being seen in the light of the I IS hypothesis, rather than its current I AM alternative. The philosophical problems that this undertaking will present will be dealt with as and when they arise. But there are two practical problems as well. And my chosen solutions to these are ones which I must spell out in advance.

The first problem is presented by the fact that my 'and' entries, my specimens, have a dual character. They are *like* alphabetical entries in an encyclopaedia in being free-standing, and of different lengths depending on significance. And they are *not like* alphabetical entries in an encyclopaedia in being presented, anyway to some extent, as a hooked-up piece of continuous prose. This is a duality which produces a certain tension concerning just what to put where as I accumulate my case across their length. On this one, I can only hope that my somewhat piecemeal solution of emphasising sometimes the self-contained nature of a single entry, sometimes the continuous nature of the whole, will be counterbalanced by the much more singular and unified presentation, in Chapter 9, of the theory I think they collectively represent. The second problem is that of there being thirty-five of them. And although some need rather less exploration than others, each needs *at the very least* a paragraph or two. What this means is that their presentation in a single chapter would make it one spectacularly long. On this one, I can only hope that my somewhat arbitrary solution of dividing their presentation between three chapters will be counterbalanced by the opportunity such division affords for some short 'where-are-we-up-to-now?' comments to be added at the beginning and the end of each of them.

And so to the experiment I want to conduct with each of my 'and' specimens.

(1) acts and omissions

Person A sees a fight in the street, pauses to consider whether or not to get involved, decides against it, and passes by on the other side. Person B then sees the fight, pauses to consider whether or not to get involved, decides that something must be done, and tries to separate the combatants. And we say that Person B has 'intervened' where Person A did not. Person B's was an act: Person A's was an omission.

Now it seems to me that this is just wrong. And that it is wrong as a function of being an argument based in an I AM conception of a person rather than an I IS conception of a person. Only an I AM consciousness, one conceiving of selves as agents who act upon, or refrain from acting upon, an out-there entity called the world, could conceive of the 'acts *and* omissions' pair in such a way. An I IS consciousness, one conceiving of selves as always and necessarily acting *in* the world because it is actions which actually *constitute* the world, would see an omission as—just another act. Person A 'intervened' just as surely as did Person B: that is, just as surely but differently.

In a very simple sense, the I AM mistake is to imagine that a person can do nothing. But that is not possible. Our decisions to act in particular ways are constrained doubly: first by our capacity (we cannot do what we cannot do); second by our incapacity (we cannot do nothing). And one can only make that 'doing nothing' mistake if one has an image of oneself as some sort of spectator of an on-going pageant of events: events which, while I am 'in here' in myself, are by contrast 'out there' in the world. This is the image of an independent agent acting *on* the world rather than of a dependent participant acting *in* or *through* the world. At its base, it is the difference between conceptualising the self as something that *has* experience and conceptualising the self as something that *is* experience. The former (spectator) self is one posited as 'over against' that which happens —to it. The latter (participant) self is one posited as 'constitutive of' that which happens—period.

I think it is a self of the first type, an I AM self, to which Skinner was referring when, in *Beyond Freedom and Dignity* (1972), he suggested that the philosophical job of an inner 'me' is to provide an explanation of what 'I' do which does not in its turn need explaining: one which means that the explanation buck can stop with this core 'something' from which all behaviour somehow emanates. Skinner's take on this conception was that it was one which made the autonomously initiating, originating, creating entity thus posited essentially divine. And as a consequence, just as unavailable to scientific investigation, indeed just as miraculous, as any other divinity. And I think it is a self of the second type, an I IS self, to which Gadamer (1976) was referring when he suggested that our selves are not things we possess but things that happen. I take both of these characterisations to be on the right track. And I take our task of the moment to be one of coming to understand the need for a shift from the (essentially erroneous) conception of a self as an I AM miraculous entity to the (essentially correct) conception of a self as an I IS non-miraculous entity.

To aid such a conceptual shift, it might be helpful to learn to see the I AM self as just preposterous: that is, as *literally* preposterous. That which is 'pre-post-erous' is that which inverts the rational order of priorities. When we have an I AM concept of the self as the *precondition* of experience (that which *has* it when it comes along) rather than an I IS concept of the self as the *post-condition* of experience (that which *is* it as it accumulates), we have a thoroughly normal but also thoroughly preposterous conception. Our 'I' would seem to be something like a singular stand-point, a stand-point a couple of inches behind the eyes and up a bit as I once heard the philosopher Galen Strawson describe it, out from which a self which is a precondition of both acting and omitting, *acts or does not act on the world as it will*. What the I IS hypothesis suggests, by contrast, is that our 'I' is actually something like a sequence of understand-moments, understand-moments just milliseconds long as later I will argue the work of Benjamin Libet suggests, out of which a self which is a post-condition of such acting, *acts in the world willy nilly*.

This conception of the 'acts *and* omissions' duo, a conception based in a shift from I AM to I IS thinking, has hugely far-reaching consequences. In brief and for now, consider just these: that because doing nothing is not an option, it is just as surely intervening *not* to pass legislation making abortion legal as it is to pass legislation *making* abortion legal; more generally, it is just as surely intervening *not* to use new technologies such as genetic engineering as it is to *use* new technologies such as genetic engineering; more generally still, it is just as surely intervening to *maintain* an aspect of the status quo as to *change* an aspect of the status quo. Indeed, the very word 'intervene' is wrong. It suggests some sort of outside and divine capacity to as it were 'poke' the world, or not to poke it, as one chooses. But this is an illusion. A non-poke is just another poke. And the world doesn't have an 'outside'. To the extent to which an act is an act, an omission is—just another act. To adopt and adapt a word used by Don Cupitt about what he saw as the necessity to learn to live in the full knowledge that there is no personal 'outside' the world God, I too am saying that we have to learn to live 'outsidelessly': that is, in the full knowledge that there is no personal (I AM) self *over-against* the world, but instead a personal (I IS) self *constitutive-of* the world. This is what my hypothesis proposes.

But before leaving this 'acts *and* omissions' specimen, it is perhaps appropriate to say something more about the nature of acts. They have traditionally been separated from behaviours by the concept of intention: an act is of an agent; an agent is that which is guided by an intention; the nature of an act is thus determined by the intention of the agent. Thus if Person A kills a child in a motor accident and Person B kills a child by strangling it, while the 'dead child' consequences are the same, the fact that they have been brought about in two significantly different ways matters to most of the legal systems of the world. Operating as many of them do with just the three categories 'res, actiones et personae' (things, actions and persons), the 'actiones' are almost always distinguished by use of the criterion of intention: the criterion on which has rested, for example, the usual distinction between murder and manslaughter. With an I AM concept of the self as unmoved mover, as agent, this definition of an act will do: despite the fact that 'intention' is often, and at the limits probably always, a difficult concept to operationalise, if your objective is simply of the law-and-order variety, it will serve. But with an I IS concept of the self, it won't: a realisation which can lead those of a behaviourist persuasion to argue, either because they think of it as an incoherent distinction or because they think of it as an immoral distinction, that the concept of an act as opposed to just a behaviour, should be abandoned.

For myself, I think it is something other that we need. It seems to me that what a transition from I AM to I IS thinking requires here is not that we abandon the concept of an act but that we recast the concept of an act. And that we recast it in such a way that, definitionally, the concept of 'intention' gets recast with it. Let acts be the behaviours of those members of the species *Homo sapiens* whose linguistically structured experience constitutes them as persons: let intentions be the *reasons for* those acts, as given by self or others, which *reasons for them* might or might not accurately reflect *causes of them* (see 'reasons *and* causes' below). In *How To Do Things With Words* (1962), J. L. Austin mounted an argument against the idea that in using words we merely say something about the world. He claimed instead that in using words we actually do something in the world: that is, we act. I suppose what I am doing is taking that somewhat further. I am saying that we do something in the world not by using words but by being words: or more accurately, by being linguistically structured experience. And that the something we do in the world is constitute it.

(2) appearance and reality

Once I have adopted the first of the propositions of my hypothesis, the 'appearance *and* reality' problem, anyway as usually formulated, just dissolves. With reality taken to be 'all that there is', and 'appearance' becoming thus just one phenomenon among other phenomena which together constitute that 'all that there is', it becomes just a category mistake to speak of 'appearance *and* reality'. We can no more speak of 'appearance *and* reality' than of 'apples *and* fruit'. Our questions, just to *be* questions, have to be not only syntactically coherent but also semantically so: they have to make sense. And with reality defined as all that there is, a 'question' concerning appearance *in addition to* reality becomes just illegitimate: it doesn't make sense.

But this does not mean there are not further things that need to be said. For several reasons—not least because despite being an issue which just dissolves under this head, 'appearance-*and*-reality' is also one, or anyway near relative to one, which will emerge again later under my 'realism *and* anti-realism' entry—there are. So let me say just four of them here.

First, let me stress that it would be an I AM consciousness which would make such a category mistake. How things *appear* to me as compared with how they *are* in themselves is a stance redolent with what might be called a 'me-alone-and-outside' theory of the reality which is the world. By contrast, it would be an I IS consciousness which would make no such error. For such a one, how things *appear* to me, plus how things *appear* to a frog, plus how things *are* in themselves—all these things, and of course many more—would be posited as what could instead be called a 'me-along-with-all-the-rest-and-inside' theory of that same reality.

Second, let me point out that the 'all-ness' I envisage in my 'all that there is' formulation means that that which is commonly called subjective reality exists just as that which is commonly called objective reality exists. It means that the 'ontology' of my hypothesis, the set of things acknowledged by it to exist, is such that if you (or anybody else) can name it, my formulation can accommodate it. It means that since 'the limits of my language are the limits of my world' (Wittgenstein), and my language is (potentially) without limit, so also (potentially) is my world. Some among the professionals, following the celebrated argument to that effect mounted by Kant, insist that 'existence is not a predicate': that is, that 'exists' is a logical construct rather than the name of a characteristic. To illustrate this point, G. E. Moore famously gave us the three sentences 'no tame tigers growl', 'all tame tigers growl' and 'most tame tigers growl, but some do not'. And then pointed out that while each of the propositions contained in the sentences may be true and may be false, they are anyway all intelligible: something which is not the case if we substitute 'exist' for growl. Well, substitute 'is real' for 'exists', and it's a similar point I'm getting at here.

Third, let me register that the 'all-ness' I envisage in my 'all that there is' formulation quite obviously makes some comment on the current debate concerning 'social construction'. When such debate takes place among the professionals, we get a book such as John Searle's *The Construction of Social Reality* (1995), in which he characterises social constructionism, which he wishes to repudiate, as the view that reality is not only socially constructed but also able to be changed at will, by us, and whenever we feel like it. When such debate takes place in philosophical conversation, we get exchanges of this sort: A says 'X is real'; B says 'no, X is socially constructed'. All this seems to me just crazy stuff. It seems to me that all we need to understand is that that which is 'socially constructed' is fundamentally, ineluctably, palpably and urgently—'real'! And that while the mass-energy part of reality is real and is *not* socially constructed, the theory-about-mass-energy part of reality is real and *is* socially constructed. As best I can understand it, Searle's problem would seem to be that he elides *reality* with *mass/energy*: and thereby fundamentally, ineluctably, palpably and urgently queers his own pitch.

Fourth, let me claim that while what I am doing here is to some extent reclaiming a name, there's nothing remotely 'mere' about that. The answer to the question 'what's in a name?' is almost always 'a very great deal'.

And certainly is in this case. As already mentioned, 'reality' has been a concept to a considerable extent hijacked by physicists, and used by them to label what is actually only a part of a whole. In tune with a comment of Bohr's to the effect that the job of physics is not to determine the nature of reality but to establish what we can say about the physical, I want to insist that the physical is only a part of 'all that there is'. And that the reductionism which accompanies the contrary aggrandizement indulged in by many a physical scientist, far from being the basis of a scientific approach, is a sure route to that dualism which sooner or later—and usually sooner—puts the scientific enterprise on hold by staking out large chunks of reality with 'no go' signs. For to the extent to which philosophers, such as Descartes, were thorough-going mechanical materialists and reductionists, to that extent they *had* to put their miracle-working ghosts into their machines. How else could they 'animate' the damn things? How else could they come near to coping with such notions as 'love', 'imagination', 'experience', or 'beauty'? What I hope to show, by contrast, is that it is only a thorough-going monistic anti-reductionism which can even start to cope scientifically with such as these. And that to the extent to which those scientists studying the inorganic believe that theirs is the only 'reality', to that extent will they more easily be drawn to the non-scientific, even the anti-scientific and the mystical, concerning the rest: concerning, that is, the organic, and particularly the personic. They will, after all, have to tell themselves some sort of story about all those things which their reductionism tells them are not 'real': that is, all those things which, actually in no way mysterious, are the thoroughly 'real' subject matter of sciences other than their own.

To reiterate: I am suggesting that some entities exist as part of the universe, other entities exist as part of the biosphere, and yet other entities exist as part of the world. They exist as *different* entities and *differently* in each of the three domains. Light waves exist in the universe; eyes exist in the biosphere; and smiles exist in the world: but they all exist. They are all part of reality. The real universe, of which light is a constituent entity, exists; the real biosphere, of which eyes are constituent entities, exists; and the real world, of which smiles are constituent entities, exists. And as also it encompasses 'appearance', 'reality' encompasses them all.

(3) arts and sciences

Playing now on the tension—as mentioned at the beginning of this chapter—between the recording of discussion concerning the 'and' specimens in several separate, discrete, encyclopaedia-like entries, and the recording of discussion concerning the 'and' specimens as one hooked-together, continuous, essay-like piece of prose, I have made what I want to say under this head both take off from, and begin by providing some summary of, what I have just said under the last.

Constructed by the sensory systems of animals, appearances are representations of the environment of those animals. In *Homo sapiens*, such representations are constructed by visual, aural, tactile, olfactory, gustatory and kinaesthetic senses. Since reality is all that there is, and appearances are, appearances are part of reality. The representations which are some appearances correlate more closely with that which they represent: the representations which are other appearances correlate less closely with that which they represent. All appearances are instantiated in the nervous systems of those who, as it were, do the 'peering': as experience in the nervous systems of those without language; as linguistically structured experience in the nervous systems of those with language. The only extant 'peerer' with language is *Homo sapiens*. And in *Homo sapiens*, the instantiation of linguistically structured experience is that which constitutes person-ness: is that which constitutes 'I'. Hereby, a summary to date.

But the instantiation of linguistically structured experience is incremental: as later I will outline it in more detail (see Chapter 9), ipseity develops. And one of the developments which I think takes place across that incremental instantiation is a development whereby humans shift from a struggle to come to know their situation primarily through the making of what is now commonly called 'art' to a struggle to come to know their situation primarily through the making of what is now commonly called 'science'. In other words, I am suggesting that, as far as knowing is concerned, what is appropriate for this 'and' is that we replace the additive mode (art *and* science) with a developmental one (art *to* science). I am suggesting that, at base, art offers *belief about* reality: such belief being that entertained by the artist, using primarily a narrative method. But that only science can offer *knowledge of* reality: such knowledge being that constructed by a whole scientific community, using the scientific method. And that the transition from reliance primarily on the one to reliance primarily on the other involves one more working out of the development from I AM to I IS conceptions of the person.

However, because what I might loosely call 'power'—and perhaps particularly power to wield the slaughtering word!—is still very considerably embedded, even now, in literary intellectual places, I am only too aware of the wrath that I risk bringing down on my head with such a formulation. But if so, so be it. My case is that, as animal-humans (and I AM persons), our primary mode of thinking is artistic (and assimilatory). And that as human-humans (and I IS persons) our primary mode of thinking is scientific (and accommodatory). The differences are both gross and subtle.

Artistic thought gives greater credence to the mythical. It is more given to awe. It is more often animistic and immanent: more readily anthropomorphic and narrative. It more readily succumbs to the naturalistic fallacy. And to passivity. Scientific thought, by contrast, tends to be active, explanatory and curious; to be at one and the same time, both better-informed and more modest; to reject, as a consequence of greater understanding, all animism, immanentism and anthropomorphism. All of which means, I think, that it is now the scientist rather than the poet who is the more complete 'maker'. But let me stress: 'maker' *of knowledge*. As must be obvious to even the most casual reader, I love poetry: I read lots; I even try to write some. Poetry gives us—gives me—many things: my case is simply that it does not give us—give me—*knowledge*. It gives me those beliefs concerning 'how things are' which are entertained by its authors. Those beliefs concerning 'how things are' might *tally with* how they actually are. It's just that they can't *establish* how they actually are. Only science can do that. But such is the systematic misreading of what science is, that lest anybody should think that I am trying to defend, say, the crass reductionist physicist vis-à-vis, say, the deeply humane poet, there is a very important rider to add to this.

In his *Lost Bearings in English Poetry* (1977), David Holbrook tries to explain the appeal which Philip Larkin's poetry has for many of his students. He suggests that it is Larkin's air of superiority vis-à-vis 'ordinary' people, an appeal 'from the half-educated to the half-educated'—as he puts it—which makes Larkin so popular. I don't know the poetry well enough to have an informed opinion on whether or not I would agree with him. But to the extent to which there is anything in his judgment, I would suggest that the 'half' missing, in both Larkin and those to whom his poetry appeals, is the scientific 'half': the 'half' that provides scientific understanding, yes, of the inorganic and the organic, but particularly of the personic. When I claim—as I do—that *natural* science gives us a greater understanding of the universe and the biosphere than does art, I am saying that which I believe to be both true and uncontentious. But when I claim—as also I do—that what is currently called *social* science, such as there is of it, gives us a greater understanding of the world than does art, while I am again saying that which I believe to be true, I suspect that I am hardly saying that which would go unchallenged. The detailed elaboration of my case for so saying is for later. For now, I want merely to make it clear that it is the lack of a particular *sort* of scientific understanding of which, primarily, I am talking:, that is, of a scientific understanding not of the universe, not of the biosphere, but of the world,

None of which is to say that a lack of a *natural* scientific understanding of the universe and the biosphere is not also there amongst many of those educated primarily in the arts. It is such a lack, for example, which caused Jacques Monod, in *Chance and Necessity* (1970), to lament the fact that a good many distinguished intellects still then found themselves unable to agree to the proposition, or even to understand the proposition, that 'from a source of noise natural selection alone and unaided could have drawn all the music of the biosphere'. But it is another and different sort of lack, a lack of understanding of the nature of very knowledge itself, which for my money is an even bigger problem. For this is a lack which, among other things, is responsible for the recent and highly unfortunate invention of a nasty, bogey-like thing.

Given the rather huge and rather obvious benefits of our scientific advances, any anti-science camp has to be careful. It has to try to avoid giving the impression that it is callous, inhuman and troglodytic—even when it is. And it is this, it seems to me, which is the reason for the invention of the bogey: the something which, even if with arguments bearing a remarkable similarity to those used (but by 'others', of course) to attack science itself, it *is* OK to be heard to attack. Attack *science* as hubristic and reductive, and sound just ignorant. Attack scient*ism* as hubristic and reductive, and sound instead—why just 'right-on'!

What I think is involved here is nothing much more than a somewhat disreputable sleight-of-hand. In his book *Consciousness Reconsidered* (1992), Owen Flanagan called scientism 'the view that science will eventually explain whatever is natural'. If that is scientism, I can see no reason to assume that it is anything other than the best hypothesis going concerning 'whatever is natural'. Indeed I would make its definition considerably more inclusive. I would say that, if that is scientism, I can see no reason to assume that it is anything other than the best hypothesis going concerning 'whatever is other than natural' as well. 'Scientism', it seems to me, is basically a fall-guy, discreditable and sham concept invented out of fearfulness and ignorance. And its invention—or so I believe—poses a huge challenge to the scientific community that it counteract the perceived, but ill-perceived, need for such a sham.

We have to get to a situation, all of us, where anxiety about that of which we are ignorant, fear of that from which we feel excluded, can no longer mean, as for many it still does, that rationality is perceived as inhuman; structured thought is perceived as cold; analysis is perceived as insidious; and science is perceived as an alien threat. We have to get to a situation, all of us, where with Marie Curie we can know that 'nothing in life is to be feared, it is only to be understood'; and that 'now is the time to understand more, so that we may fear less'. In short, we have got to get to a situation, all of us, where science is identified with the truth-telling, democracy and integrity which are its hallmarks, rather than, as is now too often the case, with their opposites. Of course there is less than good science. I don't think the struggle against all the irrationalisms, relativisms, post-modernisms and their ilk is helped one jot or tittle by such as Peter Atkins, fellow of Lincoln College, Oxford, and lecturer in physical chemistry, opening an article (*THES* 30.9.94) with the idea that 'somewhere buried in the awesome complexity of Nature, lies the truth'. But good science, the sort that knows that truth isn't hidden until it's discovered but non-existent until it's made, is of the essence of that truth-making task in a way that art probably isn't.

It has been poets in particular, perhaps because they trail such a terrible history concerning the supposed intrinsic superiority of their craft, and arts-educated people in general, perhaps because they trail an even worse one concerning the supposed intrinsic superiority of their persons, who have so often exhibited that impoverishment which is a lack of understanding of the centrality of the scientific enterprise. And it is an impoverishment quite terrible in its implications. In the language of Walter Benjamin, we now understand that 'every monument to civilisation is also a monument to barbarism': that is, that concentration camp torturers listened to Schubert in the evenings; that most of those myriad paintings of female nudes are of persons seen as things; that as Edward Said has shown us concerning *Mansfield Park* and Antigua, much of the wealth supporting that refined sensibility of the nineteenth century novel rested in the slave plantations. In his famous master/slave dialogue, the stuntings of Hegel's master (which I will call the stuntings of primariness: see 'master *and* slave' below) are just as real as the more familiar stuntings of his slave (which I will call the stuntings of secondariness: ditto). And such stuntings have been found so much multiplied in so many artists because so many artists have (necessarily) been multiply primary. Which means that when primariness is challenged, when 'most people ignore most poetry, because / Most poetry ignores most people' (as Adrian Mitchell puts it), it is perhaps appropriate to be a little more bold in our assessment, and reassessment, of the artistic mode. But—as the poetry lover in me has to repeat—*in our assessment of it as that which can give us knowledge, not as that which can give us anything at all.*

For some important things need to be acknowledged by those still defending the artistic vis-à-vis the scientific in the search for knowledge. It needs to be acknowledged, first, that the huge claims made for the moral importance of the artistic mode and of aesthetic appreciation have never even been made for scientific understanding: that is, that the pretensions of the artistic mode have been greater. It needs to be acknowledged, second, that while the sciences have been developed by a more diverse and less interested group, most of whose members have been relatively far from centres of power, the arts have been selected as being 'the arts' by a tiny and interested group, for the most part nearer those centres, and as interested to exclude as to include: that is, that the credentials of the artistic mode have been lesser. These are differences which would be damaging enough to their supposed status if the arts had been thought to be merely entertainment. But given the really quite huge claim that they are actually to be recognised as some sort of source of universal values, they are surely differences just largely obliterative of that status. If it is even only roughly the case that what gets into the artistic canon is what some (multiply primary) elite, by virtue of its (interested) power, has put there, and what gets into the scientific text is what some (less multiply primary) elite, by virtue of its (less interested, even disinterested) scientific investigation, has put there, then it is surely not unreasonable to suggest that a serious re-think about the moral status of the two enterprises might well be called for.

For if we say that a work of art is that which does X—let's say, 'extends our human sympathies'—and then along comes that which does X just as well, or even better, it is surely entirely appropriate to re-think our understanding of the efficacy now of the artistic mode: a rethinking which is perhaps what is going on in the turmoil which is much of 'the modern arts' right now. In considering this which is the philosophy of art, of course Plato has been a key figure. In the *Republic*, what he seems to reject is not art itself but that evaluation of it which sees it as supplying knowledge. If this is what he's saying, then I am agreeing: this is the point I am making —or perhaps even, labouring. I'm saying that while art can of course give us many things, it cannot give us knowledge. It gives us its creator's eye view of how things seem to be to its creator. But while how things seem to be to its creator can of course *tally* with how they are, such seeming cannot *establish* how they are. To be established as knowledge, the validation of the scientific method is what we now understand to be necessary. If Shakespeare wants us to believe that if only we are true to ourselves it will automatically follow that we will be true to others—and that in itself is in doubt: he did, after all, put the thought into the mouth of a character in other ways presented as perhaps not particularly wise—then to acquire the status of knowledge, rather than opinion, such a claim would need to be validated as having that status. It cannot acquire that status simply by being stated: not even by being stated by such as Shakespeare.

In a way, what is really under consideration here is creativity. And what I am arguing is twofold. I am arguing that creativity develops. And I am arguing that that development is exemplified in the shift from an artistic mode to a scientific mode: or, as Althusser characterised it (see Chapter 4), in the shift from relying (unscientifically) on 'conclusions without premises' to one relying (scientifically) on 'conclusions from premises'. I listened once to Andrew Hodges, his biographer, lecturing on the work of Alan Turing. And I was interested to hear him make the point that it was somewhat paradoxical that this the most creative of human beings was more than happy to believe that creativity, far from being a mysterious and special capacity which set those who had it apart from the common herd, would almost certainly come to be shown to be a purely mechanical process. It struck me at the time—and it strikes me still—that this is a fact not really paradoxical at all. Could it not be that I AM thinking—the thinking which necessarily puts its own person ineluctably 'first': the thinking which, in its assimilative nature, is characteristic of an artistic mode—is precisely that which might be tempted to appeal to the 'mysteriousness' and 'specialness' of its own creativity? Perhaps the fully creative scientist—the one who, anyway when doing her or his science, if not necessarily always, is thinking in I IS mode—is precisely the person who would have no need of such delusion. With such protection being afforded first by the essential democracy and collectivity of science, and second by a perhaps more sophisticated understanding of just what creativity might be anyway, he or she is perhaps rather better protected from any such elitism born of ignorance.

And now, because I can already hear, even in advance of their utterance, the savage force of those 'slaughtering words', I find that I have once again, and for the last time, to stress *that I am talking about knowledge*. Whatever it is that art can give us, whatever it is that I get from the reams and reams of poetry I read, the bits and pieces of poetry I write, *it is not knowledge*. It is my claim—it is an I IS claim—that knowledge is necessarily acquired only by working in a scientific mode: only by using the scientific method. As I hope to demonstrate as I develop my argument, that knowledge which is currently called a 'third person' affair is more accurately thought of as a 'just person' affair. In I AM thinking, third persons are other than first persons. In I IS thinking, just persons are the only ones that there are.

(4) autonomy and heteronomy

In the context of his far-reaching and very influential attempt to ground moral philosophy, it was Kant who developed these correlative terms. They were terms necessary to his elaboration of his version of personhood: a version whereby he too wanted to distinguish between persons and humans. For him, humans were unlike persons in that they were not capable of being fully autonomous. Derived from the Greek words for 'self' and for 'law', being autonomous—as our Oxford source quotes Kant as expounding it—meant being a moral agent who must reject all maxims which 'cannot accord with the will's own enactment of universal law'. For example, being incapable (through an insufficiency of such will) of so rejecting, being dependent upon the full personhood of their fathers and brothers and husbands, women were to be thought of as humans but not as persons. Indeed, it was to be as late as 1878 that a Canadian court ruled that a woman was not a person: a position which, when it was over-ruled by the Privy Council, proved to be a landmark in the struggle for that equality for women which continues to this day.

For a Kantian moralist, for Lawrence Kohlberg for example, coming to full moral maturity means coming to full moral autonomy. It means doing what one does, and being able to give fully rational reasons for what one does, not because some heteronomous authority says that one should—which heteronomous authority might variously be the state, society, ones mates, the church, ones parents, others of ones higher-status relatives, God—but because ones own freely and rationally chosen principles say that one should. In this tradition, it is thus manifest that autonomy is to be prized above heteronomy.

Of course it *is* a tradition. And as such, of course it has developed. It would perhaps be difficult to find a present-day Kantian who would want to put women—at least overtly!—into Kant's non-person category. But it has developed in other ways too. In *Taking Leave of God* (1980), for example, Don Cupitt argues that any morality making a claim to validity has to be rational, consistent and impartial. But unlike many another, he does not think that such a requirement rules out the possibility of there ever being a thoroughly moral society. For he argues, using the Kantian vocabulary, that it is entirely possible to conceive of a whole society of autonomous persons. Each chooses her or his own ethic—he actually writes 'his' here, but my paraphrase gives him the benefit of any Kant-like doubt!—but to the extent to which each recognises the necessity of consistency and impartiality, presumably a state to be achieved via education, an agreed social and public morality could emerge out of their individual choices. If this sounds familiar, indeed it is: for he goes on explicitly to claim that such a society, the best kind he believes, would be a free, liberal-democratic republic.

But the point that seems crucial to me, a point patently manifest in Cupitt's formulation, is that such development of the Kantian tradition as thus far has taken place has done so from *within* the I AM paradigm, rather than from *without* it. To borrow and adapt the language of Thomas Kuhn, it has been the tinkering of 'normal-philosophy' rather than the recreation of 'revolutionary philosophy'. But if moral philosophy is to come of age, if it is to become moral science, then what it seems to me is required is a complete change of paradigm: a change which, crucially, will need to be much more deeply challenging of the liberal humanist concept of 'the subject' than seems to be dreamed of in Cupitt's formulation.

Simply, the 'tinkering' could be expressed as a steady shift of the following kind. We have moved steadily away from upholding as acceptable the *normal* stance 'I who am autonomous freely do that which I choose that I should do: you who are not autonomous non-freely do that which I choose that you should do': and steadily towards upholding as acceptable the yet still *normal* stance 'I who am autonomous freely do that which I choose that I should do: you who are autonomous freely do that which you choose that you should do'. To the extent to which this shift has taken place, considerably I think but by no means completely, that has indeed been progress. But in its course, we have also steadily accumulated all the 'relativism' problems with which we are now only too familiar. This is an accumulation which I think means that, as far as any attempt to come up with a well-grounded theory of the good by such a route is concerned, we have now run more or less completely into the philosophical buffers. Is that which is the good that which an autonomous X thinks, or that which an autonomous Y thinks? Silence. Within-paradigm tinkering would seem to have run more or less completely out of steam or sense. So what is to be done?

In the case of cosmology, anyway as argued by Kuhn, at the point at which the new and alternative Galileo-Newton model of the universe emerged, the old and mainstream Ptolemaic model had for some considerable time been veritably festooned with contradictions, contrary examples, incoherences and embarrassing lacunae. But given that a less-than-perfect model is better than no model at all—indeed, given that 'no model at all' is not even possible—it had survived in this somewhat tattered fashion precisely until the arrival of the alternative. What I am proposing, analogously, is that our current model not of the universe but of the world, and in particular of what it is that constitutes the persons of the world, is similarly now festooned. And that it is one of those festooning incoherences which is here on show in the 'autonomy *and* heteronomy' duo.

The 'new' that the I IS hypothesis proposes on this one is that we have to learn to make not just a quantitative shift *within* the I AM paradigm from the normal 'I'm autonomous but you're not' (authoritarianism) to the normal 'we're both autonomous' (democracy, but also that pesky relativism), but a qualitative shift *without* the I AM paradigm completely. As far as our 'autonomy *and* heteronomy' pair is concerned, such a shift would involve our coming to recognise that it is impossible to make any sort of sense any more of there being two sorts of person in the world: that is, 'autonomous' persons whose selves are enabled, by being somehow 'outside' that world, to have it impinge on them or not impinge on them as *they* choose; and 'heteronomous' persons whose selves are constrained, by being somehow 'inside' that world, to have it impinge on them or not impinge on them, as *it* chooses. We are *all of us* 'inside' the world. Our doings, the doings of *all of us*, actually constitute the world. If we should want to, which I for one would not, it would be possible to designate *all of us* as heteronomous: and mean by that that we are all *determined* entities which, out of our thereby resultant mutuality, constitute the world. If we should want to, which I for one would not, it would be possible to designate *all of us* as autonomous: and mean by that that we are all *determining* entities which, out of our thereby resultant mutuality, constitute the world. But what I am claiming is not possible is that we designate *some of us* as heteronomous, *some of us* as autonomous, and the shift from the one to the other as moral development. That is incoherent. From which it follows, or so I would suggest, that the way to deal with this particular 'and' pair would be to dump it as incoherent.

But, a word of caution. As later I will try to make clear, while a transition from being heteronomous to being autonomous is not an idea which will cut the mustard as a way of characterising moral development, since such development does indeed take place we are certainly in need of one that does. And as later I will also try make clear, I think the concept of 'selving' can do just that.

(5) chance and necessity

At the end of *Chance and Necessity* (1971), Jacques Monod concluded that we now know that we are probably alone in an immense universe; that it is a universe indifferent to our existence; and that we emerged from it not by any necessity but by a chance formulated as its opposite. But are the 'chance *and* necessity' duo appropriately construed as opposites? Richard Rorty is fond of insisting that the opposite of 'necessary' is not 'arbitrary' but 'contingent': which translated into Monod's language would, I think, mean that we would be required to think of the emergence of life, and in time of humankind, not as a 'chance' phenomenon but as a 'contingent' phenomenon. What, if anything, is the difference being identified here? Is it a difference that makes any difference: to anything, or to anything of substance? Was Einstein getting at the same sort of Rorty-uncomfortableness before the idea of chance with his famous 'God does not play dice' quip? If chance were to be one of the secrets of 'the Old One', as the scientist was wont to call his God, would it indeed mean, as he himself put it, that we would have to view that God not only as subtle but also as malicious?

The significance of questions such as these is for me made clearly manifest in two somewhat fashionable conundrums which echo around philosophical conversation. The one concerns the status of mathematics: crudely, is it invented or discovered? The other concerns the status of chaos theory: crudely, does it imply more or less necessity in the physical world? What particularly interests me about both of these controversies, as anyway in my experience they enter into philosophical conversation, is the utter predictability of who will think what. Know only a little about the basic philosophical stance of someone, know just whether he or she tends to be an optimist or a pessimist, and 'read off' her or his attitude to both controversies with considerable reliability. (So considerable as sometimes in itself to make the case for 'chance' being a very suspicious notion!)

Among the minority of mathematicians who take what has come to be called the Platonic view, the view that mathematics is discovered, Roger Penrose is a leading exponent. He asks how on earth a merely invented mathematics could be so effective as the language of science. And since he seems to think of such a merely invented mathematics as an arbitrary construction of the human mind, perhaps his bewilderment is intelligible. That which is arbitrary is that which is as it is as the result of chance. And it would indeed be difficult to see how an arbitrary mathematics could be so successful. But take Rorty's point seriously, see 'contingency' rather than 'arbitrariness' as the opposite of 'necessity', and all is well. Mathematics has been developed by mathematicians. Mathematicians, like all the rest of us, live in the historically contingent. And they select that which is to count as mathematics over that which is not to count as mathematics based on the practical problems of a real universe, a real biosphere, and a real world. Hardly surprising then that the real mathematics which they make organises our understandings of said universe, said biosphere and said world. How could it not? If mathematics is a symbol system created to study the patterns of the real universe, the real biosphere and the real world, it is surely not surprising that it should do so. With much that would not do the job rejected along the way, it was *made* to do so.

It is a similar point that I would want to make concerning chaos theory. Over the last twenty years or so, we have come to understand that what were previously thought to be random events, events in whose occurrence there was no pattern, are actually events with very simple underlying causes, put in motion by the following of a few very simple rules. What this means is that we have found more order where previously there was thought to be less. Fine. And then some clot went and called the resulting theory 'chaos theory'. And at one (very fell) swoop indeed, it seemed that all the pessimists on the planet were united, in philosophical conversation, into a doom-and-gloom chorus of what we did not know and could never know because it was just too mysteriously 'chaotic' for us! But anyway in this manifestation, 'chaos' implies the very opposite of mayhem or anarchy. It denotes an irregular motion that arises from fully deterministic forces. Chaotic dynamics are dynamics very orderly in their apparent disorder. When the previously hidden structure exhibited by chaotic systems is uncovered, it makes understanding, prediction and control more not less possible. It tells us when to expect the unpredictable, and when not to. But from listening to such angst-laden conversation, you certainly would not think so.

What both of these stories illustrate, I think, is a deep-seated, blood-laden, gut-driven, totally irrational and visceral terror: a terror so riven into the psyches of so many people as to be one of the main variables, if not *the* main variable, structuring (or even '*de*structuring'!) all discussion of that level of reality which I am calling the personic. Writing in the *Times Higher Educational Supplement* (21.3.97), David King, the then editor of the journal *Genethics*, perfectly exemplifies the terror that I mean. His subject was cloning. What he found 'really disturbing' about it was that a never-ending drive to find the uniform, the efficient, the controllable might come to dominate even something so central to the human condition as reproduction. What he hoped was that he was not alone in thinking that some element of chance concerning our origins is a fundamental part of that condition. And there it is. The terror. The unspoken prayer of so much of philosophical conversation would seem to be: 'please God, let there be that which I cannot understand; that which nobody can understand, and never will; that which is not determined; that which is random; that which is unscientific; that which is chaotic; that which is just down to chance. And if there can't, please let me not know that there can't. Let me hide my shivering self in uncertainty quantum mechanical principles, actions at distances, pixie-dust in the synapses, superpositions, micro-tubules, inexplicable genius, *anywhere*, so long as I don't need to know that we can know. Dear God, let me not know that we can know!'

Having written such as the last few sentences, my immediate and cautious reaction is to back-peddle; to deprecate somewhat; to begin this my next paragraph with a 'perhaps I exaggerate' sort of phrase. But I won't. Because I don't. I think this terror is as real and as huge and as hugely debilitating as I also think it is thoroughly misplaced. It is, of course, the terror of determinism: a terror which will be more fully explored below (see 'free will *and* determinism'). All that I want to register for now is that this very real terror of determinism can present, quite often presents, as a veritable addiction to chance. But wait a minute. Chance is also the enemy *of the enemy* of determinism. Things brought about by chance are not things determined: good. But they are not things brought about by free agents exercising their free-will either. Not so good. In this fraught and frightening territory, it would seem that the enemy of my enemy is not necessarily my friend. It seems to me that what we are encountering here is what Erich Fromm so memorably called 'the fear of freedom'. For using if not Hegel's conception then at least Hegel's language, with freedom construed as 'the recognition of necessity'—and this is how I *will* construe it—fear of that recognition is indeed fear of freedom. And we know we are encountering it when, for all that the liberal humanist lauds free-will over determinism, let push comes to shove and he or she will laud 'chance' over both of them.

It seems to me, by contrast, that far from being lauded, chance is a concept we would do better to abandon, anyway in this 'chance *and* necessity' context. For if we are not careful, so called 'chance' events get construed, totally unjustifiably, as events without causes. And in being so construed, they open the doors, anyway in philosophical conversation, to various and assorted magics. We would do better, I think, to construe them as events caused not by that which is (in principle) unidentifiable, but by that which is (as of now) unidentified. And to help us do so, we could do worse than heed the advice of the mathematician Ian Stewart. In a piece entitled 'Think Maths' (*New Scientist*, 30.11.96), he is thinking about 'emergence'. He suggests, first, that philosophers have a tendency to talk about it as if it involves something a-causal, when all it really means is that what he calls 'the chain of causality' becomes more complicated than (yet) we can fathom. And he adds, second, that even human behaviour might be in accordance with mathematical rules, but because those rules are so mind-bogglingly complicated we do not (yet) have the mathematics which could express them. But to be able to listen to such advice—or anyway to hear it—we would need a conception as to just what a person is which is considerably different from any of those with which we operate at the moment. Such a conception would have to be one which did not permit that element of chance, as nourished by a fear of necessity, to creep in to its formulation. And I am suggesting that it is just such a conception of a person which is at last becoming available to us.

The I AM conception of a person sees it as an uncaused cause: a prime mover and shaker; a free agent, with a free will. It might murder its mother, but there is no necessity about that: it might water the roses instead. Who knows? It is free. There is a fundamental uncertainty about what it will do next. And given that 'these days' there is even a fundamental uncertainty in physics of all places—how often does philosophical conversation run down *this* groove!—how much more must we expect to find it in waffley old psychology. I am suggesting that this whole line of thought is just wrong. Perhaps some of the simplest evidence that it is wrong comes from some of the simplest, everyday behaviour. Is there anybody who at least once a lifetime, if not once a day, has not been fully aware of a behaviour which he or she would like to change but cannot? What I am suggesting is that when we come to be able to compare the nano-level realities of those situations where we could not change our behaviour with the nano-level realities of those situations where we could, what we will discover is not a failure of free-will in the first case and a triumph of free-will in the second—whatever that could be made to mean—but a thorough-going determination in both cases. The first situation will be found to have one determining past, the second another. And it will be those different pasts, both of them determining but determining differently, which will account for the difference.

There is no escape either from any of all this—supposing we should (wrongly) think that we need one—in the celebrated 'uncertainties' of physics: or as Ian Glynn so succinctly puts it in *An Anatomy of Thought*

(1999), 'Heisenberg won't help.' Glynn acknowledges, of course, that the famous 'uncertainty principle' does mean that at the level of the subatomic we must often talk only of the probability of something occurring. He accepts, of course, that we cannot predict at what point any given and particular atom of a radioactive substance will disintegrate. But he nonetheless insists that we be careful what we conclude from such realities. Even the smallest of samples of a radioactive substance consists in a huge number of such atoms. And that makes it entirely possible—witness the well-known 'half-life' measure—to say 'rather precisely' just how long it will be before half of those atoms will have disintegrated. He sees the problem, helpfully, as one of levels. And points out that at the level of middle-sized objects like members of the species *Homo sapiens*, Heisenberg's uncertainty can just be left out of account. And even if it couldn't—he adds for good measure—that would give us no comfort. For if there's anything more discomforting than thinking that all our actions have been predetermined by our genes and our history, then surely it is 'to believe that that is not quite true because there is, in addition, an element of chance.' Precisely!

What we need here, as again I will try to provide in more detail under my 'freedom *and* determinism' head, is not an escape from determinism (there isn't one), but a conception of a person which embraces determinism. The I IS conception of a person does just that. By positing a person not as that which *has* experience but as that which *is* experience—or rather, not as that which *has* linguistically structured experience but as that which *is* linguistically structured experience—it is a conception which banishes chance and recognises necessity. As it is also a conception, which later I hope to show, which thereby makes available the possibility of freedom.

On 'chance *and* necessity' then, what we have to recognise is that because they instantiate an incommensurate pairing, they need to be just dumped altogether from philosophical conversation. 'Contingency *and* necessity' might have something to offer. 'Chance *and* necessity', I suggest, does not.

(6) egoism and altruism

The insistence, in the I IS hypothesis, on the significance of the organic/personic divide carries, as one of its consequences, a challenge to an alternative hypothesis which killed George Price. For those who are interested, the details of this sad death are to be found in the first chapter of Andrew Brown's book *The Darwin Wars* (1999). Suffice it here to say that Price, a theoretical biologist, killed himself because he thought he had found proof of original sin, and could not bear it. His proof was his reformulation of a set of equations which, while they showed that altruism of a sort could flourish within the constraints of evolutionary theory, showed also that it was always and necessarily only what has come to be called '*apparent* altruism': that is, an altruism such as the 'reciprocal' variety proposed by Trivers which is actually only informed or prudent self interest. Price's reaction to this was extreme. He abandoned his atheism, adopted a passionate Christianity, and lived, in near madness, an existence which he took to require of him that he sell all he had and give to the poor. When almost all his resources were gone, he died in squalor, by his own hand, in a London squat. And his body was identified by his colleague and fellow theoretical biologist, Bill Hamilton.

His was a formulation of the 'egoism *and* altruism' problem which while it does not drive everyone who holds it first mad and then to their deaths, nonetheless racks many an other as well. And for the same reason as it racked him. What's more, it's entirely possible to argue that where it does so, it is racking, maddening and even killing precisely the good. For it is precisely the good for whom the positing of a relentless and inescapable selfishness running through all organic existence, and thus through human organic existence along with the rest, is so destructive, primarily of hope. In today's Darwinism-takes-all climate, this is a particularly pressing problem. And it seems to me that, particularly at the level of philosophical conversation where it is perhaps most rampant, it is vital that it be tackled. If the inevitability of a 'devil-take-the-hindmost' selfishness were a conclusion necessary of acceptance, then wishing it were otherwise would of course be just pointless. But it is my contention that it is a conclusion *in no way* necessary of acceptance. Before however trying to demonstrate that it's not, it is first necessary to rule out of court two arguments which purport to do the job but in fact do no such thing.

Before Richard Dawkins wrote his unfortunately titled book *The Selfish Gene* (1976), it was the already-mentioned Bill Hamilton who had established what since has become the professional consensus among evolutionary scientists: that 'group selection', the idea to which so many non-professional 'altruists' want to cling, is just an explanatory non-starter. The fact that group selection would enable some sort of case to be made for a genuine altruism does not make group selection true. It is almost certainly not true. Indeed, it seems to me that once George Williams had made the 'sex-ratio' point back in 1966, the point that in so many species group selection would predict many more females than males which was manifestly not the case, the argument should have been dead. And that once somebody else had pointed out that, even if it had existed, group selection would just as significantly mean 'selfish groups' (and thus between-selfish-group conflict) as it would mean 'altruistic individuals', it should have been buried. The fact that it wasn't, and in some few places still isn't, seems to me to have rather more to do with ideology than with science. It seems to me that those who so much wanted it to be true, who spoke of it in tones that suggested it was in some way a morally superior notion which was being trampled underfoot by wickedly selfish not to say capitalist ideas, were just wrong. Perhaps mainly as 'soft' natural scientists, or liberal-and-left social scientists, they wanted it to be true because they somehow thought that only if nature could be thought to be 'cuddly and hospitable', as somebody put it, could humans be defended as 'cuddly and hospitable' too. But that is just daft. To try to reclaim altruism, as if it were the source of morality, by resurrecting group selection, is like trying to reclaim the soul, as if *it* were the source of morality, by resurrecting vital spirits. Nature is what nature is. Culture is what culture is. If nature is red in tooth and claw, it in no way follows that culture is *necessarily and therefore* crimson too. The group selection hypothesis is almost certainly not true. And the usefulness of my hypothesis is that the 'real altruism' side of the argument does not *need* it to be true. At the very least, it is surely always a good idea to need to believe that with which the evidence tallies, as opposed to that with which, spectacularly, it does not.

The second argument to abandon is what might be called 'the argument from niceness'. In the not-very-edifying 'Darwin wars' between the Gouldians (supporters of Stephen Jay Gould) and the Dawkinsians (supporters of Richard Dawkins), there has been, among social scientists of liberal-and-left persuasion, what I take to have been an essentially *political* need to back Gould. His position seemed to be so much less 'atomistic' and 'socio-biological' (boo words, both of them). To some, I think he just seemed to be a somewhat pleasanter human being: less rude about religious believers, for example. And his position seemed to offer a way out of the inevitability of the dire 'selfishness' hole. But I think what we need here is a strong dose of Tony-Bennism. We need to insist that it is not the personalities but the ideas that matter. And in terms of the ideas, it seems to me that a responsible argument for the possibility of genuine altruism among humans has to deal with the (apparently) more hostile-to-the-cause Dawkinsians rather than the (apparently) less hostile-to-the-cause Gouldians—because *about their biology* the former are just right. There is a considerable consensus among professional biologists that the essence of selfish-gene biology is correct: that is, that it has more of the truth on its side than any credible alternative. It therefore behoves those of us who want to argue the case for genuine altruism at the level of the human to deal with *that* version of evolutionary theory, rather than with others apparently more attractive to our cause. For as was long ago pointed out by Gramsci, while any fool can deal with her or his theoretical opponent at her or his weakest, can make a case against the social efficacy of Christianity for example by considering the life and work of the Borgias rather than the life and work of Desmond Tutu, the real task is to deal with ones theoretical opponent at her or his strongest. It is for such a reason that I, at any rate, will approach 'the science wars' by trying to deal with the arguments of the majority Dawkinsians rather than the minority Gouldians. And I shall suggest that to deal with them means two things. It means to accept their ideas as powerfully explanatory of the (human and otherwise) organic. And it means to reject their ideas as just irrelevant to the (only human) personic. Let me now try to spell out *that* position more fully.

Those who choose headlines in newspapers and titles for books have often a lot to answer for. When it had become a best-seller, one commentator pointed out that *The Naked Ape* (1967) might have sold rather fewer copies with the rather more accurate title *The Relatively-Hairless Hominoid.* And I would suggest that the fate of *The Selfish Gene* (1976) might similarly have been different if it had been rather more accurately (because rather less animistically) called *The Survival-oriented Gene.* Put perhaps a little strangely, my point is that because genes don't have selves to 'ish'—take 'ish' to mean 'act in the interests of', and I can stop using my speech marks!—they just can't be 'self-ish'. We are dealing here with yet another category mistake. And it's a nice illustration of just how widespread is the illegitimate elision of the natural and the moral in our culture (see 'natural *and* moral' below) that a phrase such as 'the selfish gene' can become so ubiquitous with so little realisation of its scientific absurdity. In general usage, the dictionary allows 'self' to apply to the individuality or essence of both a person and a thing. But in scientific usage—and it's science we're trying to do here—'self' is a word in the technical vocabulary of the sciences currently called social. And that means that it is at best unhelpful, and at worst just sloppy—an unhelpfulness or sloppiness no doubt made that much easier to indulge in by the little less than contempt so often shown by natural scientists towards social scientists—for the former to use it, *in a scientific context,* so inaccurately. So how do we rescue the situation?

Step one: excise the concept 'self' from the natural sciences of the organic, and respect its technical usage in what I am calling the social/moral sciences of the personic.

Step two: recognise that in those sciences, a self is essentially an entity which develops. A *Homo sapiens* is born self-less; a self starts to emerge at around about 18 months of age; and that self then develops across a lifetime. What this means is twofold. Firstly it means that to describe anything other than a human person as self-ish is just metaphorical. But secondly it means that to describe a human person as self-ish is just definitional. Persons exist as one or another type of self. They have one or another type of self to ish. And ish it they will: necessarily because just definitionally.

Step three: understand that ishing your self, being self-ish, not only in no way necessarily *precludes* being genuinely altruistic, but in one way necessarily *entails* being genuinely altruistic. For if there is a type of self whose own interests are such that when it ishes them it necessarily ishes the interests of all other selves as well, then for such a self, far from altruism being precluded by its necessary selfishness, altruism is actually entailed by its necessary selfishness.

Step four: understand that though rare and under-developed, such selves are not only logically possible but actually existent. To see them, the essential move is fully to recognise the significance of the distinction between the organic and the personic. And to recognise it not religiously but scientifically.

We have had a religious recognition of the distinction, one in terms of such magical things as souls and spirits, for a long time. But souls and spirits no longer compel. For what is certainly a steadily rising number of persons, probably a rapidly rising number of persons, the evidence against their existing at all is now thought amply sufficient to constitute knowledge. But like all varieties of the stuff, this is knowledge which gives us not only new answers to old questions, but also new questions. And that's fine. Except that being in such a 'new-questions-but-not-yet-any-answers' situation for any length of time tends to mean that one, the other, or both of two things starts to happen. From some quarters come answers to the new questions which are thoroughly inadequate answers. From other quarters come refusals to recognise the very existence of what are only 'so-called' new questions. And in the case of souls and spirits, examples of both such strategies are readily on offer. Entailing as they do not only the emergence of the God-shaped hole but also the need to try to plug it, knowledge claims positing the non-existence of souls or spirits—or rather, their existence only as figments of our imaginations—are claims which open the way to both sorts of unhelpfulness. From some quarters, such as from evolutionary psychology for example, comes the thoroughly inadequate answer that with the demise of the distinction between the organic and the personic *based in magic*, the distinction between the organic and the personic *based in anything*, the very distinction itself, has also gone: and by reduction. From other quarters, such as from the various forms of new age mysticism, comes a refusal even to recognise that anything has changed: comes an insistence that souls and spirits are as alive and as well as ever they were, and are safely ensconced at Stonehenge or some such place. By contrast with both of such stances, my claims are threefold.

Contra those of a traditionally religious disposition, those who want to cling on to what are essentially now magical versions of the distinction between the organic and the personic, I want to claim that no such versions are any longer viable. Contra those with inclinations towards evolutionary psychology, those who want to replace such magical versions of the distinction with no distinction, I want to claim that the disappearance of its magical versions in no way entails the disappearance of the distinction itself. And contra those who want to deny the very possibility of science in such places, I want to claim that what we need is a scientific version of the distinction: one which will see the mature sciences of the organic increasingly being joined *and counter-posed* by the not-yet mature *but fundamentally different* sciences of the personic.

One way of capturing what I anyway take to be those fundamental differences is to present them in terms of interest. In such terms, what I am proposing would look something like this: that while the interests of organic entities are fundamentally that they *live*, the interests of personic entities are fundamentally that they *love*; that while the interests of organic entities can only be defended, and are therefore necessarily defended, in *competition* with the interests of other organic entities to *have* what is necessary to that *living*, the interests of personic entities can only be defended, and are therefore necessarily defended, in *co-operation* with the interests of other personic entities to *make* what is necessary to that *loving*; and that while the *genetically instantiated* competition between organic entities results in an *evolutionary* process, the *linguistically instantiated* co-operation between personic entities results in a *developmental* process. These are all differences which could not make more difference. And the difference they make to an appropriate characterisation of our 'egoism *and* altruism' pair could not be more substantial.

Picked out by their respective italicised vocabulary, there are *two* processes delineated above. There is a genetically instantiated and evolutionary process whereby organic entities compete to have what is necessary to live. There is a linguistically instantiated and developmental process whereby personic entities cooperate to make what is necessary to love. And what is happening as a result of our so-far almost complete failure to differentiate the organic and the personic in any way other than magically is that the second, as yet the less well understood, is being studied as if it were the first. As far as our 'egoism *and* altruism' duo is concerned, what that means is that a concept appropriate to a study of the first has been just 'lifted' into a study of the second.

Having been so 'lifted', it lands, of course, in a discourse saturated with I AM thinking. And what is taken to be the 'egoism' and (merely reciprocal) 'altruism' of I AM persons is then studied by analogy with the 'selfishness' of genes: that is, it is then studied a-historically (or even anti-historically) and a-scientifically (or even anti-scientifically). All of which means that a current and urgent task must be to try to develop its study from within the I IS paradigm instead. So what would that involve?

Shedding the a-historicism would involve avoiding making what in *Consciousness Reconsidered* (1992) Owen Flanagan called a 'variety of mistakes' characteristic of the person he was calling an egoist and I am calling someone thinking with an I AM conception of the self. In Flanagan's words these include believing 'that each individual starts life with a self'; believing 'that development or enhancement of a pre-existing self is possible, but that self-emergence is not'; believing that 'this original self constitutes the core of the self, our conscious control center, the source of all action plans, and the agent for whom all experiences accrue before being filed for future reference or discarded.' By contrast with such errors, he wants us to build on William James's idea that 'the self is a construct', and that 'self-emergence occurs as a person develops'. On such a conception, 'the main idea is that the self emerges as experience accrues': that is, that 'the ego is an after-the-fact construction, not a before-the-fact condition for the possibility of experience'. Or as he quotes James as putting it, the self is not 'an unchanging metaphysical entity like the Soul' but instead 'a *thought*, at each moment different from that of the last moment, but *appropriative* of the latter, together with all that the latter called its own.' These are all approaches which allow us to see that a self doesn't evolve: it develops. And eventually, as later I hope to show, that that development permits, or rather ensures, the emergence of a type of self with interests such that when it defends them it necessarily defends the interests of all other selves at the same time.

Shedding the a-scientificity would involve ceasing to think of our-selves as free-willing, individually-born Ies who are *over against* the world, and coming to think of our-selves instead as mutually determined and determining, socially-constructed Ies which are *constitutive of* the world. For us, developing rather than evolving entities, neither 'egoism' nor (only reciprocal) 'altruism', represent the limits of our possibilities. For us, the love of which we can come to see ourselves as being capable is such that, had poor George Price understood its possibility, might have been able to save him his life. It's achievement would involve our coming to understand a freedom which inheres in a very particular necessity. For that which is organic, as members of the species *Homo sapiens* are organic, it is the genes-driven necessity to live, carrying with it the necessity to compete with all other living things so to do, which is foundational. But for that which is personic, as selves are personic, this is not so. For us, it is the linguistically-driven necessity to love, carrying with it the necessity that we co-operate with all other loving things so to do, which is of the essence.

Close-packed in these last few paragraphs are some of the ideas most important to my argument. I am thoroughly aware that, as baldly stated here, I have given them as yet no sufficiently substantial explication. But as, specimen by specimen, my 'ands' experiment proceeds across this and the next two chapters, such explication will steadily accumulate. Under only this 'egoism *and* altruism' head, suffice it for now if I have shown just the following: that for a self, and only for a self, to be self-ish is just definitional; that the interests of selves develop; and that while the interests of selfish but less-developed (I AM) selves, in still trailing their animal origins, might seem to offer George Price some reason to despair, the interests of still selfish but more-developed (I IS) selves could have offered him only hope.

(7) ends and means

This is an 'and' which, in the form in which it most usually enters both philosophical conversation and academic philosophy, is derived most immediately from Kant. Although it is often only the first which is labelled as such, he actually proposed three rational principles which, taken together, were to constitute his famous 'categorical imperative'. First: 'act only according to that maxim whereby you can at the same time will that it should become a universal law'. Second: 'act in such a way that you treat humanity, whether in your own person or in the person of another, always at the same time as an end and never simply as a means'. Third: act in accordance with 'the idea of the will of every rational being as a will that legislates universal laws'. Our 'ends *and* means' dichotomy is sourced in the second of these three.

And the most important things to notice about its formulation are two little phrases: the first, 'at the same time'; the second 'never simply'. The maxim is not suggesting that we should treat others as ends *instead of* means: it is suggesting that we should treat others as ends *as well as* means. It is not proposing that we should *never* treat others as means, but 'never *simply*' as means. Thus, when I catch the bus, the driver's skills are my means of reaching my destination safely and on time: he or she is means to my end. But if the driver becomes ill on the journey, I cannot legitimately argue that since he or she is means to my end of arriving at my destination on time, hospitalisation will have to wait until I have been deposited there. Why? Because *as well as* being means to my end, he or she is also person. This, surely, is elementary stuff. So then why has this particular 'and' pair assumed the sort of high billing it quite often gets in philosophical conversation? Why have the things we do, our acts, been parcelled up to be 'means' towards other things which are called 'ends'? It is my belief that the reason is fundamentally political. But before I try to substantiate such a claim, I need first to establish a few parameters.

It is first necessary to register that there is nothing natural, given, or necessary about our acts being so parcelled. They have been, but they do not have to be: a fact illustrative of what I take to be a very marked tendency towards the naïve, in both philosophy and philosophical conversation, concerning what is natural, given and necessary. If they are even thought about at all, which frequently they are not, it is too often just assumed that the origin and history of some 'first' principle being used in argument can just be ignored. And for those who are less than sociologically sophisticated, that is certainly a danger here. For it's not that acts are just like that:

that *just naturally* they are means to ends. It's that they are means to ends given a particular take on things; from within a particular theory of things; as the result of working inside a particular paradigm about the nature of things. And what I am once again suggesting is that the take, theory or paradigm which creates 'ends *and* means' dilemmas is the one which posits an I AM self. Posit instead an I IS self, and what becomes clear is that there is just no way in which what we all do *can* be coherently parcelled out into 'ends *and* means' chunks.

It's not just that things which are ends given one objective are means given another: though they are. It's not just that the end of any given action might become the means to another end: though it might. It's not even just that linguistically structured experience does not readily break down into 'ends *and* means' chunks: though it doesn't. It's that there's a fundamental misconception involved in thinking about this part of reality, about intention if you like, in this bifurcated way: in this 'first this' (a means) 'then that' (an end) fashion. So let me illustrate what I take that misconception to be with an example. And as anyway I believe, with as difficult an example as you could hope to have to deal with if you are wanting to argue the case I am arguing.

You are in Nazi Germany. You are sheltering Jews in your attic, and the Gestapo knocks at the door. You are asked: 'are you sheltering Jews in your attic?' And the question for discussion is: is it right to lie? With 'the means' taken to be the lie, and 'the end' taken to be looking after the well-being of the persons in the attic, is it right to lie? If lying is bad, can it here be justified as means to a good end?

I want to suggest that a bad means cannot be used to justify a good end. But I want to suggest it for a reason different from that which is usually advanced. I want to suggest that it cannot be so used not because it is wrong to do so but because it is impossible to do so. It seems to me that both ends (are these good ends?) and means (are these good means?) have to be justified not the one by the other, but the both by something else. And that the 'something else' concerned is a coherent theory of the good. As of now, such a theory is spectacularly conspicuous by its almost complete absence: an absence in the face of which our 'ends-*and*-means' problem, and many another to boot, stays stubbornly insoluble. But it is my case that such a theory is available, but once again only as a consequence of the radical change of paradigm which I am proposing. Let me argue that case as it concerns this 'ends *and* means' pair in two steps: using my Gestapo example, let me first spell out that which I think would be *the right thing to do*; and second spell out that which I think would be *the theory of the good which would entail its being done.*

I do not think it would be right to lie. The human conversation fundamentally concerns a discussion of the Socratic question 'how should we live?'. It is a question which permits of no easy answer. And it is a question which, for as long as it remains unanswered, or answered less than satisfactorily, prolongs all the suffering of the world. And given that to put into that conversation that which we know is not the case is to hinder, in however small a way, the search for an answer, to lie, it seems to me, is never the better thing to do. So even in these dire straits, one does not lie. But then because answering morally illegitimate questions is never the better thing to do either, neither does one comply with the demand for an answer to the question. Instead one says something like 'you have no right to ask me that question, so I will not answer it'. Of course the result of such a stance might very well be that one ended up dead. But that in no way alters the rightness of the response. Presented with some situations, it is right, it is good, to end up dead. But only a theory of the good based in there being a fundamental difference between the organic and the personic, and thus a fundamental difference between what is good for an organic entity and what is good for a personic entity, can reveal that reality to us. And the I IS theory can so reveal.

For it is the raison d'être of only the organic entity 'human-animal' that it should live: biologically, to be dead is the ultimate failure. But for personic entities, for persons, things are different. Mine is a theory which suggests that the raison d'être of the normal personic entity 'animal-human' is that it should have: that it should *have life* is the way in which it trails its natural origins; that it should *have love* is the way in which it reaches, but faultily, for its moral possibilities. But when it has actually attained those possibilities, when it has ceased to be an I AM person and has become an I IS person, things are changed utterly. The raison d'être of the moral personic entity 'human-human' is that it should *love*: that is, that it should love its own I and all other Ies equally. And to do that, it is entirely possible that the right action is the one that results in its ceasing to exist. In the attic situation above, for example, there are, for the human-human I IS person, many Ies to love: its own I (which means not lying); the Ies of those in the attic (which means not answering the Nazi question); the Ies of the visiting Gestapo (which means not colluding with their wrongly assumed right to demand an answer); the Ies of all those other persons who come to know about, and thus learn from, the incident (which means not compromising); ultimately, the Ies of all other persons (which means understanding that for a human person, as over-against a human animal, it is *not loving* rather than *not living* which is the killer). Essentially, entities which actually consist in linguistically structured experience can be only stunted in their development when that experience is of what is not the case. And given that to love, following Erich Fromm's now famous formulation in *The Art of Loving* (1962), is actively to *pursue* the well-being of the loved, not actively to *stunt* its development, then loving equally all those Ies involves saying it like it is. With lying thus perhaps being the only act which can never be good, I think there is perhaps something like an Hegelian 'negation of negation' at work here: that is, human-animals do not have the capacity to lie, and don't; animal-humans do have the capacity to lie, and do; human-humans do have the capacity to lie, and don't.

So to finish this section by returning to the question of the political implications of the 'ends-*and*-means' pair, I would suggest that they are (simply!) these. Politics, as it has famously been defined, is about who gets what, and how. And in an I AM world the story goes something like this. If I succeed in cornering more in a situation where you corner less, I have power over you. In then institutionalising my power as 'right', as 'good', perhaps even as 'natural', I also and thereby institutionalise your powerlessness similarly. In the situation thus created I have 'legitimate' power, and the means by which to exercise it 'legitimately'. This then means that should you try to redress the balance, the means you would need to use would have to be illegitimate. And I could complain, in full-throated liberalness, that even if your ends were good, they could not justify your wicked means. Neat! Thus a crude but I think essentially accurate presentation of the ideological value of an 'ends-*and*-means' concept in the struggle to resist justice and equality. But given that there is no place for the ideological in a science of the good, the way to deal with the claim that this 'ends *and* means' duo is a conception in any way conducive of useful thinking is, I suggest, to deny it.

(8) fact and value

By far the most well-known modern presentation of this pair is that of G. E. Moore in his *Principia Ethica* (1903). Failure appropriately to characterise the distinction between the two led frequently to what he called 'the naturalistic fallacy': that is, to the assumption that an empirically verifiable feature of what is could be used to ground what ought to be. Moore, by contrast, maintained that although the laws of nature can say what is, we can from that read off exactly nothing about what ought to be. This is a stance which, although now very widely accepted in academic places, is much less readily appreciated, I think, in general philosophical conversation.

The position I want to defend is that value is of mind: to value something is to 'mind' about it in a particular way. This means that nature, being mindless, cannot be the *site* of value, still less *be* a value—whatever that could be made to mean. To the extent to which the relativists are insisting that value is of valuers, and valuers with all their wants and needs, histories and interests, are historically-situated persons, I am with them. It seems to me that that which is explanatory of value is never something about the valued, always something about the valuer. But the problem has been that, from within the I AM mode of thinking about persons, that 'something-about-the-valuer' has frequently been thought to be inaccessible to scientific understanding. And that has meant that while a spectacularly successful natural science has been coming to understand what *is* the case in the universe and the biosphere, a spectacularly unsuccessful moral philosophy has been stymied when trying to understand values concerning what *ought* to be the case in the world. On this one, what I want to suggest is that just as natural philosophy developed into natural science, so moral philosophy needs to develop into moral science: a science which can begin to replace the old and unscientific language of 'ought' with the new and scientific language of 'is'. And that such development is dependent upon a new understanding of what a person (and thus a valuer) actually is: which new understanding, as only now I appreciate, began to emerge for me from some long-time-ago polemics.

I very clearly remember the pleasure it gave me when first I managed to work out how to respond to a criticism of my then-Marxist approach to the way in which society was to be changed. The criticism was a much paraded one. If via the working through of the inevitable and invariant Marxist stages of the development of societies—primitive communism, slavery, feudalism, capitalism, socialism, communism—the world I wanted would come about anyway, why did I need to spend so much of my time and energy working away to make it happen? Why struggle to bring about the inevitable? My eventual response, the one which, having tried out various others and found them wanting, I rested in as the best, was that what was inevitable was that I, and others like me, *would* so struggle. Unrecognised as such at the time, this was probably one of the first steps I took in the direction of my current thinking. And if it was, it was a shame that I did not then know about a philosopher whose ideas, had I encountered them, would have helped me considerably in the formulation of my case.

Spinoza never says 'ought'. Throughout his *Ethics* (1677), he insists that we are free to the extent that we are rational, and refers to that which a rational and thus free person *will* do, to that which a rational and thus free person *does* do, not to what a rational and thus free person *ought to* do. What's more, the whole process is developmental: we become rational, and thus free, to the extent that we acquire rational emotions. And these are acquired not by introspection and acts of will—'I *am* being irrational, I *ought to* be rational, I *will be* rational'—but by cognitive development. Via our actions, their consequences, our experience, we slowly accumulate more appropriate ideas as to the nature of things, and it is these more appropriate ideas, more appropriate understandings, which mean not that we *ought* to behave differently but that we *do* behave differently. Amongst much else that is so liberating in such a stance, it means, of course, that blaming becomes redundant: becomes, as Spinoza himself labels it, just 'vindictive'. Which is not of course to say that we *ought* not to go in for it, but to say that as we accumulate more appropriate understandings concerning human psychology, we just *do* not go in for it. Succinctly: the way to get an 'ought' from an 'is' is to come to understand that it actually *was* an 'is' all along.

To take a simple example, consider this recently topical question. *Ought* one to use 'he' and 'his' to designate both females and males, or *ought* one to say 'he or she' and 'her or his'? Step one: get rid of the 'oughts', and just recognise instead that both practices exist. Step two: do correlation studies to see what goes with what; in other words, isolate those variables clustering more (or less) among 'he' and 'his' users than among 'he or she' and 'her or his' users. Step three: in the light of an hypothesis concerning moral development, identify the developmental stages of persons exhibiting your clusters of variables. Step four: identify thereby whether it is 'he' and 'his' users, or 'he or she' and 'her or his' users, who are morally more developed—and thus which usage, according to your hypothesis, is the one which *will* grow, rather than *ought* to grow, with moral development. Step five: wait and see what happens. If the usage identified by your hypothesis concerning moral development does in fact grow, you have validating evidence for that hypothesis: if it doesn't, you haven't! In moral science, as opposed to moral philosophy, given that no 'ought' is necessary, there's just no problem of getting it from an 'is', or from anywhere else. What was inevitable in my 'better world' story above was simply that people like me *would* and *did* and *do* struggle to achieve it.

Now it is my impression that these are understandings which, in academic places, are currently growing apace. The stance adopted by such as Paul Churchland, for example, who proposes a new taxonomy whereby moral knowledge, a mixture of theory and practice addressing the social world, and scientific knowledge, a mixture of theory and practice addressing the natural world, are to be thought of as equally capable of attaining objectivity, involves a clear refusal of the positive/normative distinction, and thus of the is/ought question, and thus of the 'moral-science-is-impossible' stance. And while there are going to be those among his academic colleagues who would disagree with him, some more some less, there are probably going to be none these days who would any longer be amazed, aghast, astonished, uncomprehending that such a proposal should be 'on the table'.

But if this is what is the case in the academy, these are understandings which I suspect are growing considerably less rapidly in more general philosophical conversation. Indeed, I would perhaps go so far as to say that it is a very widespread misunderstanding of the progress involved in coming to rely less on non-scientifically validated lived experience (common sense), more on scientifically-validated contrived experience (scientific method), which increasingly hampers our moral advance. It is what in some people produces, for example, that ill-informed hostility to psychological research which—they say—'discovers what everybody knows anyway'. What it actually does is either to validate lived experience (often), or to challenge lived experience (often too), by measuring it against the carefully controlled contrived experience of the experimental method. And especially on those occasions when 'what everybody knows anyway' (such as that hitting children does them no harm and is just good discipline) turns out to be so hideously wrong, the efficacy of the procedure becomes only too tragically clear, and the need to develop a scientific approach to the moral even more urgent.

It is my claim, yet again, that what is holding us back from developing such a scientific approach to the moral is a false I AM conception of the person. With its free-agent assumptions, fact-value distinctions, positive-normative formulations, and first-person egocentricities, it is an ever-increasing impediment to our moral development. And one of the most telling illustrations of my argument on this, a special case presented briefly here and more fully in what is to come (see Chapter 11), is to see what happens when one of the best known moral imperatives ever articulated is treated thus positively (as a fact) rather than normatively (as a value).

Much of present-day philosophical conversation about the moral would have us conceive of 'love thy neighbour as thyself' as a value: that is, as something which because it cannot be grounded empirically is just an 'ought'; I *ought* to love my neighbour as myself. My claim is that moral science needs to re-conceptualise it as a fact: that is, as something which because it can be grounded empirically is actually an 'is'; I *do* love my neighbour as myself. And because in its 'as thyself' formulation it is so powerfully a call to equality, in making now my case for its positive rather than normative status, I will be making here, but still only briefly, what will be my first direct defence of that idea.

My argument is couched in terms of a theory of our moral development: that is, of our development from being animal-human I AM persons to human-human I IS persons. As the former, the level of love we can manage *for both ourselves and our neighbours* is a lower-level love; essentially, a love stunted by what I am calling 'the having ideology'. When we experience ourselves as having entities, when we are I AM persons, we 'have' not only our experience, but also ourselves, other people, our property. Within such a paradigm, it is telling that while 'viable' means 'able to live', 'loveable' does not mean able to love. And in meaning instead 'able to be loved', this is a difference of meaning which exemplifies the way in which such a love is conceived of as something we have. We have it when somebody gives it to us or bestows it on us. Or we have it when, as commodity, it is exchanged at a price on a market. Success is when we have more of it, failure when we have less. But I will argue that even with such stunted conceptions as to what love is, it's not that we don't love our neighbours as ourselves but ought to. It's that we do love our neighbours as ourselves, but badly—in both cases. Let me spell that out a little more fully.

It is my case that as I AM persons our selves are partial: both partial in the sense of incomplete, and partial in the sense of interested. And that as persons constituted as such selves, we love *ourselves* badly (where 'badly' is operationalised as 'in a manner whereby we either acquiesce in, or strive to acquire and maintain, positions as either primary in relation to secondary others or secondary in relation to primary others'), and those *others* badly too. As I IS persons, by contrast, our self is one impartial: both impartial in the sense of complete, and impartial in the sense of disinterested. And as persons constituted as such a self, we love *others* well (where 'well' is operationalised as 'in a manner whereby we neither acquiesce in, nor strive to acquire and maintain, positions as either primary in relation to secondary others or secondary in relation to primary others, but only positions as equals with others'), and *ourselves* well also. This is an argument for equality with not an 'ought' in sight. It is a scientific argument for equality. But for now, if I do anything other than baldly state it, I get more than a little ahead of myself.

(9) freedom and determinism

In *Paradise Lost*, some of the fallen angels choose to spend their time in what I am calling philosophical conversation. Milton reports that they '... reasoned high / Of Providence, foreknowledge, will and fate, / Fix'd fate, free will, / foreknowledge absolute, / And found no end, in wand'ring mazes lost.' Some four centuries after the poem was written, 'in wand'ring mazes lost' describes very well the state it is still entirely possible to get into when trying to think coherently about this 'freedom *and* determinism' question: a question which, when it seems appropriate, I will alternatively refer to as the question concerning 'free-will'. But not only that. As to many of my readers it will already be abundantly clear, if indeed it was not at the outset, it is about *this* pair, above all others, that my hypothesis requires that I be most clear. It is therefore about *this* pair, above all others, that I will spell out my position in most detail.

A simple starting question might be to ask why it is so worrying, to so many of us, to contemplate the possibility that our actions are not free but determined. And a simple starting answer would be because it would remove the concept of responsibility, and thus the possibility of both blame and punishment. For how could we blame and punish people who were unable to do other than they did? And if we couldn't legitimately blame and punish, wouldn't 'human nature' ensure that the world would be full of out-of-control children and licensed villains? It is for this reason that, practically, the 'freedom *and* determinism' issue is so important. Though for all that, it is at the level of theory, not of practice, that a solution to the problem it poses must be sought.

Up to now, the problem represented by the 'freedom *and* determinism' pair has been just that: a problem. We have not properly understood such things. And many have been not at all sanguine that we ever would. In the *Critique of Practical Reason* (1788), for example, Kant suggested that there were three problems of metaphysics which would prove to be forever beyond the powers of the human intellect to solve: the problem of the existence (or otherwise) of God; of the existence (or otherwise) of immortality; and of the existence (or otherwise) of free will. And on this last, which is here our concern, the best he could come up with was that while determinism *does* rule in the realm of which we can have knowledge (the realm of appearances), free will *must* rule in the realm of which we can not have knowledge (the realm of 'things in themselves'). Perhaps it was an argument that did not convince even its author: in a daftly anachronistic sort of way, I suppose that I certainly hope so. For I think it is not too wide of the mark to see Kant's role in the history of the debate as being protective: as being the production of an ideology which would enable such notions as 'free will' and 'agency' still to flourish, against all the odds, in the face of the revolutionary upheavals of thought occasioned by the work of Newton, Locke and others. If ideas of the transcendental could be battened down as philosophically defensible, then perhaps the whole edifice of a consciousness-led world could be re-secured. It would be my claim that, to the extent to which that was his role, he could only do a temporary make-do-and-mend sort of job. And that as we now come more fully to understand the workings of the brain and the nature of the self, that whole traditional structure will necessarily come tumbling.

Indeed, has started to do so. From complete determinism ('hard', of the genes: 'soft', of experience and history), through all the intervening positions on a continuum, to complete free-will ('hard', of total individual responsibility: 'soft', of developing individual responsibility), philosophical conversation quite often finds someone or other to defend more or less every possible position. Indeed it is incredibly common to be able to defend two different positions at once when that seems to suit. Nowhere is this more obvious than in the 'heads-I-win, tails-you-lose' tactic whereby wherever its perpetrator's beliefs fall out on such a continuum, they fall out nearer *determinism* when explaining her or his shortcomings ('I am dishonest because my parents never checked me'), nearer *free will* when explaining the shortcomings of others ('you are dishonest because you don't care about the consequences for other people'). It would seem to be particularly easy for us to fall into adopting such a double-standard when we consider our parenting: *at one and the same time*, we blame parents who should have done this, that or the other differently (a stance which requires us to assume that *their* wills *were* free to do otherwise *then*), and exonerate ourselves from responsibility for our unpleasantnesses because they made us the way we are (a stance which requires us to assume that *our* wills are *not* free to do otherwise *now*). This is a sample of a type of special pleading so widespread that social psychologists have developed a theory with which to study it: and 'attribution theory', as it is called, has spawned a rich literature illustrating the often hilarious nuances of its widespread operation. It is of course also to be found at work when considering not shortcomings but achievements: wherever our beliefs fall out on the continuum, our opportunism finds us *nearer determinism* when explaining the achievements of others ('you were better taught than me'), *nearer free will* when explaining our own achievements ('I worked harder than you'). It would be silly to put forward as the major reason for trying to achieve a coherent theory of 'freedom *and* determinism' that it would put an end to such self-serving nastinesses. It would nonetheless be a fairly pleasing spin-off of such a theory that it would. So: what are the chances of such achievement?

Traditionally, the dualist argument for *in*determinism has been of the 'ghost in the machine' type: the type that insists that in order to explain the free will which we all feel very strongly we possess, the apparent determinism of physics has to be in some way modified. The older view was to say that such freedom was a special privilege, God-given, and only to humans. And that physics had therefore to be modified by such ill-defined things as 'souls', 'spirits', or 'minds'. But now, when 'God' and 'soul' and 'spirit' will serve as explanations for fewer and fewer people, and even 'mind' will serve only for some, and only if de-reified to 'minding' or 'what the brain does', theories about free will are in need of some not inconsiderable recasting. They need to provide some answer, for example, to what is now a famous question. Posed in *The Concept of Mind* (1949) by Gilbert Ryle, the question is this: 'if voluntary actions are those produced by an agent's "volitions" (decisions, intentions etc.), are those volitions themselves voluntary?' Ryle's own comment on his question—'If so, we get an infinite regress; if not, we get voluntary acts as the result of involuntary events'—is one which leaves it as pertinent now as ever it was. And if candidate answers are to be taken even halfway seriously, they need now to ensure that they have at least the appearance, if not the substance, of scientific respectability. Given this last constraint, it was perhaps only to be expected, though still I think unfortunate, that one search for enlightenment should have involved the mining of a very successful (if puzzling) theory from the physical sciences.

As explanation of physical reality, classical physics has, since the early part of the twentieth century, been joined by quantum mechanics: the former a theory of *macro* physical reality; the latter a theory of *micro* physical reality. Physicists agree on the determinism of the former: disagree on what some of their number call the *in*determinism of the latter. For some of those who *do* read quantum mechanics in an indeterministic way—the alternative is to assume that its determinism (and thus the way in which it can be united as a theory with classical physics) is as yet un-understood by us, possibly because we have the issues wrongly posed—here is a new and apparently scientifically respectable way of defending free will. It led the mathematician Roger Penrose and the anaesthetist Stuart Hameroff, for example, to do just that by postulating 'quantum-mechanical-indeterminacy in the microtubules'. But characterising such talk as the equivalent of postulating 'pixie-dust in the synapses', the philosopher Patricia Churchland counters—in *Brain-Wise* (2002)—that if what is wanted is some help from the laws of physics, it could be wiser to think about the law of conservation of mass/ energy. Any talk of a causal relation whereby a non-physical (because mental or free-willed) event bringing about a physical (because neuronal) change would clearly violate such a law. And anyway to date, nervous systems provide no evidence of any such violation.

But further, it seems to me that such mining of quantum mechanical indeterminacy, or the earlier but similar ransacking of the uncertainty principle of Heisenberg, to find a solution to the problems posed by the concept of free will is thoroughly unproductive. And not primarily because the principle in the one case and the theory in the other are both of them rather less safely lodged in the canons of physics than the law of conservation of mass/energy: though that is true. But essentially because it does not seem to me that indeterminacy, even were it to prove a conceptually secure notion, has anything much to do with freedom. Let me explain.

If it really were just some uncaused cause which led me, say, to murder my mother, what possible meaning could you give to the word 'freedom' to be able to argue that that made me free? Surely I would have anyway the possibility of freedom in exactly the opposite circumstance: that is, if I murder my mother as the result of some cause of which I or others can become aware, and thus have some hope of coming to control. As the philosopher Ted Honderich repeatedly points out, as far as human action is concerned, 'uncausedness' is about as far away from freedom as you can get. Of course the standard response to such an argument is to say that my action *was* caused: but by a reason. This, however, is less than helpful. With a reason construed, à la Ryle, as a volition, it can immediately be charged, à la Churchland, with an offence against the law of conservation of mass energy: a fact which, in child-speak, might be expressed as 'well, what caused the reason then?'

In *The Philosophy of Right* (1820), Hegel comments that 'the ordinary man thinks he is free if it is open to him to act as he pleases', but adds that 'his very arbitrariness implies that he is not'. If that which is conceived of as indeterminate is to be equated with that which is conceived of as arbitrary, then it seems to me that the constantly re-invented attempt to find some indeterminacy somewhere in order to protect 'free-will' is thoroughly misplaced: indeed, so misplaced as to make its very opposite that which in fact defends freedom. How come? Well, as I have just suggested, wouldn't it be out of the recognition of the necessity that some experience would result in matricide and other experience would not result in matricide that such freedom would be able not to be found but to be made? After all, if we were in a situation where the fine-grained causes of mother-killing were thoroughly understood, maybe we would then be in the position to make it totally unheard of. That would be freedom. It would be the sort of freedom born of understanding.

For consider this. Interviewed once on *Desert Island Discs*, the then Astronomer Royal, Sir Martin Rees, modestly insisted that 'it's harder to understand a frog than to understand the universe'. He's right. Frogs are considerably more complex systems than stars. And humans, with their billions of neurons in their billions of interconnections, all linguistically structured, are considerably more complex than frogs. In the face of such complexity, what are currently called the social sciences, the ones trying to make sense of matricide, are working with as yet not much access to the almost unimaginably fine-grained distinctions between one state of such a complex system and each and every other state of such a complex system. Instead, they are working with what, by comparison, are the massively gross-grained distinctions of such things as differences of social class background, or of whether or not physical punishment was experienced as a child. What seems to me to be truly amazing about working with such as-yet primitive and gross-grained variables is not the fact that pattern sometimes does *not* emerge from them, but rather that sometimes it *does*. To get the levels of correlation which we do, for example, between having violence inflicted upon you and becoming violent yourself, should perhaps be seen as just a tiny glimpse of the much more specific regularities which will emerge when our variables get down to a more fine-grained level: when they get down, for example, to the level of the detailed linguistic structuring of particular cell assemblies or neural networks.

But determinism has terrified so many so much for so long—Stuart Hampshire once went so far as to call it 'the hideous hypothesis'—that it is going to be a long haul to get such a reading of our situation generally accepted. Nonetheless, it being my position that it needs to be accepted, I was more then delighted, some fifteen years ago, to encounter the work of Benjamin Libet. It is work which now is celebrated, famous, notorious even: and much replicated. But then, when not many of the people with whom I anyway discussed things had heard of it, the most common comment I got when relating what I understood him to have shown was some version of 'I don't believe it', or even 'that's just rubbish!' An Oxford philosopher of my acquaintance, who had recently discussed the work with Libet himself, said—as I recall—that he thought 'there must be something wrong with it'. And such reactions are not at all surprising. From what I've read more recently, Libet himself seems very reluctant to draw the conclusions from his work which it seems to me that it implies. But to examine the issue more closely, let me first add one more summary (their production has been described as now a veritable 'cottage industry') of what it was that he did in his now so famous experiment.

Under the heading 'Unconscious cerebral initiative and the role of conscious will in voluntary action', his paper appeared in the journal *Behavioural and Brain Sciences* in 1985. His subjects were attached to two measuring devices: an electroencephalograph which measured the time of the onset of the 'readiness potential' in that part of the cortex which mediates hand movement; and an electromyograph which measured the time of the onset of activity in the hand muscles. They were then told that they should clench their right fist just whenever they felt like it, but should pay close introspective attention to the instant of the onset of the urge, desire or decision to perform each such clenching, and to the correlated position of a revolving spot on a clock face (indicating clock time). This meant that there arose from the experiment three 'times': the time at which the readiness potential started to fire; the time at which the decision to clench was taken; and the time at which the fist began to clench. The results were startling. In an operation that took 550 milliseconds in total, 350 milliseconds before his subjects 'willed' that they would clench their fists, the readiness potential (sometimes called the action potential, but anyway indicating that their brains were at work preparing to clench) started to fire. And 200 milliseconds after his subjects 'willed' that they would clench their fists, their fists clenched. In summary, brain potentials preparing to clench a fist were firing 350 milliseconds before the 'free-will' decision so to clench had been taken. In thinking about this remarkable result, Libet asked himself whether 'if the brain can initiate a voluntary act before the appearance of conscious intention' it therefore follows that there is 'no role for conscious function'. And he suggested that because 'conscious control can be exerted before the final motor outflow to select or control volitional outcome' there is no need for conscious initiation: it would be redundant. Why? Because 'the volitional process, initiated unconsciously, can either be consciously permitted to proceed to consummation in the motor act or be consciously vetoed'.

Of the much that there is to discuss about this experiment, it is the fact that it appears to suggest that an unconsciously initiated action proceeds to consummation *unless there is something that blocks or redirects it* which here interests me. And I want further to explore the implications of that idea by thinking about a comment on Libet's work made by the American experimental psychologist V. S. Ramachandran. He suggested that perhaps what the experiment indicated was that what we have is not 'free will' but 'free won't'. It is, of course, a witty comment. But not I think *merely* a witty comment. Indeed, I think it is a comment which, as a way of formulating what is going on in a Libet experiment, could prove highly productive of interesting explanation. And so further to examine the possibility of its doing just that, I want to offer two things which might promote such productive use. I want to offer first (and briefly) a very general observation: second (and more fully) a 'free won't' model of my own.

First, the very general observation. Could it be that what we perhaps have here is yet another example of a very widespread phenomenon: somewhat clumsily put, of the 'rule-out-what-you-don't-want-rather-than-rule-in-what-you-do-want' phenomenon. Is Libet's 'consciousness' – that which I would call 'linguistic structuring' – ruling *out* the natural in ways perhaps analogous to those in which science rules *out* wrong theories (Popper); lack of reinforcement rules *out* undesired behaviours (Skinner); evolution rules *out* less-than-fit forms (Darwin); developed moral reasoning rules *out* earlier varieties (Kohlberg)? Freedom, as opposed to free will, seems so often about *not* being able to do things: for a somewhat eclectic sample, about *not* being able to take homeopathy seriously; about *not* being able to discriminate on the basis of sex or race any more; about *not* being able to regress to childhood; about *not* being able to believe in the possibility of time-travel; about *not* being able to unknow what you know. Whenever I think about Libet's work, I register yet again the huge appropriateness of Luther's declaration 'here stand I; I cannot do otherwise' as the powerful image of freedom that surely it is. For freedom as 'the recognition of necessity' is the most powerful image of freedom I know. And for me, a crucial part of that freedom-bestowing recognition is that we do not have free will.

Second, my own 'free-won't' model. Imagine a computer screen which has a cursor 3.5 millimetres wide half way down its left hand edge. The cursor can be moved, and the task is to shift it across the screen by discrete steps so that it arrives at any one of the other possible edge positions it could occupy. Movement is initiated when the operator presses a key which causes a stream of discrete 'movers' to pass through the cursor, entering it from the right, exiting it from the left. As they thus accumulate to its left, the movers push the cursor forward in front of them. They are of different types. And by accumulating at type-specific nodes along the length of the cursor's left side, they have a differential impact on the direction of its movement, thereby making a differential contribution to the site of its eventual destination. The system runs until the cursor reaches a new edge position, then stops. Suppose now that we want to make the cursor 'free' to select both its route across the screen and its destination. To keep things simple, suppose it 'wants' to select the most direct of the many possible routes (straight ahead) to the most obvious of the many possible destinations (a point directly opposite its starting position but on the right hand side of the screen). This means, first, that it must be sensitive to the differential impact of the different types of mover. And second, that it must be equipped with a device which can either block movers or redirect movers while they are still within its boundaries: that is, while they are crossing its 3.5 millimetre width and before they pass out to the left and start to make their differential impacts on the direction of its movement. With such sensitivities and devices in place, a continuous stream of movers could pass through it, and by a mixed 'policy' of allowing some to proceed, blocking and/or redirecting others, it could 'self-move' across the screen. My guess is that programmers would have no problem in producing a visual aid which could facilitate an understanding of something like the above. And I want to suggest that, in doing so, they would be mimicking something already very much with us.

Let the cursor be a human being. Let the movers be motivators: that stream of *events in the human being* which, continuously passing back in time nanosecond by nanosecond, become the sedimented *experiences of the human being*: experiences which, completely determinedly, will carve out its route through time. Let the sensitivity to the differential impact of different motivators be the sentience of the human animal, as mediated by a genes-built nervous system. And let the device for blocking and/or redirecting motivators be the consciousness of the human person, as mediated by a language-built symbolic system. Thus would the problem of free will dissolve. During a human present which is 350 milliseconds long, as modelled by a cursor which is 3.5 millimetres wide, it's not that 'everything that ever happens *to* us happens in it' (with my emphasis, this is Nicholas Humphrey's formulation), but that what happens in it *is* us: is the continuous recreation, nanosecond by nanosecond, of the happening called 'I'. Such a present is thus both 350 milliseconds long *and* the permanent impermanence of which we are. We are not in it: we are of it. And that means, or so I suggest, that by a process of coming to determine the determinations, freedom can be wrested from determinism: *and only from determinism*.

For it seems to me that, as is often the case with such big issues, the 'freedom *and* determinism' one has tormented us only by being wrongly posed. We have worried that if determinism were true we would not be able to change the future. But it's not the future which determines, it's the past. We don't need to be able to change the future: which is just as well, because the future has not happened and will never happen. Indeed, the mistake of thinking that my freedom is something to do with my being able to change the future can be shown to *be* a mistake simply by asking how I would know when I had done it. We don't need to be able to change the future because it's the past which determines: which is just as well, because far from not being able to do so, we cannot help but change the past all the time.

Let me illustrate, if crudely. I stop now typing at my word-processor and take a sip of my water. Let us suppose that before I did so I had taken ten such sips already. I have now taken eleven. I have changed the past. In whichever direction across the screen I was heading as the experience 'water sipping', that direction has been exquisitely modified. And each time I change the past in such a way, which means continuously, I am changing that which is determining me. If you like, it is that which we push out behind us which pushes us forwards.

But to see this reality less crudely, more clearly, we have once again to problematise the apparently innocent little word 'I' which I have been using in presenting my model. If that I is conceived of, as almost always and almost everywhere it *is* conceived of, as an entity *having* experience, then dualism remains (me and my experience) and the free will problem remains with it. But if that I is conceived of not as *having* experience but just as *being* experience, then monism rules. And it's OK. When I wave my arm, the reason why we can say, and with apparently equal validity, either that a *person* is behaving or that *a state of a nervous system* is behaving, is because they are actually one and the same thing. When I think of myself (impossibly) as something over-against both 'my' body, which I can direct in various ways, and 'my' mind, which I can change when I will, then the old I AM concept of I is blocking still the possibility of a new development. An I IS concept of I, the concept of I which my hypothesis proposes, is one which understands it not as a material thing which *has* experience but as a happening which *is* experience: which is the linguistically structured experience of a *Homo sapiens*. And such a formulation is one which is making three claims.

First, that the moment by moment states of a linguistically structured nervous system actually constitute ipseity: that that is what I-ness is. From which it follows that the question as to whether it is an I or a state of a nervous system which is behaving just dissolves: with an I posited as *being* a state of a nervous system, no such question exists. And from which it also follows that the question as to whether or not I am responsible for my actions dissolves with it. With an I posited as *being* a state of a nervous system, the answer is still 'yes'. But only because the state of a nervous system which produced matricide *is* the person which murdered its mother. That which is responsible is a person alright. But it is not an I AM person conceived of as an extentional material thing who *has* this experience and *ought to have had* that one. Instead it is an I IS person conceived of as a durational happening which *is* this experience and *is not* that one.

Second, that being aware in no way *necessarily* involves being an agent; and that being an agent in no way *at all* involves being an originator. Indeed the very word 'agency' should alert us to its non-originator status. Agents are entities which act as the expression of an origination which is other and elsewhere. And the agents which are persons act, ultimately, as the expression of an origination which is the sum total of the linguistically structured experience of all their kind.

Third, that what is needed is a recognition. Unrecognised, not re-cognised, as being a determined singular entity, the linguistically structured experience which is the singular human person can *have* 'free will', but only as an illusion. Recognised, re-cognised, as being a mutually determined and determining entity among other mutually determined and determining entities which together constitute the world, the linguistically structured experience which is the whole human community can *be* 'free', but for real. For with persons posited as both awarenesses of, and nodal points in, a linguistic web which is the world, mutual awareness, via the quintessentially collective process of coming collectively to determine the determinations, offers the possibility of that freedom which is the recognition of necessity.

But only of course to everybody. We have always known that one person cannot be equal. We must now know that one person cannot be free either. Such a recognition, involving as it does our coming to understand that the attempt to identify our humanity with our capacity to anticipate the future is fundamentally misplaced, and that ours is a humanity much more importantly to be identified with our capacity to anticipate the past, is of course in its infancy. A major problem in being able to get very much further with its development at the moment, I think, is our current inability to formulate the concept 'information' without ambiguity. It is a commonplace of current biology that information *in the sense of the instructions necessary to the making of a body or an organism* is that which is somehow carried genetically. It is no way such a commonplace, but by analogy I am suggesting it, that information *in the sense of the instructions necessary to the making of a mind or person or I* is that which is somehow carried linguistically. In both cases, it is that 'somehow' which we need to understand.

Such understanding might well prove hard to come by. But at least the search for it, already of course well underway, will protect us against mysterianism. For being essentially emergent entities, Ies are subject to all the confusions, further to be explored in Chapter 9, seemingly inherent to the concept of 'emergence'. Seeing it as a process which somehow disrupts causality, we are inclined to allow emergence to be posited as that which permits answers little short of the magical to be given to difficult questions. By contrast, an hypothesis which sees the emergent entity 'I' as the sequential linguistically structured states of nervous systems is one which requires us to posit this particular emergence in what I think is a far more productive way: that is, as a process thoroughly causal, but with a causality so complex that we do not (yet) have the capacity to fathom it. As I have already mentioned, the mathematician Ian Stewart repeatedly suggests that behaviour maybe only seems non-determined because its determinations are so many and so small that we do not (yet) have a mathematics to cope with them. In time I suspect that we will.

The outlining of these three claims ends what I want to say, anyway directly and anyway for now, about what I take to be the implications of the Libet experiment for our understanding in general, and for my hypothesis in particular. But before concluding my remarks on the importance of this perhaps my most important 'and' specimen, I want quickly to register just a few of the implications of the position I have outlined.

First, it has implications for how we think about the present: put another way, implications for how we think about our situation vis-à-vis a celebrated rodent. The animal concerned is a mouse: that much quoted 'wee, sleekit, cow'rin, tim'rous' mouse of Rabbie Burns. Addressing this small creature, whose 'best laid plans' he had inadvertently caused to 'gang astray' by accidentally ploughing up its nest, the poet offers consolation: 'Still thou art blest, compar'd wi me! / The present only toucheth thee: / But, och! I backwards cast my e'e, / On prospects drear! / An forward, tho' I canna see, / I guess and fear!'. He is suggesting that while the mouse lives only in a continuous present, he must live in more demanding places: always with the knowledge of a past, some parts of which he regrets; always with the anticipation of a future, some parts of which he fears. I know what he means. It is frequently claimed that he is writing about the essence of 'the human condition'. But there is, nonetheless, an important sense in which an alternative view is even more significant.

It is unfortunate that, in order to point to that alternative, I will have somewhat to misuse the language. As is often the case, and as is the case here, I will have to use space language—'boundary', 'leaving', 'entering', 'line'—in a time context. For now I will just use it. But when later I am considering my 'time *and* space' specimen, I will try to defend my contention that it is directly as a consequence of our having been stuck in (extentional)

I AM conceptions of the self for so long that what would be a more appropriate (durational) I IS vocabulary is not yet available. So in what is for now the absence of appropriate 'time' language, let me stumble along (!) in its 'space' substitute as best I can.

Ever leaving the former, ever entering the latter, both creatures, man and mouse, live at a boundary between past and future. But it is a boundary of different 'width' for the pair of them. In an important sense, the present for the mouse is almost non-existent: think of it as being at a boundary *line* between past and future, which like all lines has no width. Whereas for the human, what Libet's work suggests, I think, is that what for the mouse is a boundary *line* becomes for the human a boundary *area*: that is, an area which is 350 milliseconds 'wide'. However wide the mouse's boundary—a question the answer to which would need to be established empirically—it would at all events seem to be the case that it is narrow enough to mean that, within its range, there is next to no room for, next to no capacity for, the struggle to *recognise* necessity, and thus to *make* freedom. For the human, by contrast, its 350-millisecond-wide affair, a present which is both a permanent impermanence and all that there is, would seem to have been made wide enough—as I suggest, by its linguistic structuring—as not only to *permit* but actually to *require* that, within its range, a more or less continuous struggle to recognise necessity, and thus to make freedom, will take place. Both creatures are determined. It is absolutely vital to understand that were there chance or randomness rather than determination, there would be absolutely no possibility of freedom being able to be made, at any level, by any creature, ever. But in ways both to some extent already understood, to some extent as yet unexplained, what seems to have happened at the level of the human is the development of the capacity to *recognise* the determination, and in the 350 millisecond gap between its starting to play itself out in the firing of a 'readiness' potential, and its consummation in the musculature, to inhibit and/or redirect its flow.

Second, it has implications for how we think about the question of changing the world. People impatient with various of the injustices and nonsenses of the world sometimes ask, usually rhetorically and sometimes in some anguish, how we can change it: the world, that is. But in the light of these 'freedom *and* determinism' issues, it can be seen, in an important sense, to be the wrong question to ask. The world actually *being* that which we do, everything that everybody does *being* the world, the question we really need to ask is how we can actually help changing it—simply by that doing. As I have already indicated, and will argue further in Chapter 11, this is a conception which makes the whole enterprise of changing the world quintessentially collective. When I is understood as the linguistically structured experience of *each* human being, and that experience is quintessentially of the linguistically structured experience of each *other* human being, there is no freedom-via-the-recognition-of-necessity for one without a freedom-via-the-recognition-of-necessity for all. The process of *making* better one-and-all worlds, collective to the core, consists in an ever more complete accommodation of an ever growing percentage of persons to an ever better understood world. The steady accumulation of such freedom can only become complete for one when it is complete for all: the position I take Marx to have been adopting with his insistence that the free development of each and of all are co-determinate.

Third, it has implications for how we think about child development. When the Jesuits suggested that, if they had the care of a child until he or she was seven years old, then they were likely to have formed its basic mind-set for life, they were giving expression to a view still very much with us: that is, to the view that it is quintessentially *early* experience which is differentially important in human development. My position would spell that out by suggesting that experience is durational, and durational units in sequence behave exponentially. Thinking of my cursor model, when 'a lifetime' consists of one durational unit, all determination is in one direction, and thus 100 per cent of the outcome is explained by it. When it consists of two such units, the influence of each, at 50 per cent of the whole, is halved; when of four, halved again—and so on. In this manner is the impact of each additional durational unit drastically reduced. This is one of the reasons offered by my model as to why early experience is likely to be crucial. The other is the state of our 350 millisecond 'struggle band' in the early years of life. In infants it is non-existent: in small children it is not yet fully functioning. Which means that in those early years, natural determination can roll through the existence of such little creatures relatively undeterred, relatively unchecked. Indeed it would be a specific and no doubt eventually testable prediction of the I IS hypothesis that there will be a neurological and linguistic correlate of the 'give-me-a-child-until-he-or-she-is-seven' stance which will be able to be related to 'struggle band' development.

Fourth, it has implications for how we think about punishment. When we feed the new conception of 'I', the I IS conception of 'I', into the question: 'am I responsible for my actions?', the answer is an unequivocal 'yes'. But it is a 'yes' which has to be understood in the context of an appropriate answer to a prior question as to what this 'I' (which is responsible) is. An I which is the linguistically structured experience of a *Homo sapiens* is an I which *is* its experience rather than an I which *has* its experience. As such, it is determined by its past because it is constituted by its past. To say 'I am responsible for my actions' is to say the same thing as '"the linguistically structured experience of Ann Long" is responsible for my actions' *because 'I' and 'the linguistically structured experience of Ann Long' are one and the same thing*. Substantially then, the determinism I am defending most certainly *does* make the 'punishment' aspect of responsibility totally indefensible: substantially *does* insist that given that our determination goes 'all the way down'—as some among the professionals might put it—Skinner was bang-on right in his campaign against punishment as a method of shaping behaviour. The essence of his operant conditioning was that while being *reinforced* for *desired* behaviours made them *more* likely to be repeated, being *punished* for *undesired* behaviours did not make them *less* likely to be repeated. His campaign was thus positive ('it doesn't work') rather than normative ('it's morally unacceptable'). But for a conception of morality which wants to rid itself of 'oughts', as mine does, this is indeed 'a difference which makes no difference'.

Fifth, it has implications for how we now think about 'the will'. In *The Birth and Death of Meaning* (1968), Becker described four stages in the emergence of a will more free: the direct attractive or aversive responses of micro-organisms; the classically and then operantly conditioned responses of creatures with nervous systems; the making of spontaneous connections using such signs as chimps use when banging on oil-drums to register status; and the use of fully semiotic systems—of which natural language is by far the most significant—whose meaning is validated only on the basis of consensus. This, or something like it, is what I am talking about. And it means that when Hannah Arendt sees the will as 'the organ of the future', then if she means by that what I think she does, I am with her. I am with her to the extent to which she means that freedom as the recognition of necessity is available only to persons; that complete freedom as the recognition of complete necessity is available only to complete persons; that complete persons are those who have developed the sorts of selves which recognise the necessity of loving to personic flourishing; and that the point (moment) at which such recognition becomes habitual for all of us will be the point (moment) at which her 'organ of the future' will have come into its own.

And sixth, it has implications for how we think about what previously I have referred to as the terror of determinism. In *Darwin's Dangerous Idea* (1995), Daniel Dennett drew attention to the religious longings of those who try to defend a mind-first, top-down image of the place of humans in the cosmos. He quoted, in particular, Stephen Jay Gould (1989) who 'has certainly seen his task as one with cosmic implications, something that is especially clear in the epiphanies about the Burgess Shale in *Wonderful Life*'. Dennett characterises such yearnings as being those of secular humanists who 'sometimes think that Humanity cannot be special enough to matter if it is the product of merely algorithmic processes.' As they contemplate Darwin's dangerous idea, he says, what they really want is some 'radical contingency': something that 'will give the *mind* some elbow room, so it can *act*, and be *responsible* for its own destiny, instead of being the mere effect of a mindless cascade of mechanical processes!'. I think that in essence he is right. It seems to me that the Gouldian position in 'the science wars' is fundamentally a product of the (totally false) notion that you can't get something more special from something less special—when in fact you can't get it from anywhere else. But the irony of the present situation is that, in the same way as water could only become known to fish if they could emerge from it, so the very reason why our determination is just becoming known to us is because we are just starting to be able to see how we might emerge from it. All of which means that the veritable terror (and thus huge hostility) that the thought of such determination arouses in so many people—secular humanists along with many of the rest—is arising, *and could only arise*, just as it is becoming unnecessary. As I hope to show in later chapters, those 'religious yearnings', totally legitimate it seems to me, can only be satisfied by a complete and determinist re-think as to the nature of the personic.

And so, a summary. We have thought of 'free-will' as a once-and-for-all and always-already phenomenon: a gift given and passively accepted, whole and complete, at the origins of our humanity. I am suggesting that we must now think of 'free-won't' as a little-by-little and never-yet phenomenon: a prize taken and actively wrested, in steady increments, across the development of our humanity. And those are changes more easily seen if we go vertical:

the old free will is	*the new free won't is*
once-and-for-all	little-by-little
always-already	never-yet
a gift	a prize
given	taken
passively accepted	actively wrested
whole and complete	in steady increments
at the origins of our humanity	across the development of our humanity

When Hegel characterised freedom as 'the recognition of necessity', then even as I suspect that I have taken his phrase to places he did not at all intend that it should go, I think he was offering a characterisation which chimes just splendidly with my I AM to I IS formulation. Freedom so conceived is not discovered where previously it had been hidden: it is made where previously it had not been. It is made when we come to understand what in *Beyond the Chains of Illusion* (1962) Eric Fromm called 'the passions and interests which direct us behind our backs'. In full the relevant quote is:

> His [Marx's] position is very much the same as Spinoza's: we are determined by forces outside our conscious selves, and by passions and interests which direct us behind our backs. Inasmuch as this is the case, we are not free. *But* we can emerge from this bondage and enlarge the realm of freedom by becoming fully aware of reality, and hence of necessity, by giving up illusions, and by transforming ourselves from somnambulistic, unfree, determined, dependent, passive persons into awakened, aware, active, independent ones. Both for Spinoza and for Marx the aim of life is liberation from bondage, and the way to this aim is the overcoming of illusions and the full use of our active powers.

My copy of the book, very old, has the 'behind our backs' phrase underlined, and the marginal note: 'being directed behind our backs is a good image—try to develop it.' And my hope is that, with my I AM to I IS reading of this 'freedom *and* determinism' pair, I've done just that. For when Pascal quipped that, had Cleopatra's nose been a little longer, the whole course of human history would have been different, it has only so far been possible—for him as much as for his readers—to take such comment as humour. To get us all to that level of understanding where it can be appreciated that while it is *not* the case that the length of a nose determines the peace-or-war fate of a people, it nonetheless *is* the case that every single material thing and happening making up each discrete past 'then' aggregately determines every single material thing and happening making up each discrete succeeding 'then' *up to and including 'now'* will be no easy undertaking. But it is my contention that it is an undertaking on the successful outcome of which depends the whole possibility of our ever making better worlds.

(10) good and evil

The philosopher's stone of the alchemists was a substance which could turn base metals into gold. Of course it turned out to be an imaginary substance, but not before centuries had been given to the fruitless task of trying to identify it. The philosophers stone of the philosophers themselves, or anyway according to Kant, was whatever it was which would produce in us the disposition to be good. And it is widely believed that, given the fallen thing that is 'human nature', the search for this stone too is equally hopeless. Much of twentieth century thinking about the good was dominated by the return, initially in G. E. Moore's *Principia Ethica* (1903), of David Hume's insistence, long ignored, that it was not possible to derive the 'ought' of the good from the 'is' of the merely existent. And that was a resurrection which did nothing to challenge such hopelessness. Leading, as it did, to both the moral scepticism which maintains that we can't say anything true about what is good, and the moral relativism which maintains that moral statements are all relative to particular individual and social standpoints, it leant support instead first to an existentialist response which told us that philosophy could not tell us how to live (so we were 'free'!), and then to a postmodernist response which told us that there was no escape from being relativists about the good, as about everything else.

None of which did anything, however, to quell a considerable yearning. For widespread in philosophical conversation, but present as well among the professionals, there remained an ache that it might be otherwise. Thus writing in the *Times Higher Educational Supplement* (5.11.99), for example, the moral philosopher Mary Warnock was to be found lamenting 'the current widely perceived loss of any sense of shared values', and stressing the urgency of 'rebuilding such a sense' by 'appealing to our common humanity.' My response to such lament would be that while the yearning is appropriate enough, for as long as we defend societies in which some of us have more of 'our common humanity' than do others, then any such 'appeal' will fall on ears which cannot afford to hear it. Moral theories which offer some criterion by which to judge the goodness of an action are sometimes called 'substantive': thus in offering 'the extent to which an action promotes pleasure' as its criterion, utilitarianism is a substantive theory. In such terms, mine too is a substantive theory. With 'the extent to which an action promotes that which is essential to loving' as its criterion, it will suggest—in the slightly funny language of some among the professionals—that 'for every action X, X is morally good if and only if it is such as to promote that which is essential to loving'. And in attempting so to recast the 'good-*and*-evil' pair that we are saved from the resignedness bred by relativism, that we can go on to think about just what it is which is essential to loving, it is my claim that the first step is to rid ourselves entirely of the second term: that is, of 'evil'. It is therefore with 'the case against the coherence of the concept of evil' that I want to begin.

Tabloid headlines notwithstanding, any good dictionary will confirm that the old meaning of the word 'evil' is now obsolete. That old meaning drew upon essentially mediaeval Christian theology. In Western Europe, the post-Greek but pre-Enlightenment period was dominated by a Christian theory of morality which, anyway at the level of the pew, anyway at the level of what I am calling philosophical conversation, had it this way: that that which was good was that which was in accordance with the will of God; and that that which was evil was that which was in accordance with the will of the devil. Both were supernatural powers. But anyway by comparison with God, the devil died a relatively early death theologically. After all, how could God be thought of as all powerful if that charismatic fallen angel was allowed to harass and thwart him at every turn? He had to go. But although in more sophisticated theological places he did, in the guise of 'profound immorality, wickedness and depravity, especially when regarded as a supernatural force' [O.E.D.], his almost eponymous 'evil' lingered. And even now, when in all conscience it has come to mean nothing much more than 'bad', or maybe 'very bad', there are some—witness those tabloid headlines: witness the outpourings of some contemporary politicians—who indulge themselves, from time to time, in the type of 'evil' hysteria which does nothing much more than further undermine their already fragile credibility.

In those more sophisticated places, both what used to be called 'moral evil' (the sins of humans, such as torture and murder) and what used to be called 'natural evil' (the what-if-not-sins of God, such as earthquakes and floods), are for the most part now theorised differently. And the tradition of such theorising with which I would want to identify myself, one which coheres with that abandonment of the concept of evil which I am suggesting is a necessary first step in rethinking this 'good *and* evil' pair, is that which recognises an essential asymmetry.

The best-known Christian source of the asymmetrical approach is that which develops across the writings of Dionysius the Areopagite, St Augustine and St Thomas Aquinas. In its fully-fledged version, it argues that evil is not a presence but an absence: that the evil is nothing but the absence of the good. This is a position which can quite readily be interpreted to be part of a tradition which, extending back at least as far as to Socrates, is perhaps the mainstream one in Western philosophy. Its most usual form is that which suggests that, in thinking about what we were wont to call 'evil', we are actually dealing with the privation of something. Kant's version was to explain evil in terms of a privation of the will. A bad act was not the result of somehow *having* within one that which was evil, but of *not having* within one (anyway enough of) that which was good: specifically, a good will. Thus within a Kantian system, 'do evil', let alone 'do radical evil', as he called it, could not possibly be adopted as a guide to human action. To think that it could would require that we associate people with that 'devilishness' which he specifically denied as any sort of appropriate description of human behaviour. Thus does 'evil', but not of course 'good', disappear from philosophical contemplation. And good riddance! Such an understanding means that in much the same way as both the 'hell' member of the 'heaven *and* hell' duo and the 'devil' member of the 'God *and* the devil' pair have largely disappeared, so too must the 'evil' member of the 'good *and* evil' pair follow them down. But leaving—what?

In contrast to the relativists, who if they too drop the 'evil' member of the pair (as probably most of them do) would need to replace the remaining 'good' pluralistically, would need to talk in terms of 'goods' of various culturally-specific shapes and sizes, some of them mutually incompatible, what I want to suggest is that what is left is 'good *and* pre-good': or getting them better into sequence, 'pre-good *and* good'. This is the universalist position I seek to defend: specifically here; more generally throughout the book. It means that I have to shed the notion of evil: but there I would seem to be on safe enough ground because almost everyone else seems to shed it too. But it also means that, anyway as far as persons are concerned, I have to shed its softer stable-mate 'bad' as well. So how can that be done?

Working still from within what I have called the asymmetrical tradition, badness is to be thought of as a defect: things that are bad fail to come up to some standard of what we think such things should be like. Deck chairs that are bad won't fold; oranges that are bad are not edible; people that are bad are not honest. With such a usage, if a hidden entity were to ask: 'Am I good or bad?', and then give us a list of its characteristics, we would first have to discover what it was—deck chair? orange? person?—before we could offer an opinion as to whether its proffered list made it 'good' or 'bad'. But such a usage of the idea is not careful enough for clarity. I am suggesting that the 'folk-philosophy' which permits us to talk so needs to be toughened up in such a way that deck chairs and oranges are recognised as conceptually distinct from persons. I am suggesting that while a less-than-good deck chair (a non-developmental entity) can be 'bad', a less-than-good person (a developmental entity) can only be 'not yet good'. Which means that, for persons, it's not just that the pair 'good *and* evil' is out because, in its primitively additive mode, it does not make clear that 'evil' is not the positive opposite of 'good'—as it isn't. It's that the pair 'good *and* bad' is out as well because, even in a more sophisticatedly additive mode, it does not make it clear that 'good' has *no* opposite of *any* kind. For persons, whatever is the 'evil-like' or 'bad-like' word we pair with 'good' must be paired with it not symmetrically as an opposite, but asymmetrically as a precursor.

The asymmetry is there because of a time-irreversible, developmental relationship. It is there because the process of making the good is essentially just that: a process. In *P.N. Review* (vol. 18, no 4, March/April 1992), Laura (Riding) Jackson suggests that 'probably language and the idea of the Good are co-original'. This seems to me to be on the right track. In the terms of the I IS hypothesis, using its terminology, that which is natural, flowing along evolved lines of least resistance (but with language and the emerging 350 millisecond gap ensuring that such resistance is ever growing), is that which is mindless: does not mind; is not capable of the good. That which is not natural but moral, struggling along developing lines of more resistance (but with language and the emerging 350 millisecond gap ensuring that such resistance is ever diminishing), is that which is mindful: minds; is capable of the good. And just as Galileo came to understand that the not-still is the not-yet-stopped and/or not-yet-redirected natural motional, so we need to come to understand that the not-moral is the not-yet-inhibited and/or not-yet-redirected natural *e*motional. Amongst many other revolutionary things, such understanding would enable us to see that while praise (of the good) is entirely appropriate, blame (for the bad) is entirely ridiculous.

Denying the existence of 'evil' or of 'bad' is not just being generally cuddly about the human. It's not just warming to the frequently observed 'terrible temptation' to do good, or claiming that, like Stevie Smith's cat, we humans love to 'gallop about' doing the stuff. It's not just feeling with the observation of Charles Lamb that 'the greatest pleasure I know is to do a good thing by stealth': for after all, he goes on to add 'and have it found out by accident'! No, this is a fundamental and vital theoretical point concerning the nature of persons, and the consequent nature of 'the good': a point which, when it has been understood, can undermine such ready cynicism, and can challenge the way in which 'do-gooding' has come to carry its special and peculiar meaning. It is a point which, by taking off in turn from three distinctly different moral stances, each much in evidence today, I will try now to identify more closely.

Stance One is the *there-are-many-moral-truths* stance. It is adopted, for example, by Alan H. Goldman in his book *Moral Knowledge* (1989). He argues for what he takes to be a realist position concerning moral judgements, suggesting that, like other judgements, they are indeed either true or false. But he argues that 'true' in this context does not mean 'matching reality' but a combination of two other things: it means being internally coherent, and being 'coherent with social aims and norms of individual psychology'. He adds that although it is normally important to us to think of truth as singular, with moral truth we have to learn otherwise. We have to learn that there are many moral truths: that moral truth is 'multiplex'. And that means, he says, that there will be as many true moral systems as pass the test of coherence. Using *this* stance as my starting point, I would be saying: yes, fine. But the test of coherence is a pretty rigorous test: so rigorous, in fact, that only one moral system passes it.

Stance Two is the *there-are-no-moral-truths* stance. It is adopted, for example, by Bernard Williams in his book *Ethics and the Limits of Philosophy* (1985). If I have understood him aright, he seems to have been chiefly concerned to separate ethics and science: to argue that moral reasoning and scientific reasoning are essentially different. He takes the scientific enquiry to be one where 'there should ideally be convergence on an answer, where the best explanation of the convergence involves the idea that the answer represents how things are'. But in the case of an ethical enquiry, he argues that while 'convergence on an answer' is only 'unlikely', convergence on an answer 'because it represents how things are' is impossible. Along with many another, he would seem to be claiming that there is no 'how-things-are' in morals. Using *this* stance as my starting point, I would be claiming, along I think with Spinoza, that there is.

Stance Three is the *there-is-one-moral-truth* stance. It is one very widely adopted by sundry biological reductionists. Dismissive of what they see as the total mess which is moral philosophy, and eager for a scientifically respectable theory of the good, such people base themselves in evolutionary theory. What is good is that which promotes the flourishing of the biological entity *Homo sapiens*. And there is no need to consider any other sort of human entity, any 'self' or 'person' or 'soul' or such, for such things just don't exist. As anything other than the 'kin selection' of Hamilton, or the 'reciprocal altruism' of Trivers, evolutionary altruism does not exist and could not exist. This means that the one moral truth that any theory of 'the good' must respect is that of biological flourishing. And that means that anything which requires us to 'love our neighbours as Ourselves' or 'act only on that maxim which we could will to become universal law' is just a moral non-starter. Using *this* stance as my starting point, I would be agreeing that there is one moral truth, but suggesting that it concerns the flourishing not of the organic entity *Homo sapiens* but of the (thoroughly existent) personic entity 'self'; agreeing that moral philosophy is a mess and that we need a scientifically respectable theory of the good, but suggesting that the science concerned is not biological science but moral science; and agreeing that evolutionary altruism, as anything other than the 'kin selection' of Hamilton or the 'reciprocal altruism' of Trivers, does not exist and could not exist, but suggesting that that is a fact just irrelevant to the question of whether or not moral altruism exists.

This third is a stance which, unlike the other two, needs further explication. It needs it, first, because as a consequence of the recent ascendancy of 'evolutionary psychology', some version of it, in philosophical conversation anyway, tends to predominate. And it needs it, second, because the position on it which is inherent to my hypothesis is both somewhat singular and very important to its defence. Let me therefore pursue it a bit further.

It would be a total non-sequitur—and is, for it is often perpetrated—to suggest that because evolutionary altruism *does not* exist in the biosphere therefore moral altruism *can not* exist in the world. As I have already argued, (see 'egoism *and* altruism') such 'arguments' are just not worth wasting time on: there is absolutely no 'therefore' about it. It is, however, a totally other matter to suggest that *as a matter of independent fact* moral altruism *does not* exist in the world. I *am* suggesting that. But I am suggesting it not because evolutionary altruism does not exist in the biosphere—which is true, but irrelevant. I am suggesting it because it can be shown not to exist by a quite independent examination of that world *in its own right*. And what's more, what was in the earlier 'egoism *and* altruism' discussion referred to as the 'cuddly and hospitable' view of humans does not need it to exist. On the basis of an 'I AM to I IS' hypothesis as to the nature of ipseity, human persons can be shown to be capable of the highest imaginable acts of moral goodness without a drop of anything which need remotely be thought of as altruism in sight. So let me finish my comments on my 'good *and* evil' specimen by spelling out briefly just how.

First: conceptualise 'a person' as the linguistically structured experience of a *Homo sapiens*. Second: conceptualise 'a self' as a stage in the development of such a person. Third: conceptualise 'being selfish' as a defining characteristic of any such self. Fourth: conceptualise 'selving' as the process in a person by which there is a developmental shift across a sequential series of emergent selves, each with emergently new and emergently different interests. Fifth: conceptualise that sequential series as one in which, when its earlier I AM types look after their own steadily developing but not yet fully developed animal-human interests (which definitionally they do: there is no moral altruism), those interests are such that, in looking after them, they thereby and necessarily fail to look after the interests of other selves. Sixth: conceptualise that sequential series as one in which, when its final I IS type looks after its own fully developed human-human interests (which definitionally it does: there is no moral altruism), those *new and different interests* are such that, in looking after them, it thereby and necessarily succeeds in looking after the interests of other selves too. Seventh: conceptualise the whole process not as one which we *ought* to move through but one which we actually *do* move through.

With these seven moves in place, 'moral altruism', conceived of as putting the interests of another self ahead of the interests of your own self, becomes redundant: as any part of any attempt to defend the possibility of human goodness, just redundant. And it is because I am suggesting that these moves are *precisely* those it is appropriate to make that I am arguing that the problem supposedly entailed by the 'good-*and*-evil' pair is a non-problem. The 'evil' is non-existent. And the 'good' is that which, out of the 'not-yet-good', is *made*, slowly and steadily *made*, across the selving process. In turning now to my next specimen, it is a further elaboration of the nature of that very process which, neatly, can be appropriate.

(11) I and thou

In introducing this pair in Chapter 4, I suggested that, in philosophical conversation, as opposed to in Buber's own usage, what usually goes on under this head is either discussion of the 'appearance *and* reality' and 'fact *and* value' pairs (concerning first person versus third-person views of what is), or discussion of the 'egoism *and* altruism' pair (concerning degrees of genetic relatedness as a source of morality). Since those are all 'ands' in their own rights, as it were, and therefore considered elsewhere, I will use this 'I *and* thou' specimen to consider a different issue. The critical distinction for Buber between an I *and* it' relationship and an 'I *and* thou' relationship is not that the first is always between a person and a thing, the second between a person and another person (or God). Although such is usually the case, it is not what is critical. What is critical is that where the first never involves what he calls 'the whole person', the second always does. And it is perhaps worth exploring how, if at all, his distinction tallies (or doesn't) with that which I am making between I AM and I IS persons.

Using my terminology, Buber's 'I' is clearly an I AM: its Other is a 'thou' not an 'I'. And in the context of my hypothesis, the status of an I AM as a whole person is problematic. In one sense, the sense in which wholeness pertains to unity over fragmentation, clearly all persons are whole persons: given its delineation of a person as essentially a state of a nervous system, which system definitionally cannot be in two states at the same time, mine is an hypothesis which puts considerable emphasis on the unity of the person as an all-or-nothing event. But in another sense, the sense in which wholeness pertains to completeness over incompleteness, things are different. With my essentially developmental approach, I AMs are not whole (in the sense of complete) persons: there is another developmental stage, the I IS stage, available to them. And as a further explication of what that stage consists in, I want to take this 'I-*and*-thou' opportunity to look specifically, but very briefly, at two stage theories of development, the one of Erik Erikson and the other of Lawrence Kohlberg, to illustrate the way in which there is already in the psychological literature some small exploration of what I am thinking of as 'selving towards the moral'.

Erikson's stage theory often goes under the heading of 'the eight ages of man', and is probably most easily understood via a simplified, graphic presentation (see Table 2). My purpose here is in no way to provide a detailed explication of Erikson's theory: for anyone who should want it, his own classic text *Childhood and Society* (1950), or any number of psychology textbooks, can provide that. Instead I want to no more than suggest, and that via the briefest of outlines, a possible similarity between Erikson's person at stage 8 (positive) with my I IS self.

Stage	Age	Psychosocial Crisis	Radius of Significant Relationships
1	0–1	basic trust v. basic mistrust	mother or mother figure
2	1–3	autonomy v. shame or doubt	parents
3	3–6	initiative v guilt	basic family
4	7–12	industry v inferiority	neighbourhood and school
5	12–18	identity v role confusion	peer group
6	20s	intimacy v isolation	sexual partners
7	30–50s	generativity v stagnation	friends and workmates
8	50+	ego integrity v despair	humankind

Table 2: The Eight Ages of Man [after Erikson (1950)]

Erikson thought that each of his eight stages, genetically based and thus the same in all cultures, was characterised by a developmental psychosocial crisis such that its outcome, somewhere on a continuum from positive to negative, determined the advantage (or disadvantage) with which the person entered the next stage. These were not, however, once-and-for-all outcomes: thus although the main thing that the Stage 1 infant learns is trust (positive) or mistrust (negative), and the resulting behavioural biases will certainly present as strong tendencies towards further biases thereafter, nonetheless there is some leeway. First, each of the terms in each of his pairs of developmental duos represents only the extreme of what is, in fact, a continuum. And second, even for those entering a subsequent stage disadvantaged by being biased toward the negative end of the previous one, or even all previous ones, all is not lost. Erikson thought it was entirely possible to make up for early negative experience later on: though decreasingly so as the person concerned got older.

With that as the barest of explications, let me look now at his Stage 8 (positive) persons: that is, persons of perhaps 50 or more whose development has been so spectacularly positive that they achieve complete (!) 'ego integrity'. What, according to Erikson, would such persons look like? By working out some coherent theory as to 'what it was all about', they would have made sense of their lives. Because they understood the inevitabilities of things, they would not be racked by 'if only' questions. Because they understood something about development, their own and that of others, they would no longer blame those others, perhaps particularly parents, for doing what they could only do. They would have no fear of death. And the radius of their significant relationships would be sufficient to encompass the whole of humanity: of all their kind.

Here are selves at least somewhat akin to my I IS variety. There is of course much in Erikson, more fully considered, with which I would be out of sympathy: even in this thumbnail sketch, it is clear that his *genetic* determinism, though not his determinism as such, would not be mine. But especially in what anyway would be my reading of that last point, of the fact that he sees such folk as those who, in the end, have come to identify with, and thus be personally concerned about, the lot of all their kind, at least embryonically here they are: my I IS selves. And the fact that Erikson thought such persons to be extremely rare also conforms to the grain of my approach: as long, that is, as one also whispers—'so far'.

Erikson's is thus a theory which can offer my hypothesis a useful conceptual tool. For a reworking of his 'radius of significant relationships' idea can give me a powerful way of characterising the I AM to I IS shift I am proposing. In bringing steadily more and more people into that radius, each succeeding I AM self, steadily less partial (in both senses of the word) but none the less still partial (in both senses of the word), can be seen as a *quantitative* advance upon each preceding I AM self. And in bringing finally all people into that radius, the final I IS self, no longer partial (in either sense of the word) can be seen as a *qualitative* advance upon all preceding I AM selves. Indeed, so useful is this formulation that it will be one to which I will several times return across the rest of my book.

Lawrence Kohlberg's stage theory (see Table 3), is both different and similar. It is sometimes presented, erroneously I think, as a theory of moral development. It isn't. It's a theory of the development of moral reasoning. Clearly grounded in Kant's version of the 'autonomy *and* heteronomy' dichotomy—that is, clearly of a liberal-humanist ilk—it is interested in the reasons which people offer for those of their behaviours which have ethical implications. And again, it is a simplified graphic presentation which fits my bill. Kohlberg argued that the development of moral reasoning was contingent upon cognitive development. He made his study of it by presenting his subjects with stories, of which there were nine in all, which involved moral conflict in each of ten areas which he considered universally significant to humankind: punishment, property, law, affection, authority, life, liberty, distributive justice, truth and sex. What he was interested in was not the content of the decisions which his subjects made when he presented them with his 'dilemma' stories, but in the reasons that they gave for those decisions. It was these reasons which he then classified as falling first into one of three levels, and then, there being two stages at each level, into one of six stages.

Level	Stage	Orientation	Description
1: Pre-Conventional	1	Punishment and Obedience	*what is wrong is what is punishable*
	2	Instrumental-Relativist	*what is right is what is rewarded*
2: Conventional	3	Interpersonal Concordance	*good behaviour is what pleases others*
	4	Law and Order	*being good is doing one's duty*
3: Post-Conventional	5	Social Contract-Legalistic	*what is right is a matter of personal values and opinions*
	6	Universal-Ethical Principle	*what is right is based on self-chosen abstract ethical principles*

Table 3: Kohlberg's Stages [after Gross (1992)]

As with all stage theories, while the ages at which people reach the levels and stages vary, it is the sequence of development which is claimed to be invariant: nobody who has not in the past given a Level 2 Stage 3 reason for a behaviour will leap straight to giving a Level 3 Stage 6 one first. Indeed, so sparsely populated was his Level 3 Stage 6 category that at a later stage of his studies—he began his work in 1955—he abandoned Stage 6 altogether as being beyond the reach of almost all human beings. But when, however rarely, such were to be found, this is what, according to him, they would look like. Their conscience would be their own: neither the opinions of others nor the laws of the land would determine for them what they would do. And the principles by which they lived would be those of justice and equality. The evidence of their existence might be thin on the ground, but you would know one alright when you met her or him.

If there is much in Erikson with which my hypothesis would not be consonant, there is even more in Kohlberg. His theory is thoroughly individualistic. His Level 3 Stage 6 characters come by their most highly developed moral reasoning via self-chosen individual reflection, quintessentially of their own free wills. And that is an approach which could not be more different from that intrinsic to my own hypothesis. But at least to the extent to which his identification of such folk is based in their adherence to principles, particularly to principles of justice and equality, here also perhaps almost lurk incipient I IS selves. If they do, I like to think that their lurking stance is consequent upon their waiting for a more adequate theory to permit their more complete emergence!

I see stage theories such as these as among the first attempts of psychological science to articulate in some way what I am calling the selving process. I don't make reference to them here because I want to defend anything of their detail: I don't expect much of it to survive when what will probably be technological developments permit more close-grained study of such things. I make reference to them only because I think they help me to articulate what I mean by the process I am calling 'selving'. In Chapter 11, I will consider it further. But for now, what I want to make clear is that, suggestive as they may be, what my hypothesis raises are issues not even yet dreamed of, or so it would seem, in the philosophies of such as Erikson or Kohlberg. And this is why.

An I understood as the linguistically structured experience of a *Homo sapiens* is an I which develops not individually but in an interlocutory web of other such Ies. Each interlocutor in the human conversation is 'con-scious' within *and only within* the web: 'knows with' each other interlocutor within *and only within* the web. Each interlocutor develops a 'con-science' within *and only within* the web: a 'knowing with' each other interlocutor within *and only within* the web. Each interlocutor is thus a joint bearer, with all the other interlocutors, of the representations and discourses of the web. When the interlocutors are I AMs, their discourse is skewed by each experiencing herself or himself as first: the web is pulled this way and that by interest and partiality; and the pressure to selve ones way out of such stresses remains. With the gradual emergence of I IS persons, things change. When the interlocutors are I ISes, their discourse is the 'straight-for-straight' stuff of just persons: a shift which means that the thrashings of the web are stilled to disinterest and impartiality; and the selving process is freed to reach its completion. This is what my hypothesis is proposing. And that means that mine is therefore not just another stage-developmental theory—though it is such. It's instead a stage-developmental theory in which what is being presented as its final stage is one qualitatively different from any of the others. And not only any of the others as so far proffered by the psychological literature, but any of the others it would be *possible* to proffer from within a paradigm in which the Other of 'I' was still only 'thou'. Buber's language is still that of the originating first person: of one still partial in the sense of interested, and partial in the sense of incomplete. And as such, it is language still less than capable of giving voice to that level of loving which becomes available, or so I will be suggesting, to I IS persons. Persons for the most part still of the future, they will be whole. Where Rimbaud knew that 'I *am* an Other', they will more importantly know that 'an Other *is* an I'.

As in the previous paragraph, I have now several times used my invented 'Ies' word as the plural of 'I'. But especially in this 'Buber' context, I think it maybe needs some further comment than already I have given it. In I AM language, there is no plural of 'I': that is, in I AM language there is no 'world without Others'. The first person plural names I *and* others as 'we'; the second person plural names *only* others as 'you'; and the third person plural names *only* others as 'they'. A plural with *no* other, a plural with just the equality of 'I and I', does not exist. And it is because the I IS hypothesis needs such a plural that I have invented 'Ies'. I have had to invent so because I am trying to show how a new understanding of the real nature of 'I', and its 'Ies' plural, will change our understandings utterly.

The sixteenth and seventeenth century revolution in thought which created the modern physical sciences of the inorganic, the revolution associated with the names of such as Copernicus and Galileo and Newton, involved our coming to understand that the earth is no 'first' planet at the centre of the universe. The nineteenth and twentieth century revolution in thought which created the modern biological sciences of the organic, the revolution associated with the names of such as Darwin and Wallace, Mendel, and Crick and Watson, involved our coming to understand that *Homo sapiens* is no 'first' creature at the centre of the biosphere. It is my contention that the third and coming revolution in thought which will create the moral sciences of the personic will involve our coming to understand that there is no 'first' person at the centre of the world. The full significance of each of the previous revolutions was hard enough to grasp: indeed the full significance of particularly the second is still not fully grasped. But it is also my contention that the full significance of this last, the revolution which will involve our coming to understand that there is no I AM first person, will be almost immeasurably harder. And that when it finally and fully penetrates, when it is fully understood that in a 'world without Others' Buber's 'I *and* thou' must cede to 'I *and* I', the consequences will be both world-changing and mind-blowing: literally. Of which, more later.

* * *

So I come to the end of my first 'experiment' chapter. By now you will no doubt be getting the hang of my approach: first demonstrate the way in which an I AM reading of 'I' will not let us solve the additive problems which the particular duo represents; then show how an alternative I IS reading can rid us of them. It is an approach which will continue, specimen by specimen, across Chapters 7 and 8 as well. Inevitably, I think, this will involve me in some repetition. But precisely because, to many people, my case is likely to be more than a little counter-intuitive, that may be no bad thing. For you never know: what has perhaps not lodged as significant via its consideration under an 'acts *and* omissions' head might yet lodge as significant under a 'nature *and* culture' or 'time *and* space' head instead. So although it has, of course, already occurred to you that you might now skip the next two chapters and move straight to Chapter 9, I hope that nonetheless you will instead persevere.

Chapter Seven

EXPERIMENT TWO

From 'identity and relationship' to 'mind and body'

This chapter deals with the second set of my 'and' specimens: with the next ten of them. And once again my aim will be to point up the difference made to an understanding of each of the pairs by situating it within a changed conception of ipseity. Once again I will be arguing that, within the I AM understanding of the nature of a person, all that is coherently available is an additive reading: 'there's this *and* there's that'. And that it is a shift to an I IS understanding which can offer one of two alternative readings, each of them more informative than the postulation of mere 'and-ness'. Such a shift can either facilitate the replacement of the additive mode with a more highly differentiated subordinative alternative: that is, demonstrate the way in which the two members of the duo can be more informatively related, the one to the other. Or it can show that because a juxtaposition of the pair concerned does not carve conceptual space at any significant joint at all, it needs to be just junked. Which then is that which applies to my next ten specimens?

(12) identity and relationship

In academia, this pair would seem to identify issues which figure, in some way or other, in almost all branches of philosophy: from the esoteric reaches of the philosophy of mathematics, right over to the hurly-burly of political philosophy. In the former, Paul Benacerraf is to be found arguing in the *Journal of Philosophy*—'What numbers could not be' (1965) and 'Mathematical truth' (1975)— that what is significant about numbers is the relation in which they stand the one to the other, rather than what they in some sense 'are' in themselves. In the latter, all and sundry seem to want to tackle the vexed question of the way in which individual identities and societal relationships are to be articulated. And this is a range of reference which figures also in philosophical conversation. In both places, what most discussion of this 'identity *and* relationship' duo essentially boils down to is discussion of the problem of how to think about change. Is it that reality consists in change-initiating identities embedded in thereby-changed relationships, or in change-initiating relationships embedding thereby-changed identities? Until we can answer these questions, all we can do, with this 'and' specimen as with all the others, is to say 'there's this *and* there's that'. Where what we need is a mode of thought analytical because subordinative, until we can answer these questions all we've got is one descriptive because additive.

It seems to me that the way to remove this particular bit of primitive 'and-ness' from philosophical conversation is to hang on to that vital 'change' word. Whichever way the 'identity-*and*-relationship' cookie crumbles, *change* is the order of the day. Nodding in the directions respectively of Heraclites, Leibniz and Hume—and no doubt many an other besides—we need to hang on to ideas such as that you can't step into the same river once; that no two things can be exactly the same in every way; and that identity over time requires changelessness, and is therefore impossible. We need to hang on to the understanding that all those identities that we are sometimes inclined to consider changeless and eternal are simply relatively-long-maintained moments in a sequence of moments: or, as Woody Allen puts it, to the understanding that 'tradition is simply the illusion of permanence'.

This, of course, is the type of stance which 'dialectics', in its variety of forms, has struggled to defend. By taking change as the norm, and apparent stability as just apparent, dialecticians are wont to think of the basic units of reality as being not things (identities) but relations-between-things (relationships). And the great strength of such an approach is that it does succeed in getting some sort of purchase on the ubiquity of change. Things become processes. Instead of change being seen as something which (from outside them) happens to things, it is seen as something which (from inside them) constitutes things. In perhaps the best-known version of this approach, the traditional Marxist formulation has been that motion (change) is the 'mode of existence' of matter (things). It seems to me that while this is an approach which can be a real aid to thought, it is not without its problems: in particular, the problem that it can lead to the shedding of one reification (in the form of *the* identity, or *the* matter) only to acquire another (in the form of *the* relationship, or *the* motion). There would seem to be some almost irresistible tendency in human thought towards 'entification'—to coin a word. But while it is a tendency incredibly useful, indeed probably essential, it is not without its dangers.

*Id*entities are entities. And in my experience, entities are for many of us troublesome things to pin down. For a start, *an* entity, singular, is of course always actually plural. It is always actually a collection of things: on a grand scale planets, stars (and the rest) as the entities we call solar systems; on a small scale protons, neutrons, electrons (and the rest) as the entities we call atoms. To become an entity, such collections have to cohere in some way. And given that the degree of coherence has to be enough to distinguish the entity (one collection of things) from its environment (another collection of things), the boundary between the two, between entity and environment, becomes critical. But there are different ways of conceptualising a boundary. It can be thought of as *actual*: that is, as a function of some sort of skin, within (inside) which the entity is enclosed. Or it can be thought of as *virtual*: that is, as a function of some sort of patterning, without (outside) which the environment is open. When thinking in the virtual manner, which is the 'dialectical' or 'process' manner, both the *type* of coherence necessary to entification and the *degree* of coherence necessary to entification will vary with purpose. The type of coherence—of what *sort* is it?—will vary with purpose: to list just some possible examples, it might be spatial coherence, temporal coherence, social coherence or functional coherence which is what matters. And the degree of coherence—what is *enough* of it?—will vary with purpose: to use a social coherence example, the entity 'family' will variously include anything from just 'mum, dad and the kids' to 'mum, dad, the kids, grandma, granddad, the cousins, the dog—and Uncle Tom Cobbley and all'.

These are some of the problems of entifying. We need to remember that turning verbs into nouns, turning processes into things, turning relationships into identities, is just fine: as long as we remember something. We have to remember that such nouns, things and identities are 'capta' not data. We have to remember, for some examples, that they are matters of agreement and convention (what *is* a nation?); matters of more-or-less rather than all-or-none (what *is* a marriage?); and matters of constant change (what *is* a shore-line?).

But entifying also has other and to some extent contrary problems. For if in order to protect ourselves against the entification of things (good!) we instead go in for the entification of relationships-between-things (bad!) we are hardly further forward. It seems to me that the dialectician is almost certainly right if he or she argues that the entification of things is a considerably greater danger than the entification of relationships-between-things, and thus that the stress on relationship, motion and process is progress upon the stress on identity, matter and thing. Nonetheless, it also seems to me that we could do with a formulation of these issues which would protect us against the entification of either: that is, which would enable us to understand this 'identity *and* relationship' pair in such a way as to entify neither of its constituent parts. And in once again suggesting that it is our erroneous I AM thinking which puts such understanding beyond our ken, I am once again claiming that the I IS hypothesis is one which can offer such a way. Let me spell that out by reference to what I take to be a three-stage process.

First, when I AM thinking is at its most pervasive, 'I' is thought of as an unchanging fixed point of reference, and we get a formulation of the 'identity *and* relationship' duo which is both wrong and dangerous. It is wrong because any such fixed and homuncular 'I' is an illusion. It is dangerous because, in giving false priority to unchanging 'identity' in this absolutely basic case, we are all the more inclined to give it, by extension, in others. And if we do, we tend to lose almost all conception of the ubiquity of process and change.

Second, when the grip of I AM thinking on our consciousnesses is beginning to weaken, when the I AM God-person is fading and developments in psychology and neurobiology are beginning to raise questions about its I AM human-person creator as well, we get a more sophisticated (because dialectical) formulation of the 'identity *and* relationship' duo, which is both less wrong and less dangerous. It is less wrong because notions of a fixed and homuncular 'I' come to be reformulated historically and developmentally: change at the level of the person can now be acknowledged. It is less dangerous because, in giving no priority to 'identity' vis-à-vis 'relationship' again in this basic case, we are less inclined to give it, again by extension, in others: and change at other levels too can now be acknowledged. And with such dialectical development, we get a much better conception of the ubiquity of process and change.

But it seems to me that there is, nonetheless, another step to take. In order fully to avoid not just the entification which is *always* involved in prioritising 'identity' over 'relationship' (non-dialectical thinking) but also the entification which is *sometimes* involved in prioritising 'relationship' over 'identity' (dialectical thinking), we are in need, or so I claim, of another development in thought. It's a development which is a particularly important ingredient of my hypothesis. And what it consists in is a recognition of the all too frequently overlooked fact that change changes as well.

Change is indeed ubiquitous: as soon as there is something there is change. Only if there were nothing would there be no change. But change changes as well. When the question is whether it is change-initiating identities which are embedded in thereby-changed relationships, or change-initiating relationships which embed thereby-changed identities, it's not just that we have to understand that both identities and relationships change: it's also that we have to understand that change changes too. In the language of the I IS hypothesis, change changes both *between* its levels and *within* its levels. Between level-two reality and level-three reality, the change which at the level of the organic is evolutionary is not the same as the change which at the level of the personic is developmental. Within the former, the discrete combinatorial system, based in DNA, which powers all organic change genetically was itself selected, in a process of evolutionary change, from among other replicating systems. And within the latter, it would be a prediction of my hypothesis that the discrete combinatorial system, based in a feature of language as yet largely unidentified, which powers all personic change will be found to have been itself created, in a process of developmental change, from among other making systems.

So: to accommodate not only changing identities and changing relationships but also changing change, perhaps our formulation should be—and this would be the third stage of the three-stage process I referred to earlier—that the mode of existence of changing identities is changing relationships, *and* the mode of existence of changing relationships is changing identities. This is a formulation which gives reality an intrinsic 'monality'. It solves the problem presented by this particular 'and' by reading it as *monal* (identity-*and*-relationship) rather than dual (identity *and* relationship). And it would mean that what we might call 'the directional question'—is change in identity producing change in relationship, or is change in relationship producing change in identity?—is to be solved by pushing it down to the level of specifics. To illustrate my meaning here, and at the same time provide an introduction to my next-but-one 'and' specimen as well, let me go to the level of one such specific, by way of illustration, and give some first consideration to 'what is changing what' in the case of individuals and societies.

The now very widely used phrase 'always already' is traced sometimes to Heidegger's 'always already there' notion of our being 'thrown' into the world, and sometimes to Derrida's 'always already' (toujours déjà). Whichever, both were anyway pre-dated, as no doubt by many an other, by Malthus: his version was to speak of the young coming into 'a world already possessed'. It is precisely because it is an idea so obviously important that it is also an idea so variously parented. To get at yet another source, when Durkheim argues that all concepts are derived from experience—there are no innate ideas—he adds that that experience is 'collective': that is, for each new human, a public and shared world is there —'always already'.

What I want to suggest is that, however it is expressed, this 'always already' idea is not unproblematic. Thinking without it can certainly get us into difficulties. But thinking with it allows us easily to make at least three sorts of error as well. We make a first sort of error if we think of it as suggesting that our identities are only 'always already' *constrained by* social relationships of which we were not the originators, rather than also 'always already' *enabled by* social relationships of which we were not the originators: as someone once put it, it is possible for something to be at one and the same time both an obstacle to doing something and a necessary condition to its being done at all – as is gravity to high-jumping. We make a second sort of error if we forget that that which 'always-already' *is* is actually shorthand for various combinations of that which 'always-already' *is coming to be* with that which 'always-already' *is ceasing to be*. And we make a third sort of error if we allow 'always already' notions to befuddle our thinking about social reality: to persuade us, for example, that since William James suggests that the self is based in 'individual experience', and George Herbert Mead suggests that it is 'socially constructed', they are therefore at odds with each other, when in fact, if we think of 'individual experience' as being primarily of that which has been 'socially constructed', they are easily combined. To protect ourselves from such errors, it seems to me that we need to think of 'always already' as *always* being accompanied by 'always anew'. Let me illustrate this – to conclude this section – by reference to the work of Roy Bhaskar.

In his *Reclaiming Reality* (1989), he suggests that 'the relations into which people enter pre-exist the individuals who enter them, and whose activity reproduces or transforms them; so they are themselves structures'. It seems to me that this sort of formulation is literally just wrong: and even metaphorically somewhat misleading. It is not 'the relations into which people enter' which pre-exist them: it is *other* and *similar* relations which pre-exist them. Literally, before people enter them, the relations *they enter* don't exist. 'General relations' are not Platonic forms, free-floating structures or ideal types which are merely *exemplified by* 'particular and actual relations'. Just as there is not an ideal duck, an essence of duck-ness, to which actual ducks correspond more or less well, so there are not ideal relations, essences of relations, to which actual relations correspond more or less well. There are only actual ducks, exhibiting variation, and actual relations, doing the same. And 'enter', of course, is also wrong. People don't *enter* (note: a *space* word) relations with other people: they *make* (note: a *time* word) relations with other people. As so often, surely the bottom line is the *making*. We *make* both the identities and the relationships between them – and both for our purposes. But it is only when we have shifted fully from I AM to I IS thinking that we can register the full significance of the fact that – to echo Marx – we can make them, and thus ourselves, out of only the ingredients to hand: that is, out of only that linguistically structured experience of which we consist. And only together.

(13) ideology and science

As indicated already in Chapter 4, it is the Marxist notion of ideology, ideology as 'necessary false consciousness' as Adorno described it, which has perhaps been the most thoroughly elaborated. But despite that elaboration, there are still problems. In particular, what is often called not *a* problem of ideology but *the* problem of ideology is this: if, as is much more widely accepted than just from within Marxist scholarship, ideologies are those systems of ideas which legitimate the domination of some over-against others, then why do they so often hold? Various authors have tried clearly to formulate this question. Rousseau argued that if the strong were to maintain their position vis-à-vis the weak, then what they had to do was to transform force into power and obedience into duty: in effect, construct ideologies. But he is not at all clear as to why such constructions should so often be successful. Don't they require subordinate people to go along with things which are not in their interests? And if they do, then why do they go along with them? In *The Mass Psychology of Fascism* (1933), Wilhelm Reich is similarly puzzled. 'What has to be explained' he writes 'is not the fact that the man who is hungry steals or the fact that the man who is exploited strikes, but why the majority of those who are hungry *don't* steal and why the majority of those who are exploited *don't* strike.' In his book *On Voluntary Servitude* (1996), these are the sorts of formulations of the problem which Michael Rosen explores. If ideology is to fulfil the function of maintaining domination without relying on coercion, then surely it goes against the interests of both the subordinate individual and the subordinate groups of which he or she is a member. So how come 'ideological consciousness' is often so readily embraced? The I IS hypothesis—or so I would claim—can neatly untie this Gordian knot. By experimenting with what *its* formulations would mean as far as ideology is concerned, we can get—or so it seems to me—not only a clearer distinction between ideology and science than is usually presented, but also a solution to this 'voluntary servitude' problem. Let me spell out both things.

On what an ideology is: yes, wherever there are unequal power relations there will be ideologies; but no, they are not systems of ideas which legitimate the domination of some over-against others. They are instead systems of ideas which seek to legitimate the differential distribution of power. An ideology embodies partiality. It is a set of ideas promoting the interest of those *whether primary or secondary* who seek to maintain that partiality against the disinterest of those *whether primary or secondary* who seek to remove that partiality. Those expressing the ideas aim, consciously or unconsciously, not to buttress primariness against secondariness, but to buttress the primary-secondary divide against equality. The struggle thus becomes that to replace 'the ideology of the normal' (in which normal persons, whether primary or secondary, accept primariness and secondariness in the name of 'naturalness', 'inevitability', 'efficiency', or the like) with 'the science of the moral' (in which moral persons, no longer either primary or secondary, have successfully challenged primariness and secondariness in the name of equality).

This means that when we see it as a problem that subordinate people accept forms of thought that go against their own interests, we see it that way because, in failing to ask 'interests *as what*?' or 'interests *as whom*?', we fail to separate out our categories of thought appropriately. In the terms of the I IS hypothesis, for animal-humans (normal persons with I AM selves at various stages of development), such forms of thought are *not* those that go against their interests because their interests are those of animal-humans: while for human-humans (moral persons with I IS selves), such forms of thought *are* those that go against their interests because their interests have become those of human-humans. Inequality *is* in the interests of animal-humans: I AM persons who are not yet developed to the level of disinterestedness. It's just that it's *not* in the interests of human-humans: I IS persons who are. Animal-human normal persons and human-human moral persons *are just different persons.* Where animal-human normal persons have still the interests of the interested, human-human moral persons have now the interests of the disinterested. And that is a difference making all the difference in the world.

For it means that there just *is* no 'problem of ideology'. To shed it as an illusory problem, a non-existent problem, we have only to hold in our minds four distinct propositions: (1) that when primary people believe in the appropriateness of their primariness (and thus other people's secondariness), when they are thus still animal-humans rather than human-humans, theirs is a belief which is in their interests *as animal-humans*; (2) that when secondary people believe in the appropriateness of their secondariness (and thus other people's primariness), when they are thus still animal-humans rather than human-humans, theirs is a belief which is in their interests *as animal-humans*; (3) that when primary people no longer believe in the appropriateness of their primariness (and thus other people's secondariness), when they are thus human-humans, theirs is a belief which is in their interests *as human-humans*; and (4) that when secondary people no longer believe in the appropriateness of their secondariness (and thus other people's primariness), when they are thus human-humans, theirs is a belief which is in their interests *as human-humans*. And to register, yet again, that it is only the failure properly to distinguish first the natural from the cultural, then (from within the cultural) the normal from the moral, which has made us think that while (1) and (4) might well be the case, (2) and (3) are surely not. Putting I IS terminology into the debate, they *all* turn out to be the case. Putting I IS terminology into the debate, while 'self interest' is a given, is of selves just definitional, it is not just that the *type* of self whose interests are being looked after *changes*, but that the *level* of self whose interests are being looked after *develops*. And that development, a struggling out of 'the having ideology', as later I will argue it, is that which constitutes the making of a self with the interests of the disinterested.

What I am suggesting then is that *ideological* consciousness is normal: it is that which, at current levels of human development, is '*almost*-always-already' in place. And that what really needs explaining is how *scientific* consciousness can come to be accepted in its stead. Such explanation, I am suggesting, lies in the normal-to-moral shift I am describing: a shift *powered by* the process of selving across persons. This means that when I experiment by substituting I IS terminology into that of Rousseau—where he argues, as outlined above, that the strong have to transform force into power and obedience into duty if they are to succeed in maintaining their position vis-à-vis the weak—I come to the conclusion that he's got things back to front. I am arguing that strong-weak relations, what I am calling primary-secondary relations, come into existence always already structured in such a way that both the primary and the secondary see both the power of the primary and the duty of the secondary as 'only natural'. They see it that way because, in origin, it *is* that way: in origin, it *is* 'only natural'. It's just that what we have to learn to do is to see the 'only' as clearly as we see the 'natural'. And to understand that the process of struggling fully *out of* primary-secondary relations requires, among other things, that both the primary and the secondary come to recognise (*re*-cognise) the power of the primary as 'only' force, the duty of the secondary as 'only' obedience, and both as stunting of the fully human relationship they might develop the capacity to make.

To do this we need more science, particularly more personic science. But there is a problem. In *An Intelligent Person's Guide to Philosophy* (1996), Roger Scruton identifies the two key problems of our present human condition as being the loss of agreed certainties and the arrogant and presumptuous self-importance of science. And in so doing, he articulates what is a very-widely accepted anti-science stance which therefore needs serious challenge. Of Scruton's claim, for example, we must ask 'who agreed the agreed certainties when they were agreed?' And when the answer has revealed the ideologies of the interested, ideologies of course often propagated precisely by the arrogant, presumptuous and self-important, we must develop in their place the sciences of the moral. Common sense is not enough. Indeed, common sense might almost be *defined*, in anyway many of its manifestations, as unconsciously articulated ideology. It is moral *science* that we need. And to develop moral *science*, we have first to root out two less-than-helpful diversions along the way. We have to stop trying to discover it, ready-made, at the level of the physical sciences of the inorganic—because that just produces ideologies. And we have to stop trying to discover it, ready-made, at the level of the biological sciences of the organic—because that just produces more ideologies. Let me illustrate what I mean here with two examples of an ideology-making process in operation, as it mines concepts from each of these two inappropriate levels.

Considering, first, the mining of the *physical* sciences, no physical principle has been more extensively used for ideological purposes in what are currently called the social sciences than the uncertainty principle of Heisenberg. No matter that all the way through it has been a principle in quantum mechanics; no matter that, even there, it is now in question: it has been just impossible for the ideologues of doom to keep their hands off it. Because it appeared to be suggesting that it is not possible to measure the momentum of a particle and the position of a particle at the same time, it was immediately fair game. Anything which told us that some things were just 'not possible' had to be worth broadcasting. After all, if physicists could not be certain (which such doomsayers so wrongly took to mean 'scientific') in their terrain, how much more impossible must it be for so-called social scientists to be certain (ditto) in theirs. No matter that, in the physics where it belongs, it is probably wiser to see the uncertainty principle as saying that a quantum particle does not *have*, at one and the same time, a position and a momentum. Such niceties are just ignored. We have to *stop* ignoring them.

Turning, second, to the mining of the biological sciences, a principle carrying like if not equal ideological freight is that enunciated in the nineteenth century by the German biologist Ernst Haekel as 'ontogeny repeats phylogeny'. Haekel noticed that a growing mammalian embryo first acquires structures looking remarkably like gills which only later go on to develop into jaws. Since, in evolutionary terms, the gills of creatures living in the sea preceded the jaws of creatures living on land, he suggested that the individual embryo was retracing the evolutionary stages through which its pre-mammalian ancestors had moved. The ideological use of such an idea is obvious. For those who want to present us as 'only animals really', what a gift to be able to add that we are not even only *higher* animals either: that in our individual development, each of us is fish 'en route' as it were! The fact that, anyway as Haekel had it, the so called 'principle' is wrong does not seem to matter at all. What embryonic development actually retraces, and that only in part, is the evolution of earlier *embryos* rather than the evolution of earlier fully-fledged animals: when a sequence of instructions to make gills in fish is built upon to become a sequence of instructions to make jaws in mammals, it will of course result in the earlier stages of both sequences being common. But such niceties faze the ideologues hardly at all. If they want to find a deep and mysterious way of accounting for mammalian embryos making gills and fins 'en route' to making jaws and legs, then find one they will. And the fact that we now understand that the tinkerer which is evolution, as opposed to the designer which is the human engineer, just cannot do any better than cut and paste, botch and bungle, as it blindly comes upon not the best way of making legs from scratch but an OK way of evolving legs from fins, seems not to impress. Ideologues don't really *want* scientific explanation: consciously or unconsciously, they just *want* it to be the case that our gills and fins should be 'deeply significant' even 'mysterious'. We have to *stop* so wanting.

If these are the sorts of approaches which, in making the moral sciences, are at best unhelpful and at worst mischievous—and I am suggesting that they are—then what instead is the approach that we need? I am suggesting that by far the most important ingredient of any alternative approach is the getting of a much firmer purchase on the nature of personic development: that is, on the nature of a shift from I AM to I IS thinking. To which end—and remembering that mine is a formulation whereby organic entities struggle to live, but personic entities struggle to love—a summary of the 'ideology *and* science' duo, as informed by such a purchase, would look I think something like this.

The earlier stages of the development of the personic are characterised by I AM thinking. The I AM mode is a dualist mode: the I is represented as that which *has* experience. The I AM mode is an ideological mode: it represents the interests of first-person I AM persons as being over-against those of all Others. I AM persons *have* experience; they seek to *have* in competition with all Others; above all, they seek to *have* love. In short: they are the animal-human, normal persons, who by incorrectly construing the entity 'person', understand their condition ideologically rather than scientifically. By contrast, the final stage of the development of the personic is characterised by I IS thinking. The I IS mode is a monist mode: the I is represented as that which *is* experience. The I IS mode is a scientific mode: it represents the interests of just-person I IS persons as being over-against those of no Others. I IS persons *are* experience; they seek to *be* in cooperation with all Others; above all, they seek to *be* loving. In short: they are the human-human, moral persons, who by correctly construing the entity 'person', understand their condition scientifically rather than ideologically. And these being the differences between all the earlier stages of the personic when persons are construed ideologically, and the final stage of the personic when persons are construed scientifically, what I am suggesting is that our progress consists in a developmental shift across such stages, from the ideological to the scientific.

In the last quarter of the last century, there were those who proclaimed the 'end of ideology', meaning primarily the end of Marxism. But of course, as C. Wright Mills pointed out, 'the notion of the end of ideology is itself an ideology'. Indeed. It was a notion which had something of the air of magic about it: something of the stuff of wish fulfilment. As if saying something loudly enough could make it so, it sounded like some wild and wonderful attempt to foreclose on the further development of the human enterprise by just stating it to be completed. What I am suggesting, to the contrary, is that ideology will only end when there is science to take its place: that is, that our 'ideology *and* science' pair are a developmental duo. And that as far as that development is concerned, the I IS hypothesis represents a theoretical position which enables the difference between the two to be formulated reasonably precisely. Normal ideology is that which nurtures an I AM system of ideas which seeks to legitimate having, in competition with others also needing to have, as the basis of what is a lower-level (I AM) reading of the personic. And moral science is that which nurtures an I IS system of ideas which seeks to legitimate being, in cooperation with others also wanting to be, as the basis of what is a higher-level (I IS) reading of the personic.

(14) individual and society

Further to some first discussion of this 'and' a few pages back, when in considering the 'identity *and* relationship' duo I used it as an example, I should begin here by registering that although it is a contentious issue across all the social sciences, it is specifically in sociology that it has been most extensively debated. Indeed, I think it could be claimed that the single most important philosophical question facing sociologists is that concerning the most appropriate way to formulate the relationship between the two members of this particular pair. In his book *The Sociological Revolution: From the Enlightenment to the Global Age* (1998), Richard Kilminster convincingly argues that the establishment of sociology as an independent academic discipline—in the nineteenth century with the work of Comte, Spencer and the like—was a turn away from philosophy's concern with the timeless and the abstract, and towards the new discipline's concern with the socially-rooted and the historically specific. But that does not mean that there are not philosophical issues of some magnitude still to be tackled within the now well-established sociological tradition. And among them, the one represented by this 'and' pair is undoubtedly central.

There have been many philosophical enterprises searching for a way of bridging, reconciling, uniting in one discourse, the individual and the social. All those mid-twentieth century attempts to mesh the contributions of Marx and Freud, for example, were of this ilk. But short of any such meshing being theoretically compelling—and I would argue that we are still way short of any such meshing being theoretically compelling—it seems to me that stances which, in various ways, see individuals as primary are probably more widespread than their opposites. Even for those who would have not a good word to say about the policies she tried to situate in such an outlook, Margaret Thatcher's now infamous remark that 'there's no such thing as society, only individuals and their families' is a stance with considerable 'common sense' appeal. In philosophical conversation anyway, it seems in some ways just obvious that first there are individuals, who then together make societies.

Even in academically sophisticated places, the stance is not unknown. When developing his case against a Marxist position, for example, F. R. Leavis was wont to argue that anyone who knew anything at all about literature could not for a moment escape the truth that human life exists only individually, and that it is therefore only in individuals that society exists. And although much less common, it can even be found in academically sophisticated places overtly concerned with the social. Thus in *Growing Artificial Societies: Social Science from the Bottom Up* (1996), for example, Joshua Epstein and Robert Axtell, then of the Brookings Institute, reported on their attempt to re-run history on their computers. They created artificial societies consisting in thousands of independent 'agents'—which in true Thatcher style represented *either* individual people *or* individual families!—and then put them onto what in effect was an electronic 'map' to have another (and maybe different) go at making history. Their built-in assumption of independence in the agents, the assumption that as individuals we freely act and that those actions, aggregated, give us the societies we have thus freely chosen, represents if not a mainstream position then certainly a considerable position even within what might loosely be called the human sciences. To get at something different, we need to turn to the discipline where, at the very least, this is an assumption that has to be justified rather than just stated. That is, we have to turn to sociology.

I think it was in *Central Problems in Social Theory* (1979) that Anthony Giddens first articulated his key idea of 'structuration'. He argued that social structure and individual action are inextricably linked: a safe but not particularly illuminating position to take, given that his discipline was struggling not to acknowledge a linkage, which to most seemed just self evident, but to conceptualise it. He suggested that we can all draw on our personal experience to transform the structures that structure us—at least to some extent. What he seemed to be saying was that individuals both make social forms and are made by social forms. Societies reproduce themselves by socialising individuals into cultural scripts, yes. But for one thing, because social structures are only ever reproduced through the practices of social 'actors', there have always been scripts which offered some opportunity to improvise and ad lib. And for another, the capacity to improvise and ad lib has considerably increased with the development of a modernity which enhances such 'reflexivity'. Relationships between actors and structures, once heavily biased towards the latter, are becoming relationships between actors and structures, now increasingly biased towards the former. This, as I understand it, is what he seeks to capture with his 'structuration' concept. Social structures are to be understood as 'the unacknowledged conditions and the unanticipated consequences' of human actions. On the one hand, he insists that the classical nineteenth century sociologists were wrong to regard social structures as almost 'forces of nature': on the other, he also insists that Margaret Thatcher was wrong in her famous quip. [In the light of such apparently impeccable 'middle-of-the-road-ism', it should perhaps have come as no surprise to have found him a key adviser to Tony Blair in the latter's search for 'the third way'.]

His 'structuration' concept, or so it seems to me, indicates quite clearly what he is going to be against. He is not going to like—and he doesn't—all those theories which suggest that social structures exist in some way independent of human agency: good. He is not going to like—and he doesn't—the attempted take-over of the sociological by the evolutionary: good. And he is not going to like—and he doesn't—any type of positivism: good. All of which means that I anyway turn to such a thinker with keen anticipation. But he does not deliver.

Or rather: he delivers—but not boldly enough, not tightly enough. His insistence that social structures, though of course very powerful and influential, are nonetheless 'what we do in and through them all the time' reminds me of nothing so much as Richard Dawkins' last sentences in *The Selfish Gene* where he insists that though 'built as gene machines and cultured as meme machines' we nonetheless 'have the power to turn against our creators' because 'alone on earth' we are able to 'rebel against the tyranny of the selfish replicators'. In both cases, what I *feel* is—hooray! But also in both cases, what I *think* is—well, I just think! And there's the rub. There seems to be something fundamentally unsatisfactory, fundamentally fudged, about both formulations: something that leaves me anyway with the distinct sound of goalposts being moved ringing in my philosophical ears. If what we do can make what makes us, what is left of the understanding that it makes us? If alone on earth we can rebel against that which builds us as machine, what is left of the understanding that it builds us as machine? Try as I might, I can't get away from the more than nagging suspicion that the answer has to be 'nothing'.

But that, of course, is only from within a paradigm which fails to problematise the 'we' and the 'us' of such formulations. From within a paradigm which *did* so problematise, which *did* insist on enquiring into the nature of the 'we' that would be doing the making and the 'we' that would be doing the rebelling, and offered some reformulations, things could perhaps be different. Reformulations such as neither Giddens nor Dawkins came up with could perhaps be on offer. And since the I IS hypothesis is one which precisely *specifies* such reformulations, it is worth considering how it would work out in this context. Let me try to convince you that it works out quite well.

What is wrong with the Giddens approach as it stands is that, yet again, a less than bold enough idea as to what 'I' or 'we' can be made to mean, a less than tight enough argument as to how 'I' or 'we' is to be conceptualised, queers his pitch. By contrast, the I IS approach requires that yet again we experiment with an alternative conception of both: an alternative conception which I think can yield a less fudged solution to the problem of the relationship between 'individual *and* society'. But before I can try to make clear just how, I must first spell out a problem. It seems to me that there is something very wrong with the way in which, traditionally anyway, the issues are posed. And what that means is that it's not just that I want to question the various answers to the question as to how individuals and societies are related. It's that first I want to question the very question.

As traditionally it has been posed, the question has been: what is the relationship between the individual and society? That is, there have been assumed to be two entities, and the problem has been appropriately to formulate the relationship between them. I think we have to begin by ditching that question. We have to begin by understanding that it is not that there is an entity 'individual' and an entity 'society', which entities *have* a relationship which it is our task to articulate. Rather, it's that the relationships between the entities 'individuals' are that which actually constitute the entity 'society'. With such a re-formulation, society becomes not the totality of individuals within it, but the totality of relationships *between* individuals within it. And the strengths of such a re-formulation—or so it seems to me—are at least threefold.

First, it challenges an inappropriate reductionism. The intuitive appeal of Mrs. Thatcher's 'nothing but' approach—society as 'nothing but' the individuals who compose it—is challenged. There are individuals, and there are relationships between individuals. And society is the sum total not of the former but of the latter.

Second, it challenges that reductionism without introducing any element of the mischievous mysteriousness which is so often associated with non-reductive formulations. Society is not 'nothing but' the sum total of the individuals within it. But the 'something more' that it is, a 'something more' which makes it instead the sum total of the relationships between those individuals within it, gives it an emergent property of a totally transparent nature. In no way mysterious, it is thus totally amenable to scientific investigation. In initial and nursery language, for example, that investigation might start from something like this:

1 individual:	inconceivable (metaphorically and literally!)
2 individuals:	1 relationship (the simplest 'Adam and Eve' society)
3 individuals:	3 relationships (a less simple and pivotal society)
4 individuals:	6 relationships (a more complex society)
5 individuals:	10 relationships (a yet more complex society)

And we are launched up a Pascal's-triangle expansion whereby, totally non-mysteriously, society is 'more than' the sum total of individuals within it because instead the sum total of relationships-between-individuals within it. And there just *are* more of them!

Third, it clarifies thinking about many of the different levels of the construct 'society'. Social scientists study, bounded and labelled, nested sets of relationships (societies), such as 'family', 'nation', 'class'. As traditionally they have been posed, the questions have been: what is the relationship which the individual *has* with the family, which the individual *has* with the nation, which the individual *has* with the class? I think we have to come to understand instead that it is the relationships *between* individuals which *is* the family; the relationships *between* individuals which *is* the nation; the relationships *between* individuals which *is* the class.

With the ground thus cleared by a reformulation, I can now experiment. I can now feed an I IS conception of 'I' into these formulations rather than into traditional ones, and see what difference, if any, that makes. In particular, I can see if it can begin to clarify our thinking about the *direction* of social change. For in an argument seemingly as old as argument, this has been the key issue here. Some have suggested that we need to change individuals in order to change society—which is, after all, just individuals collectively. Others have suggested that we need to change society in order to change individuals—who are, in the most celebrated version of this position, just 'the ensemble of their social relations' (Marx).

I want to propose that when we posit ourselves as I AM first persons who *have* their experience, the primary-secondary relationships thereby entailed *do* constitute the societies which 'always already' *do* shape us. They *do* shape us because, as a consequence of that posit, we don't have a theory adequate to an understanding of our only apparent situation: armed less than fully, with only ideology, we see 'through a glass, darkly'. But that when we come to posit ourselves instead as I IS just persons who *are* their experience, the equal relationships thereby entailed *will* constitute the communities which 'always anew' we *will* shape. We *will* shape them because, as a consequence of that new posit, we *will* then have a theory adequate to an understanding of what is our real situation: armed now more fully, with burgeoning moral science, we will see then 'face to face'. Put differently: to the extent to which the social structure (the structure of relationship) between persons is one of inequality, to that extent *it constitutes them;* and to the extent to which the social structure (the structure of relationship) between persons is one of equality, to that extent *they constitute it*. Perhaps it is unnecessary to add—but I will add it anyway—that to date, social structures of the second type exist for the most part only as potential: that is, only in the thought experiments of the imagination.

This is a formulation which has juxtaposed not just first-persons with just-persons, but also societies with communities. And for two reasons, I want to look more closely here at that second shift. I want to do so, first, because it is a shift vital to my argument. I want to do so, second, because I must clearly distance that argument from anything remotely like support for what is currently called 'communitarianism'. So how have the social sciences usually conceptualised the difference between societies and communities?

In *Intersubjectivity* (1996), Nick Crossley elaborated an understanding as to what constitutes a community which is I think one fairly standard among social scientists and philosophers of the social sciences. 'To be a community', he wrote, 'a population, who need not be geographically close, must share in certain understandings of themselves as a group and of their environment'. He went on to argue that they must 'derive a common identity from a shared lifeworld', and that their roles must be both 'interdependent and relationally defined', and 'mutually recognised as such'. Thus is constituted what is 'an irreducible structure of interactions, which is both self-defining and self-regulating', to be a member of which is 'to be involved in the shared understandings of the community, to have an interdependent role within it and to be recognised as such.'

Now it is not that I want particularly to challenge any of this: it's just that I want to add to it in ways which I think fundamentally deepen its significance. And to do that, I want to borrow and lean on the words of the Scottish moral philosopher John Macmurray. In *Persons in Relation*, published in 1961 but taken from his 1953 Gifford Lectures, he writes about communities in this vein:

> Any community of persons, as distinct from a mere society, is a group of individuals united in a common life, the motivation of which is positive. Like a society, a community is a group which acts together; but unlike a mere society its members are in communion with one another; they constitute a fellowship. A society whose members act together without forming a fellowship can only be constituted by a common purpose. They co-operate to achieve a purpose which each of them, in his own interest, desires to achieve, and which can only be achieved by co-operation. The relations of its members are functional; each plays his allotted part in the achievement of the common end. The society then has an organic form: it is an organisation of functions; and each member is a function of the group. A community, however, is a unity of persons as persons. It cannot be defined in functional terms, by relation to a common purpose. It is not organic in structure, and cannot be constituted or maintained by organisation, but only by the motives which sustain the personal relations of its members. It is constituted and maintained by a mutual affection. This can only mean that each member of the group is in positive personal relation to each of the others taken severally. The structure of a community is the nexus or network of the active relations of friendship between all possible pairs of its members.

This gets much closer to the significance that I want my 'community' word to carry. I would use neither 'group' nor 'individuals' in the way he does in his first sentence. I would say 'mutual love' rather than 'mutual affection'. I would be sure to make clear that his 'positive personal relation to each of the others taken severally' would for me carry no connotation of any 'face-to-face' necessity: 'active relations of friendship' would for me describe the stance each I IS person would take up towards each and every other I IS person in the world. And at the same time as stressing, with him, that community 'is not organic in structure, and cannot be constituted or maintained by organisation' I would also point out, as he does not, that the way to ensure that those structures which are 'an organisation of functions' do not preclude the development of community—as, by making each member a function of the group, they do now—is to disassociate the idea of the functional slot from the idea that some one person occupies it. But with these provisos, here is what I mean by community.

It is a quite distinct conception. And to ensure that its distinction is clear, I must immediately and carefully distance myself from a different conception with which I want no truck at all. In *Modern Philosophy* (1994), Roger Scruton writes, at its end, of his hope that by 'understanding our concepts, as they inform and are informed by our social experience, we may find a path back to the natural community'. And adds that if such a path is 'one which is closed to us', then it is vital that we know 'why that is so'. As a way of distancing myself totally from such a conception of community—fundamentally, I think, a communitarian conception of community—let me try to tell him why I think that 'that is so'.

Let me tell him, first, that a 'natural community' is best understood as a contradiction in terms. And let me tell him, second, that 'back to' anything is a nonsense. Let me offer him instead a process of development: first, various natural groups in which human-animals live as non-person, not-I creatures; second, various normal societies in which animal-humans love less than fully as having, first-persons; and third, a unified moral community in which human-humans love fully as being, just-persons. Let me suggest to him that *out from* 'natural groups', *out from* 'normal societies', 'moral communities' have to be made. They are not natural. They are thoroughly non-natural. They are moral. They have to be made. But let me also suggest to him that if it really is the thoroughly non-natural thing called 'community' that he's interested in, then he can, if he so wishes, be part of the struggle to create it. If he wants to be a part of the struggle *out of* natural groups, *out of* normal societies, *into* moral communities, that would be wonderful. But first he would need to embrace an idea which might not be much to his liking. Let me explain.

In his *Categories*, Aristotle made one of the first surviving attempts to divide the world into its fundamental parts. And concerning that part called 'God', he insisted that God is not something that *has* beauty but something that *is* beauty; not something that *has* goodness, but something that *is* goodness; not something that *has* truth, but something that *is* truth. This is the first attack, of which I anyway am aware, on what I am calling 'the having mode'. For Augustine, who comments on it, it would seem not to have gone far enough. He wanted to insist that putting anything after 'is' in sentences that begin 'God is' is just as problematic as putting anything after 'has' in sentences that begin 'God has'. To the extent to which I understand the difference between pair of them, I am with Aristotle. I too am challenging 'the having mode', and in all its manifestations. I see it as 'the ideology of ideologies': the one which feeds an almost always totally unconscious reduction of the category 'just person' to the category 'first person', and thus of the category 'community' to the category 'society'. I am suggesting that where the mode of being of persons in a normal society is the having mode, the mode of being of persons in a moral community will be the being mode. In the former, it is the having of more or less than others which leaves 'having' modal: in the latter, it is the having only equally with others which allows the transcendence of its modality in favour of being. In brief, I am suggesting that the making of moral communities is the making of equality.

The transformation is a process. It involves not the modification of the having mode, as perhaps was the project of many a socialist agenda, but the total transcendence of the having mode. It involves not coming to have all things collectively *with* others having them too, but rather coming not to have any things individually *without* others having them too. As I have already illustrated, it involves 'I *and* Thou' relationships becoming 'I *and* I' relationships: as later I will illustrate, it involves 'master *and* slave' relationships becoming 'just-person *and* just-person' relationships. It is a process which, at one and the same time, both reveals the *necessity* of making, and invents the *means* of making, that equality from which will flow the freedom which is the recognition of its necessity. And when he's thought about it all some more, who knows? Maybe Roger Scruton might want to join the ranks of those struggling to make it a reality. It would be nice to have him on board.

(15) individualism and anti-individualism

At first flush, it can seem as though the issues raised by this 'and' are the same as those raised by the previous one: that here we have just a re-run of the debate concerning 'individual *and* society', with 'individualism' being the position that stresses the primacy of the first, and 'anti-individualism' being the position that stresses the primacy of the second. And in philosophical conversation anyway, sometimes perhaps not as careful with its terminology as is academia, the actual discussions that go on under the two heads are very often elided. But it seems to me that there really are two 'ands' here rather than one. And that for philosophical conversation as much as for the professionals, there is something to be gained from respecting their distinction.

I have seen it suggested that it was Durkheim who first severed what had been thought to be the direct and obvious link between individualism as a moral value concerning the inviolability, almost sacredness, of individual rights and liberties (of which he approved), and individualism as a methodological principle of explanation in the human sciences (of which he did not). Whether or not this is its appropriate origination, this I think *is* the nature of a certain 'twoness' about individualism. And thus also, of course, about anti-individualism.

My own position is easily summarised. If 'individualism *and* anti-individualism' is used to map onto the 'internalist or externalist' debate—in my terms, is there a pre-existing entity 'person' which has experience (internalism/individualism), or a post-existing entity 'person' which is experience (externalism/anti-individualism)?—then obviously and unambiguously, mine is a stance which puts me firmly in the second camp. If, on the other hand, it is used to map onto the 'individual *and* society' methodological duo—is the relationship between the two such that various types of the first *constitute* various types of the second (and thus the methodology of the human sciences should privilege the psychological study of individuals), or various types of the second *produce* various types of the first (and thus the methodology of the human sciences should privilege the sociological study of societies)—then obviously and unambiguously, my stance puts me firmly—just in some totally other discourse. Since, in the second case, *which* totally other discourse is spelled out immediately above in the previous entry, I will comment here on only the first.

That which traditionally has been 'the left' has always had a very strong conservative streak in it. It was a conservative streak made manifest on a grand and terrible scale in the authoritarianism of Stalinism. And on a smaller but also terrible scale, it is a conservative streak made manifest in its everyday lauding of the collective over the individual: ironically, of Roger Scruton's thoroughly collective natural community over my thoroughly individual and non-natural alternative. Given such a streak, it seems to me no accident, for example, that in the wake of the debacle which was Eastern Europe, the theoretical and political malaise which has all-but overwhelmed the once proud socialist and communist traditions has manifested itself, in some places, in an attempt to align what once was the red of revolution with what now is the green of conservation.

My own approach is thoroughly other. What I am saying is that a seriously externalist stance in the 'what is a person?' debate—and with his 'ensemble of social relations' formulation, Marx was quintessentially after a seriously externalist stance—requires that we embrace rather than regret individuation. It is my contention that the individuation which cracks apart marriages, families, villages, nation states *in a capitalist context and thus with immediately difficult-to-cope-with consequences*, is the same individuation which, *precisely for that reason*, will spearhead the renewed search for an alternative to such context. And it is my hope that the I IS hypothesis will be one which will assist in a theoretical transition which I think needs to be made if this is to happen. As spelled out above, it would be a theoretical transition from the '*having-all-things-collectively-with-others*' stance of the old socialist and communist movements (which actually thrived on the maintenance of inequality), to the '*not-having-any-things-individually-without-others*' stance of what I am proposing will be the new movements of the future (which will not just thrive on, but actually be constituted by, their making of equality).

(16) is and ought

As I pointed out in Chapter 4, the case here is directly parallel with that already argued under the 'fact-*and*-value' pair. As a consequence, I will restrict myself here to just a one-paragraph addendum.

The 'you-can't-derive-an-ought-from-an-is' position is an I AM position. It suggests that an 'ought' from my (first person) point of view may not be the same as an 'ought' from your (second person) point of view, let alone an 'ought' from her or his (third person) point of view. If something 'is' the case, it only 'ought' to be otherwise given certain projects. And from an I AM position, 'projects' are always and ineluctably mine, yours, hers or his: that is, they are always and ineluctably from out *my* I AM, *your* I AM, *her* I AM or *his* I AM. From an I IS position things are different. From an I IS position, certain projects just *are* given—*by what I is.* Which means that certain projects just *will* be pursued—*as a consequence of what I is*. Such a project—'and the greatest of these is love'—is that of the *making* of love, via the *making* of equality.

(17) justice and benevolence

Attempts to escape the logic of equality have been various. And one of the oldest is that which uses the 'charity' argument. It goes something like this: that one of the 'evils' of equality is that, should it ever be accomplished, it would remove the need for benevolence. Putting it rather crudely, since benevolence is 'a good thing', that must make equality 'a bad thing'. It is, of course, a nonsense argument. And since I can come up with no better form of words to show it as such, let me again borrow and lean on those of John Macmurray, again in *Persons in Relation* (1961), where his quite splendidly decisive 'nailing' of such an undignified wriggle goes like this:

> To be generous without being just is to be generous to some at the expense of others; and so to produce a minor mutuality which is hostile to the interests of the larger community. It is to create and defend a corporate self-interest, and this destroys the universality of the moral reference. To be more than just to some and less than just to others is to be unjust to all.

Precisely! But yet still there are voices to the contrary. The majority are those of the usual, traditionally authoritarian suspects: the folk who, in the philosophy of the academy, might be labelled 'ethical individualists'. Two contemporary groups, each of which might or might not legitimately be said to fall under that rubric, are the 'virtue ethicists', and some among the feminists. The first are those seemingly attempting to create, within moral philosophy, a 'third way': a position neither deontological nor consequentialist, but based instead in a cultivation of the virtues. And to the extent to which I get their drift, it would seem to be the 'benevolence' member of our current duo, if only because it can so much more easily be construed as an individual virtue, which sits much more comfortably in their canon than does the necessarily societal and political 'justice'. The second, more worrying perhaps if only because from them one might have expected more, are those taking the 'care' side in a directly analogous and currently active 'justice-or-care' debate going on within feminist philosophy. But although each of these voices is both currently distinct, and certainly significant, it will be via a consideration of the arguments advanced more generally by mainstream 'ethical individualism' that I will outline the way in which my hypothesis would tackle our understanding of this 'justice *and* benevolence' pair. My aim will be to show them as being related essentially developmentally: to show that while I AM selves tend to rest easily in mere benevolence, their I IS successors can only insist on the moral superiority of justice.

In *Living as Equals* (1966), Ronald Dworkin contributes a piece under the heading 'Do liberty and equality conflict?' In it he suggests that ethical individualism has two foundational principles: the principle of equal value, whereby 'it is intrinsically, objectively, and equally important that human beings lead successful lives'; and the principle of special responsibility, which maintains that 'the connection between you and your life is nevertheless a special one' and that 'someone who treated his own life as having no different place in his plans and no greater call on his attention than the life of any stranger would not be a saint but a Martian.' It is this second principle, the one which allows special concern for the 'first' person, which is the problem. A person working within it could be benevolent alright: at least when it was not too costly. But could he or she be just? Is not the demand of justice precisely that the first person should *not* be of any special concern? And if that *is* the demand, is it possible of attainment?

In *Ethics and the Limits of Philosophy* (1989), Bernard Williams argues for the validity of the second principle. He suggests that a first person who took seriously the demand of impartiality would suffer a crisis of identity: put simply, would not have enough of the stuff adequately to look after her or his self, never mind her or his dependants, if any. There are, of course, perfectly respectable traditions which challenge him. Contrary to Aristotle, who thought that concern for others could only extend to those within ones own city state, the Stoics were amongst the first to argue that we reached our highest moral development only when we were concerned for all rational beings. The whole of the Christian tradition invites us to love our neighbours as our selves, tells us that our neighbours are everybody, and very specifically requires of us that we be perfect, even as our father in heaven is perfect. And for the non-religious, Tom Paine was speaking for many an other when he claimed that his country was the world, and his religion doing good. Even Edmund Burke, whose 'little platoon' argument is so often quoted to defend our first-person partialities, actually said something to the contrary: actually suggested that 'to love the little platoon we belong to in society is ... the first link in the series by which we proceed toward a love of our country *and of mankind* (my italics)'. Nonetheless, these and many other voices notwithstanding, the position adopted by Bernard Williams is one so seemingly congruent with common-sense as certainly to need careful consideration. And in trying now to give it such, I will at the same time make my case for an alternative: a case, famously recognised as such by the Archbishop of Recife, which establishes the total inadequacy of benevolence ('when I give to the poor they call me a saint') in the face of the counter-claim of justice ('when I ask why there are any poor, they call me a communist'). It is a case directly contributory to the main theme of my book: that the only fully moral way to defend those persons who are weak and vulnerable is to make them strong and not-vulnerable; that those persons who have less than others are made weak and vulnerable by having less; that those persons who have more than others are made weak and vulnerable by having more; and that thus the only fully moral way to defend all persons against weakness and vulnerability is by the making of equality. So: is it in fact the case that 'someone who treated his own life as having no different place in his plans and no greater call on his attention than the life of any stranger would not be a saint but a Martian'?

I think not. I think that such a one would be neither saint nor Martian but just human: not an animal-human but a human-human; not a normal person but a moral person; not a first person but a just person; not an I AM person but an I IS person—all as previously defined. It seems to me that where the 'principle of special responsibility' is in error is in its seemingly unconscious adoption of a seemingly ubiquitous dualism: a dualism embedded in the ideology whereby 'you' is one entity and 'your life' is another, which 'you' then 'have'. The formulation whereby someone (a male, it seems) is able to treat something called 'his own life' in *either* one way *or* the other is just incoherent. It is indeed the case that someone who treated her or his own life as having no different place in her or his plans, and no greater call on her or his attention than the life of any stranger, would be odd: but no more odd than someone who didn't. For in being dualist entities consisting of (1) an I that AM and (2) a life that I 'has', both are exceedingly odd: so exceedingly odd indeed as not to be capable of existence. That which actually exists, or so I am arguing, is something significantly other: or rather, two 'somethings' significantly other. First, there are (definitionally) selfish I AM persons whose selves are so constituted that when they (definitionally) pursue their I AM interests, those partial interests privilege themselves *as first persons* over against others as second and third. Second, there are (definitionally) selfish I IS persons whose selves are so constituted that when they (definitionally) pursue their I IS interests, those impartial interests privilege all selves *as just persons* over against nobody. And it is perhaps worth pointing out that implicit here is yet another example of the way in which freedom has nothing much to do with choice. Because benevolence is interested and partial, it informs part of the behaviour of I AM persons interestedly, partially, and sometimes: because justice is disinterested and impartial, it informs all of the behaviour of I IS persons disinterestedly, impartially and always.

I want now to consider four issues which arise I think from this application of my hypothesis to the 'justice *and* benevolence' pair: to further explore the inadequacy of the 'benevolence' member of the duo; to further make the case that its 'justice' counterpart involves equality; to have a first go at explaining why I think those who have more than others are as equally stunted (but differently) as those who have less—a point which will be further elaborated below (see 'master *and* slave'); and to make a first defence of my stance on these questions against the post-modernist charge that they constitute a grand narrative, and can elicit therefore only incredulity.

Our *Oxford Companion* source of this specimen points out that 'benevolence', in a philosophical context, has changed its meaning. Where once it might have taken the form of 'compassion, mercy, kindness or generosity', it is now more narrowly understood to refer to acts of charity whereby 'some benefit is freely bestowed by one individual with a surplus on another who is in need'. So what then is wrong with it, in either its broader or its narrower guise, for the I IS moral person? Two things, I think.

First, it celebrates giving. And giving can be conceived of as a good only when some are more needy and some are less needy. Quite apart from all the reasonably devastating work done by anthropologists and sociologists, Mauss and others, on the power-relationships inherent in the gift—in brief, money is power and the desire to give money can sublimate, probably necessarily sublimates, the desire to gain some control over those to whom one gives—there is the more general question of *re*-distribution to consider. In modern liberal democracies, those forms of governance that post-modernists want to tell us are the best of which we will ever be capable, the extent to which progressive income tax take precedence over regressive purchase tax is sometimes taken to act as some sort of rough and ready measure of their 'justice'. But quite apart from the fact that, even on such a puny measure, so many of such democracies are found to be so severely wanting, this is anyway rather more like 'compulsory benevolence' than it is like justice. And benevolence, compulsory or otherwise, is not enough. *That* it is not enough is made abundantly clear in a much commented upon (but actually only apparent) contradiction in the work of Adam Smith. It has often been noticed that while he eventually chose 'sympathy' as the unifying principle for his ideas on morality, as outlined in *The Theory of Moral Sentiments* (1759), he nonetheless put all that emphasis on 'self-interest' as the unifying principle for his ideas on economics, as outlined in *The Wealth of Nations* (1776). But his texts do not contradict. It's just that his was a sympathy, and where it is still found (which is in many places) remains a sympathy, just fundamentally flawed by its lack of imaginative reach. Compassion, which the poet Milosz so memorably calls 'that ache of the imagination', requires us now to use a rather more developed imagination than was available, it would seem, to Smith: an imagination which, in better understanding the full force of the words of whoever it was who commented on the laws of France forbidding the rich and the poor equally from sleeping out under bridges, reaches for a rather more developed understanding of what justice requires. I will be suggesting that what justice now requires is not a (very partial anyway) *re*-distribution of that which has initially been distributed unequally, but an initial and equal *distribution* in the first place. And that is a stance very far indeed from one which can remain for long comfortable with mere 'giving'.

Second, it encapsulates loyalty. Giving being essentially a piecemeal affair—I give to Oxfam (but not Amnesty), you give to your mother (but not your father), she gives to the Women's Institute (but not the RNIB)—it is essentially a celebration of our individual and personal loyalties. And if I was inclined to accept the language of traditional religion—which, of course, I am not—I might go as far as to suggest that loyalty to an individual Other is best thought of as some sort of sin: to argue that loyalty, whether to individual or corporate persons, is in some very important sense just immoral. Fundamentally bestial things can be done, have been done, in the name of loyalty: think only of that 'loyalty' to their husbands which kept the wives of concentration camp guards silent, even as they suspected. Let us hope that as, happily, such a flawed concept dies, they are being done rather less often today. For my partner right or wrong, my firm right or wrong, my sex right or wrong, my race right or wrong, my anything right or wrong—each is as fundamentally immoral as the 'my country right or wrong' from which they all derive. Where morality is about doing right rather than wrong, support for anything 'right or wrong' is more or less definitionally just immoral. And because it is, it seems to me that the only loyalty whose claim is a moral claim is loyalty to that which is right; to that which is just; to that which is good. To justice.

Now this is an approach which has far reaching, and sometimes startling, consequences. Obviously it turns things such as sexism and racism into immoral stances, but with that there would be widespread agreement. But it goes much further. For some much more contentious examples, it turns marriage into an increasingly immoral institution, so that we can rejoice that it is dying on its feet; it turns company loyalty into betrayal, so that we can be pleased when it gets harder to come by and whistle-blowers start to flourish; it turns patriotism into a 'sin', so that we can delight in its growing irrelevance to more and more people as globalisation spreads. Some of these examples will be further explored in later chapters. For now, suffice it to register that Macmurray's 'minor mutualities' are indeed 'hostile to the interests of the larger community', do indeed 'destroy the universality of the moral reference'. And that that 'benevolence' which involves us being 'more than just to some and less than just to others' does indeed involve us in being 'unjust to all'. Which brings me to the second issue I want to consider: that of justice *as* equality.

It seems to me that 'justice *as* equality' is a quite spectacularly paradoxical idea. It is spectacularly paradoxical because, at one and the same time, it is both widely accepted as obvious, and widely derided as impossible. From the childhood chant of 'it's not fair' to all the versions of the Golden Rule; from Kant's 'categorical imperative' to Rawlsian decision-making from behind 'the veil of ignorance'; from Mill's 'principle of perfect equality' to Walt Whitman's 'By God! I will accept nothing which all cannot have their counterpart of on the same terms': from so many other such dreams of the good to so many other such visions of the just, it would seem that the world has known for a very long time that it is only relationships of equality which are truly morally defensible. And yet. From Forster's less famous quip that 'all men are equal—all men, that is to say, who possess umbrellas' to Orwell's more famous one that while 'all animals are equal' some are 'more equal than others'; from Samuel Johnson's jaundiced dictum that 'it is better that some should be unhappy than that none should be happy, which would be the case in a general state of equality' to Barrie's cynical comment, put into the mouth of a servant character, that while 'His Lordship may compel us to be equal upstairs, there will never be equality in the servants hall': from so many other such weary acceptances of the tawdry to so many other such collusions with the second-rate, we seem to have wanted, at the same time, to pour scorn on and deride that which, at another level, we know is our highest endeavour. Such a schizophrenic take on a single notion might be thought likely to produce a widespread angst. But to protect us, we would seem also to have developed a spectacular capacity to deal with the cognitive dissonance which might otherwise be thought likely to overwhelm us on these issues. We use sundry 'defence mechanisms'.

There's the 'don't-expect-too-much-of-us-we're-only-animals' defence mechanism: currently called evolutionary psychology. There's the 'anyway-equality-would-be-boring' defence mechanism: wielded by people who also wield the 'ends-don't-justify-means' mantra when it suits. Just lately, the old favourite called 'equality-would-not-be-fair-to-the-talented' has been joined by a young upstart called 'and-anyway-don't-both-me-with-grand-narratives-of-any-sort'. Why on earth do we need to be *so* defended against such an *attractive* idea?

Let me suggest that it is precisely because it is so attractive that we have needed to be so tightly armoured against it. This is a point well captured, I think, in a phrase from Peter Sloterdijk, the German philosopher who in 1987 produced his first major work under the evocative title *The Critique of Cynical Reason*. He refers to cynicism as 'enlightened false consciousness'. And borrowing the phrase—if not the rest of his stance—it seems to me that the cynic, who is after all he or she who is most likely to be found using all the 'save-me-from-equality' defence mechanisms listed above, is one of the stereotypical characters of our times: someone whose 'enlightenment' means that he or she can no longer accept traditional grand narratives, but whose 'false consciousness' means that he or she can neither begin to formulate any new ones. If something is at one and the same time both quite beautiful to contemplate longingly (if only in moments when the censor is off-guard), and quite impossible to theorise coherently (not least because that version of 'false consciousness' which sees primariness as a benefit is the one most likely to afflict the would-be theorists), then what else can one do but defend oneself against it? As we have long known, terrible indeed is the temptation of the good.

The third issue arising from the application of my hypotheses to a conceptualisation of this 'justice *and* benevolence' pair is one which, because it raises a very counter-intuitive aspect of my position, needs much fuller consideration than it will get here: a fuller consideration which it will be given under the 'master *and* slave' head below. For now let me just register that it is a critical part of my position that the benevolent stance adopted by (some of) those who have, towards (some of) those who have not, is not singly compromised but compromised doubly. It is certainly compromised because its partialities create those 'minor mutualities' which are 'hostile to the interests of the larger community', 'destroy the universality of the moral reference', and mean that we become 'more than just to some and less than just to others': again. But it is also compromised because, in doing so, we necessarily become 'unjust to all' *where 'all' means exactly what it says.*

To spell that out: it is my position that when we tinker with inequalities via benevolence, our injustice is as much to those who have as it is to those who have not. My position is that, in considering the development of the human-human out of the animal-human, of moral persons out of normal persons, the primary (those who have more of something perceived as a good by both primary and secondary) are as stunted as the secondary (those who have less of something perceived as a good by both primary and secondary). And this is not just for reasons that are anyway widely recognised to be valid. It is not just because, as David Hare has commented, the last people to be liberated from empire are the imperialists. It is not just because, as in *The Hidden Injuries of Class* (1972) Sennett and Cobb point out, 'the strong are constricted within the circle of action which maintains their power'. It is not just because the fact that John Locke had a hand in writing the Carolina slave constitution, and Roger Scruton, to this day, calls women 'he' without a qualm, pretty clearly indicates a certain stunting. It is not just because 'the emancipation of the working class' forgot the emancipation of the ruling class—and any other class you care to mention. And it is not just because life lived on a pedestal stunts by clearly giving rather less room to move, and anyway encouraging the striking of poses. These are things we many of us understand already. What is not so well understood, or so I claim, is this.

When academic philosophers are considering issues of equality, there is among them a fairly widely accepted idea that two questions need to be taken into account: first, what is to be regarded as a good; second, how is it best distributed—equally or unequally. Working *within* the constraints of such a paradigm, my position would of course be that in a situation where some persons have more of 'good X' and other persons have less of 'good X', those 'some persons' are as disbenefitted by having more as those 'other persons' are disbenefitted by having less. For me it would therefore clearly follow that, to the extent to which 'good X' *can* be re-distributed, to that extent *all* persons would benefit from its re-distribution. And to the question as to whether there is any particular good which could be substituted for 'good X' which would make this statement untrue, my answer would be that there is not: that both secondary and primary people—that is, literally, all of us—are stunted *by the nature of the unequal relations between us*. But I would also want to claim that we need to move beyond the constraints of such a paradigm, and recognise the existence of one 'good' *which is only brought into very existence* by equality. It's not that a good called 'well-being' exists *unequally* distributed where it would be better if it existed *equally* distributed. It's that a good called 'well-being' *is only brought in to very existence* by equality: is a property *emergent with* the emergence of equality. This is the line on equality which later, in Chapter 11, I will argue more fully.

The fourth and final issue I want here to consider is to make a first defence of my position regarding the relationship between 'justice *and* benevolence' against the post-modernist charge that it constitutes 'a grand narrative': that is, 'a grand (equality) narrative'. This is a charge to which I plead totally, utterly and completely guilty. Although later (see 'stories *and* explanations') I will want to make the even more outrageous claim that 'narrative' is perhaps not strong enough, and that 'explanation' might better fill the bill, for now let me plead totally, utterly and completely guilty *at the very least* to a 'grand (equality) narrative'. And in order to start to make my case that such is exactly that of which we are in need, let me consider a formulation of these issues put forward by Foucault, as it contrasts with my own.

Perhaps cynically, Foucault suggested that there is no resistance to power which is not in itself a bid for power. He's wrong. But anyway to my knowledge, the I IS hypothesis is the only one which can make his error manifest. I would argue that he's wrong because, despite his promising insistence that power is a network of forces rather than an 'object' that can be 'seized', he nonetheless goes on to posit the struggle as one which means that those who are ruled must escape from the 'will-to-power' of those who rule them. In this way, it seems to me, he fails to give appropriate weight to three other vectors in his 'network of forces': staying within his language-style, to the vector 'will-to-power' of those who are ruled—which he himself has after all identified in his 'bid' formulation; to the vector 'will-to-powerlessness' of those who rule; and to the vector 'will-to-powerlessness' of those who are ruled. The result of this is that his formulation is ultimately one which flops back into positing a line-up whereby those who are powerful (and trying to hang on to their power) are on one side, and those who are powerless (but trying to take power from the powerful) are on the other.

Mine, by contrast, is a formulation which sees those of both the powerful and the powerless who *accept* the inequalities between them (and are thus trying to hang on to either their power or their powerlessness) on one side, and those of both the powerful and the powerless who *challenge* the inequalities between them (and are thus trying to shed either their power or their powerlessness) on the other. This is a formulation, crucially, which allows that 'resistance to power' which is no way a 'bid for power' to emerge. It is the resistance which comes, from some among both the powerful and the powerless, for equality. And as such, it is that resistance, for example, which gives the lie to the quite shabby anti-equality defence mechanism which is being used by those who make the 'politics of envy' charge against egalitarians.

With a thrust which can be once again emancipatory, it is such a formulation which permits the emergence of a new grand narrative. For given that a grand narrative is just a theory, however ill-formed, unconscious and ideological, concerning the nature of the human condition, it is just philosophically incoherent to claim to be without one. That there is thought to be no defensible grand narrative just *is* a grand narrative: the grand narrative of postmodernity. That liberal democracy has triumphed over all comers and we are therefore at the end of grand narratives just *is* a grand narrative: the grand narrative of the 'end-of-history' brigade. That because 'the first great experiment in running a modern economy without relying on avarice and anxiety has failed' we should therefore 'give up the attempt forever' (the words are Gerry Cohen's from his 1995 book *Self-ownership, Freedom and Equality*, and the stance is one he opposes) just *is* a grand narrative: the grand narrative of reaction. We all of us, and necessarily, think from within a grand narrative of some shape or form. What we are in need of is a way of constructing a grand narrative of disinterest: that is, a theory well-formed, conscious and scientific concerning the human condition. And I see no reason to apologise for trying to make some contribution towards its formulation.

Our theories about the inorganic stuff of the universe and the organic stuff of the biosphere have involved, I suggest, the steady creation of a *natural*-scientific enterprise from within which an ever-more-inclusively-expanded 'some' have been increasingly able to stand ever more equal and thus ever more disinterested before the evidence: which evidence is anyway easier to see, concerning as it does that which is essentially 'Other'. And it is my contention that our theories about the personic non-stuff of the world will involve a process both very similar and distinctly different. It will similarly involve the steady creation of a *moral*-scientific enterprise from within which an ever-more-inclusively-expanded 'some' will be increasingly able to stand ever more equal and thus ever more disinterested before the evidence. But because that evidence, concerning as it does that which essentially is 'us', will be that much harder to see, it will differently involve not just an ever-more-inclusively-expanded 'some' but an ever-more-closely-approached 'all' to come to be able increasingly to stand ever more equal and thus ever more disinterested before the evidence. I am therefore defending a 'grand (equality) narrative' as the only one which will permit the *making* of that structure already well-identified as essential to a universalist stance: the structure known as 'the view from nowhere'. I am arguing that our new 'view from nowhere', that of which we are so desperately in need to defend moral absolutes, can only be made out of a 'view from everywhere'. Only such a view will allow us to create a well-formed and scientific theory not of the universe, not of the biosphere, but of the world. And the creation of equality is fundamental to the disinterestedness necessary to its making.

As introduced into the sciences of the organic in 1973 by Maynard-Smith and Price, the concept of an evolutionarily stable strategy (ESS) was of a strategy such that, if every member of a group of animals were to use it, the group could not be successfully invaded by a competing strategy. By analogy, I want to try to introduce into what I am calling the sciences of the personic the concept of what could be labelled a developmentally stable strategy (DSS), being a strategy such that, if every person in a community were to use it, the community could not be successfully invaded by a competing strategy. I am suggesting that it is of course an empirical matter to determine what such a strategy might be. But that just as in evolutionary game theory, 'demand half' was the only ESS, and from behind Rawls's original veil of ignorance, 'go for equal shares' was (in my language not his) the only DSS, so I am hypothesising 'make-more-and-more-equality' as a candidate strategy. I am also suggesting that when such a task had been completed, it would mean that a society of normal animal-human persons dealing sometimes in 'benevolence' would have transformed itself into a community of moral human-human persons dealing always in 'justice'. And that the way therefore to conceive of the relationship between the individual items in this 'justice *and* benevolence' specimen is as I AM precursor (benevolence) and I IS successor (justice).

(18) knowledge and science

The issue here, very much akin to that already discussed under the 'art *and* science' pair, is whether or not there can be knowledge which is other than scientific knowledge. Can one legitimately modify the noun 'knowledge' with some adjective other than 'scientific'—'moral', say, or 'literary' or 'psychological' or 'what you will'—which would mean that one had identified some clearly other-than-scientific *type* of the stuff. As always, this is the cue for somebody to observe that it all depends on what is meant on the one hand by 'knowledge' and on the other by 'science'. Of course: so far so obvious. But going no doubt for the jugular, I am going to insist that, since the way in which this particular 'and' actually *functions* in philosophical conversation involves the 'knowledge' member of the pair acting as banner behind which an 'anti-science' stance assembles itself, it is only appropriate to read the 'knowledge *and* science' pair in the light of that fact. In so doing, I will of course be defending the credentials of science. But that does not mean that I am not very well aware that the *pro*-science lobby is capable of gratuitous daftnesses. It is. Indeed, to make it quite clear what are the sorts of things I am definitely *not* trying to defend, let me register, if briefly, just three such.

First, there is the daftness of those natural scientists who want to swallow up what are currently called social sciences. They seem to think, these frequently militant ones, that the observations, hypotheses, experiments and theories of the sciences studying the universe and the biosphere can just be lifted wholesale into a study of the world. This was the thrust of Rutherford's previously quoted quip that everything which wasn't physics was stamp collecting, as it is also of Francis Crick's perhaps less frequently quoted remark that all sciences which are not natural sciences are merely 'social work'. This is precisely the sort of ignorance and arrogance which causes those rightly repelled by it to rush for 'knowledge' (or even 'wisdom'!) as some sort of antidote. Damaging doubly the very enterprise it seeks to promote, its ignorance makes natural science appear ridiculous (which it isn't): and its arrogance makes moral science appear impossible (which it isn't either). A recent example of such unhelpfulness comes (yet again!) from Edward O. Wilson. In *Consilience* (1998), he looks forward to a situation in which 'the humanities and social sciences shrink to specialised branches of biology; history, biography, and fiction are the protocols of human ethology; and anthropology and sociology together constitute the sociobiology of a single primate species.' I would happily agree that 'the sociobiology of a single primate species' is a totally legitimate object of study. And totally legitimate whichever of the many 'single primate species' which have evolved across the last seventy million years or so is involved, up to and including the most recently arrived *Homo sapiens*. It's just that such studies would not be studies of persons. To study a *Homo sapiens* is not to study a person; to study a group of *Homo sapienses* is not to study a society or a community of persons. And to bring this point into relief, think only of the notorious 'nature *and* nurture' duo, which in a later entry will be the subject of more detailed consideration. When discussing many aspects of the human, notoriously when discussing intelligence for example, it is a commonplace of philosophical conversation to argue that because 'nature' and 'nurture' (taken together) offer a whole (100%) explanation of whatever is under discussion, therefore 'nature' and 'nurture' (taken separately) must offer each a partial (80% : 20% maybe, or 50% : 50%) explanation of whatever is under discussion. Sophisticates of course will always add that it will be very difficult, if not impossible, to disentangle the actual proportions contributed by each factor. What I am suggesting is that, on the contrary, it will be very easy. The proportions are going to be 100% : 100%. And that is not mathematical illiteracy, but just a recognition that 'the human' here is two different things. One hundred per cent of the explanation of one thing (a *Homo sapiens*) is going to be a 'nature' explanation. And one hundred per cent of the explanation of another thing (a person) is going to be a 'nurture' explanation.

Second, there is the daftness of those scientists who, while (for the most part) appropriately trying to defend realism against anti-realism, nonetheless become totally inappropriately enmeshed in a quite indefensible attack on what is sometimes called 'social constructionism'. Such a one is Lewis Wolpert. In *The Unnatural Nature of Science* (1993), he attacks the position adopted by, among others, Barry Barnes and David Bloor of the so-called 'Edinburgh School', who elaborate and defend what they call the 'Strong Programme in the Sociology of Knowledge'. Wolpert suggests that 'those who hold to the Strong Programme believe that all knowledge is essentially a social construct, and so all science [good or bad] merits the same attention.' But this is surely just silly: a prime example of the sort of thing that can both easily and dangerously get *pro*-science positions a very bad name. For isn't it just a curious non-sequitur to argue that because all knowledge is socially constructed, it is therefore not possible to distinguish between that of it which is well constructed and that of it which is less well constructed? Of course science is socially constructed. Who else could it be made *by* other than by socially situated persons? What else could it be made *with* other than with socially constructed observations, hypotheses, experiments and theories? And where else could it be made *from* other than from within social structures? It has been the detailed working out of such relatively obvious understandings as these that has been the considerable contribution of sociologists to the study of science. And the fact that some of them—*some of them* only—go on totally unnecessarily to draw wholesale relativist conclusions from such study is no excuse for the would-be scientific to indulge themselves in a bout of sociology-bashing. It seems to me that two points need to be made about the way in which it is possible to distinguish between that of socially constructed science which is *well* constructed, and that of socially constructed science which is *less well* constructed: the first quite widely accepted as important; the second more specifically just mine.

The widely accepted point is that since every *new* discovery in science both leans on and supports every *already-existing* discovery in science, and the resulting consistency of the whole therefore puts pretty firm limitations on what can be added to the whole without causing chaos, then by and large the socially-constructed science which is *well* constructed is that which does *not* create such chaos, and the socially-constructed science which is *badly* constructed is that which does. But the point specific to my own argument would go even further. Because the societies which construct science are 'sets of relationships between persons', it is possible to study which are the features of such sets of relationships which *promote* the making of science, and which are the features of such sets of relationships which *hinder* the making of science. Take, for example, the generation of hypotheses. Their necessarily social genesis will of course throw light on which are the circumstances of social construction which correlate *well* with the generation of successful hypotheses, and which are the circumstances of social construction which correlate *less well* with the generation of successful hypotheses. And it would be a prediction of my hypothesis that what such correlative studies would reveal would be the following. They would reveal that to the extent to which the sets of relationships between persons are such as to constitute *societies* (entities consisting in sets of *unequal* relationships, and persons thus more *made by* than *making of* those relationships), the sciences there constructed will have a higher ideological content. And that to the extent to which the sets of relationships between persons are such as to constitute *communities* (entities consisting in sets of *equal* relationships, and persons thus more *making of* than *made by* those relationships), the sciences there constructed will have a lower ideological content. This is a formulation which accounts well for much: for the relatively late emergence of science; for its much-discussed 'why Europe?' characteristics, as explored, for example, in Basalla (1968); for its association with Protestantism; and for its association with democracy. And because both the type and the level of equality needed is so much more considerable when constructing the more difficult sciences, of which the moral sciences will be the most difficult of the lot, it also makes sense of its earlier application to, and greater success so far in, first the study of the universe vis-à-vis the study of the biosphere, second the study of the biosphere vis-à-vis the study of the world.

Third, last, and rather more briefly, there is the daftness of those scientists who do their own cause no favours by seeming to lack some sense of proportion. These range from the lack of such a sense among those with larger than appropriate egos to the lack of such a sense among those with smaller than appropriate visions. As an amusing example of the first, it was said about Fred Hoyle, with what truth I don't know, that he never forgave either the scientific community or the universe for rejecting the steady-state theory. And as examples of the second, how about Comte suggesting, in *Cours de Philosophie Positive* (1825), that there were obviously going to be some things, such as what the stars are made of, which scientists would never be able to fathom just a few years before the German physicist Von Fraunhofer used a spectroscope to analyse sunlight and tell us they were made of hydrogen. Or, more recently, the highly respected embryologist David Salter announcing, in an article in *Science* in 1984, that 'the cloning of mammals by simple nuclear transfer is biologically impossible' just a few years before the arrival of Dolly the sheep?

My three examples of scientific daftnesses—being predatory on the territory of others, being philosophically unsophisticated, getting things out of proportion—are just my three examples. No doubt there are others. But in listing them briefly and then moving on, I want to insist that it is no more sensible to 'knock' the contribution of science for these sorts of reasons than it would be to judge the contribution of Christianity by the activities of the Borgias. If we really want to question the validity of something, it behoves us to question it at its strongest (which is difficult) rather than at its daffiest (which is easy). It will be my case that, at its strongest, science just *is* knowledge, knowledge just *is* scientific. And that if they were to contain any sort of knowledge at all, things called 'psychological knowledge', 'literary knowledge', 'moral knowledge' or 'what-you-will knowledge' would need to be 'psychological science', 'literary science', 'moral science' or 'what-you-will science'. What this amounts to is that the way to deal with the 'knowledge *and* science' pair is to collapse the first into the second: or rather, to think in Enlightenment terms of a development whereby 'pre-scientific belief ' is progressively replaced by 'scientific knowledge'. To spell out what I think such a project involves, I will first try to outline what I take to be the limitations of pre-scientific belief; then look at its more dangerous stable-mate 'anti-science ideology'; and then conclude this section with a first reference to that moral knowledge which, to be any sort of knowledge at all, will involve a development from 'moral philosophy' to 'moral science'. This last is an idea which will be more extensively covered in Chapter 11.

It is not always altogether obvious just how to separate out pre-scientific belief and anti-scientific ideology. But for my rather elementary purposes, I will adopt a simple but I hope not simple-minded demarcation. I will take pre-scientific beliefs to be those which pre-date scientific knowledge, and anti-scientific ideologies to be those that are contemporary with, but in opposition to, scientific knowledge. That this will be a demarcation which will not always demarcate—astrology, for example, *was* a pre-scientific belief then, *is* an anti-scientific ideology now—is not I think significant to the points I want here to make. Here, and by way only of example, I want to consider just four of the more obvious limitations of pre-scientific belief: first, that it is not amenable to development; second, that it is not formulated clearly enough to be investigated; third, that it pays scant attention to fine-grained often statistical nuance; and fourth, that it rather too readily succours its close but more dangerous offspring 'anti-scientific ideology'—to which, then, I will turn my attention.

To illustrate, first, the way in which pre-scientific belief is not amenable to development, consider combustion. If you believe that fire is a gift of the gods, and/or one of the four basic elements, there is nowhere much else to go with such an idea: you just believe it, and tell stories based on it. But if, by contrast, you have about fire even a faulty hypothesis, a hypothesis such as the phlogiston hypothesis, for example, which dominated eighteenth century chemistry, then development becomes possible. As is well known, the phlogiston hypothesis explained combustion by positing a substance, phlogiston, which was expelled from a burning material. And for Joseph Priestley, one of the consequences of working within such an idea was that when what he had in fact done was to isolate oxygen, he could only comprehend its doing by postulating something he called 'dephlogistonated air'. Nonetheless, the fact that his success very quickly opened the way to the idea that, rather than expulsion, it was actually combination, with oxygen, which was the basis of combustion, was in no small part due to his scientific approach: truth emerging more readily from carefully formulated error than from ill-formulated confusion, his work meant that the issues involved had been so much more clearly spelled out.

It is such ill-formulation of pre-scientific beliefs which, as my second example, is another way of drawing attention to their severe limitations. By a too-easy use of extravagant metaphor, a too-lazy reliance on unexplored concepts, they can often lead to that obfuscating confusion which sees strange 'mysteries' where there are only everyday ignorances. Thus Wordsworth could tell us that 'every flower enjoys the air it breathes', and be just wrong. Thus, according to one of his critics, T. S. Eliot could 'at any given time—speak with equal firmness and dignity on both sides of almost any question, and with no realisation of the difficulties in which he is involved'. Thus will all true 'mysterians' lament the work of Michael Howe, professor of psychology at Exeter, who in his book *Genius Explained* (1999) challenges all the myth-making around the 'genius' concept to reveal people much the same as everybody else, if a bit more hard-working and single-minded. It is frequently the poet who goes for the mysterious. When such a one sees, say, 'the ravens flock / Around the dying deer', he or she will tell fanciful and often immanent stories as to what is going on. The scientist, by contrast, seeing the same thing, wants to try to explain. And in this particular 'flocking' case, has been spectacularly successful in doing so. In 1986, Craig Reynolds, of the software company 'Symbolics', created 'boids'. They were computer-birds, programmed to follow just the three simple rules when in company: try to go in the same direction as the rest; don't bump into them; aim instead for their average position. And now some twenty years later, the follow-up work flowing from his seminal research means that 'flocking' can now be described with as much accuracy as can the movement of molecules in gases and liquids —no doubt to the chagrin of many a mystery-monger,

The third limitation of pre-scientific beliefs, their paying of scant attention to fine-grained and often statistical nuance, is much discussed at the moment in the context of lamentations from scientists about our collective inability to think clearly about risk. Pre-scientific thinking is common-sense thinking. And while Gramsci was bang-on right to point out, in the *Prison Notebooks*, that 'common sense is not something rigid and immobile, but is continually transforming itself, enriching itself with scientific ideas and with philosophical opinions which have entered ordinary life', it is nonetheless the case that statistical ideas in general, and ideas concerning risk in particular, enter pre-scientific 'ordinary life' only with considerable difficulty, and only in part. In a much quoted experimental example of this reality, Tversky and Kahneman (1982) presented their undergraduate subjects with an imaginary character, described him as 'a shy man with a passion for detail', and asked them whether he was more likely to be a farmer or a librarian. A sizeable majority of them said 'librarian', regardless of the fact that, in the relevant population, there were so many more farmers than librarians that he was much more likely to be a farmer. It would be my contention that such errors of thought, endemic as I think they are to pre-scientific belief systems, are everyday events for almost all of us. (I would be delighted—but also surprised!—if the eagle-eyed could not find some of them in this book.)

Such—among others—are the limitations of pre-scientific thinking. But while the negative limitations of the pre-scientific can frustrate understandings, the positive hostilities of the anti-scientific can totally undermine them. In turning now to try to understand such hostilities, I want first to present three examples of them, and then suggest the way in which I think that my hypothesis as to the changing nature of 'I' can be useful in trying to combat them.

First, and this one rather obviously is just a more pernicious version of the pre-scientific belief point made above concerning statistics and risk, there is an anti-scientific ideology concerning certainty: one which (often) gets its (many) adherents into a rare old pickle. On the one hand, there is nothing they want *more* than certainty: and it is a weakness of science that, when what they want is absolute proof or certainty, all it can offer is validity or quantified risk. On the other hand, there is nothing they wants *less* than certainty: for where science offers necessity and determinism, what they really want is freedom and uncertainty. I sometimes idly speculate that for anti-science, 'perfection' would be a world in which the absence of absolute 'proof-of-safety'—and the presence, of course, of the ubiquitous 'precautionary principle'—would mean that contraceptive pills would be banned: while yet the presence of total 'freedom-to-choose', and thus total responsibility for ones actions, would mean that teenage girls could rightly be blamed for not taking them. (And there are countless other absurdities for which this one can stand as template.) Anti-science gets into this sort of pickle through its desire to get what Dennett calls 'radical contingency' in the place of 'a mindless cascade of mechanical processes'. It is my belief, to the contrary, that such 'radical contingency' is just not available. And that the positing of it, the positing of 'the radical contingency ideology', if you like, can only therefore be dangerous. Wanting to be certain that things cannot be certain is, like all irrationalisms, necessarily dangerous.

Second comes what I will call the 'spiritually corrosive' anti-science ideology. I take the name of this one from *Understanding the Present: Science and the Soul of Modern Man* (1992) by Brian Appleyard. Rarely to be found missing when brickbats are to be thrown at science, Appleyard's characterisation of it as 'spiritually corrosive' sees it as 'burning away ancient authorities and traditions'. Of course this is not the sort of stuff to be carelessly indulged in by those functioning from within the more rarefied atmosphere of academe. But even there, concerning the idea that science will come to account also for the human, no less a figure than Isaiah Berlin could protest that it was 'one of the most grotesque claims ever made': because, no doubt, it totally failed to come to grips with something not too far away from that 'spirit' which, for Appleyard, was being 'corroded'.

Third comes the 'we'll never know' anti-science ideology: the one so well illustrated in John Horgan's book *The Undiscovered Mind* (1999). Part of his argument was that even though governments devote considerable resources to brain research; even though there has been a considerable proliferation in the number of neuroscience journals and neuroscience conferences; even though we are from time to time promised the key which will unlock our understanding of aggression, of depression, even of consciousness: despite all these things, nothing to rival even Freud has yet appeared, and the problem is that 'they never get there'. He elaborates on this stance by detailing all the considerable disagreements that exist among such researchers—connectionists, Darwinian psychologists, dynamicists—and can only hope that neuroscience might be delivered from all this by the arrival of a 'genius' of some sort, his example being one who could do for neuroscience what Murray Gell-Mann did for particle physics. He does not think this likely. Since, compared with neuroscience, particle physics is child's play, we might well have to accept that it may never have that unifying insight which can bring order out of what is currently chaos. After all, nobody has even come close to understanding such 'core mysteries' as those of free will and consciousness.

It's all there: the only hope a genius; the problem not just a mystery but, heaven help us, 'a core mystery'; the author yet another of those comically and cosmically self-centred individuals for whom 'never' seems to mean nothing much more than 'not in my lifetime'. Such a one is also, of course, quite adept at using the terms 'brain' and 'mind' in quite spectacularly sloppily defined and differentiated ways. And even when he does come near to holding them clearly separate, and adopting the very common position of saying that to understand the brain is one thing but to understand the mind quite another, he simply fails to recognise that while the two quite different entities do indeed need two quite different understandings, *both* of those understandings need to be scientific. We need a scientific (natural scientific) understanding of the brain: and on that one we are moving at an ever increasing rate of knots. And we need a scientific (moral scientific) understanding of the mind: but on that one, we have hardly left port. The main reason for such tardiness in the second case, I am claiming, is a failure to appreciate just what science is, combined with a veritable terror of determinism. And it will only be when we can come to see science as the only possible basis for knowledge, and determinism as the only possible source of freedom, that we will be able to find 'lift-off' into the creation of the moral sciences.

For the old dichotomy whereby there was 'subjective' (knowledge) and 'objective' (science) is surely just incoherent. It is incoherent because it requires us to posit the 'I', the subject who is being subjective, in I AM terms. And if we do that with respect to its (obviously subjective?) knowledge, it is not at all obvious why we should not do it with respect to its (why objective?) science. However, if we instead posit the 'I' in I IS terms, think of it as the linguistically structured experience of a *Homo sapiens*, it becomes (literally!) as objective as we need it to be. The objectivity which is thereby made consists not in *any* consensus made by persons—quite huge majorities can turn out to be spectacularly wrong—but in a *particular* consensus made by persons: one using a particular objectivity-making method. Scientific method is a way of *making* objectivity. And to make it completely, the I AM to I IS shift, to which and finally I once again turn, is what we need.

Making objectivity is no small undertaking. Writing in the *Journal of Consciousness Studies* (Vol. 6, No. 8-9, 1999), Guy Claxton, of the University of Bristol School of Education, drew attention to the fact that 'the self-perpetuating nature of belief systems is blindingly obvious, except when one happens to be inside them'. He used as his examples one from 'then', and one from 'now. Then: 'to a culture that is inside one belief system, it is simply a fact that the crops failed because the gods were displeased, and there will be plenty of "evidence" that makes this "fact" self-evident'. And now: 'in one dominant worldview, it is obvious that hunger and malnutrition in Africa and Asia are caused by a shortage of food', so that it takes 'the meticulous research of an economics Nobel laureate, Amartya Sen, to show that this obviousness is in fact ill-founded'. His conclusion was that it is especially when 'the dominant interpretation seems to offer its interpreters comfort or privilege' that it takes 'sharp, disinterested observation to force a reluctant acknowledgement that the way things seem is not the way they are'.

Indeed. And it is my claim that his 'sharp, disinterested observation' can fully challenge his 'dominant interpretation' in only one circumstance. It can be only when no *interpreters* are dominant that we can be certain that no *interpretation* is dominant as a consequence. It can only be when no *interpretation* is dominant as a consequence that disinterest becomes available. And it can only be when disinterest is available that anti-science ideologies can be exposed for what they are. At its best, scientific practice already provides us with some sort of model of the workings of disinterest. But what I am suggesting is that this is a model in need of further development. Specifically, it is my claim that the very societies (sets of relationships between persons) which were necessary to the *emergence* of science *then* have become inimical to the *development* of science *now*. And that if the level of the disinterest exemplified in current practice is to develop, if 'at its best' is to become 'habitually', it is once again our conception of 'I' which must shift.

The need for such a shift can be nicely expressed in the language of Hegel. Following his use of the word, we can first of all say that if 'aufhebung' is to occur, if the disinterest already available to us is to be both carried to its next level of development *and* transcended by that very carrying, then it is such a shift that is required. And then we can add that the very nature of such a shift can be expressed in classic 'negation-of-negation' terms. Let me briefly spell that out like this.

First there were societies, complex and sophisticated civilisations such as those of ancient China or India, in which, unless as something it was a good idea to try to suppress or to try to loose, a self was hardly posited at all. Hinduism, Buddhism, Taoism, Confucianism were the belief systems which carried—and carry—such a stance. And what I am suggesting is that when the earliest 'I AM' versions of 'I' first emerged, when the choices were to cultivate it (gingerly) or suppress it (decidedly), theirs was the latter reaction. Of course the full explanation of any such 'decision' would take us, in classic Marxist terms, deep into considerations of the different economic bases pertaining at different times in different cultures. But such 'full explanation', important though it undoubtedly would be, is way beyond what is necessary here. What it is necessary here to register is that, anyway in its more superstructural manifestations, here is one of the most important sources of a major intellectual divide: that is, of the divide between one set of cultures which did spawn the scientific method and another set of cultures which did not. For science needs its 'I'. The scientific method, as so far it has developed, has needed an observing 'I', separate from that which it observes; an hypothesising 'I', confident enough to put its self in the place of God's self, and hazard a guess as to how he might have put things together; an experimenting 'I', with the nerve to dissect, probe, investigate his handiwork; and a theorising 'I', willing to postulate laws in accordance with which even 'He' would have to behave.

So, and by negation second, it was in the societies of 'the West', the societies fed by the later systems of ideas to be found in Judaism, Christianity and Islam, the societies in which *dualistic* I AM first-person selves were posited as 'over-against' that which was 'Other', that the scientific method which we now know developed. Such I AM selves were *very* successfully posited as 'over-against' the universe: because it was seen as totally 'Other', it was also seen as totally open to objective investigation, and the sciences of the inorganic flourished spectacularly. They were also *quite* successfully posited as 'over-against' the biosphere: because it was seen as considerably 'Other', it was also seen as considerably open to objective investigation, and the sciences of the organic have flourished somewhat. But they have been *not at all* successfully posited as 'over-against' the world: because it has been seen as hardly at all 'Other', it was also been seen as hardly at all open to objective investigation, and the sciences of the personic have not flourished at all.

To complete our Hegelian trio, to get us to the third stage of the development of thinking scientifically, the 'negation-of-negation' which now we need, or so I contend, is that whereby the positing of *monistic* 'I IS' just person selves, constitutive of a world *without* 'Others', will enable that extension of the scientific into the realms of the personic which will at last permit the flourishing also of the so-called 'social', better-called 'moral', sciences. It is my case that in circumstances in which the possibility of such flourishing just appears over the horizon as real, what is needed is that the old I AM idea that the essence of the human is competition and inequality be no longer cognised as just the (relatively innocent) pre-scientific belief which up to now it has been, but instead come to be *re*cognised as the (thoroughly dangerous) anti-scientific ideology which now it is becoming.

In concluding this section, I want, as promised, to make a first reference to that moral knowledge, which to be any sort of knowledge at all will involve a development from 'moral philosophy' to 'moral science', which is the main topic for consideration in Chapter 11. I will there be trying to develop what I take to be Paul Churchland's position. He would seem to be suggesting that while scientific knowledge is an inextricable mix of *praxis* and *theoria* focused on the *natural* world, moral knowledge is an inextricable mix of *praxis* and *theoria* focused on the *social* world. My own position will be similar, but I will be using a different, and I think more appropriate, vocabulary. I will be suggesting that while natural science *already consists in* theories concerning the inorganic stuff of the universe and the organic stuff on the earth, moral science *will come to consist in* theories concerning the personic non-stuff of the world. In terms of my current 'and' pair, in terms of 'knowledge *and* science', that will mean that moral *science*, the only source of moral *knowledge*, will have come of age.

(19) liberty and equality

When Walter Bagehot insisted that there is no method by which men can be both free and equal, he was succinctly summarising what has been the orthodox position on the relationship between these two for some time. Perhaps the best known recent exponent of this orthodoxy has been Isaiah Berlin. In fashionably pluralist vein, he has argued that there is not one 'good' but many 'goods'. And that there are some goods (for example, 'liberty') which are incompatible with some other goods (for example, 'equality'). This is a reality, he suggests, which means—sadly, of course—that we have to choose between them: the more of the one, the less of the other. And it perhaps comes as no very great surprise that, from such a primary source, it is 'equality', it is argued, which must draw the short straw. But Walter Bagehot had insisted on something else as well. He had also said that one of the greatest pains to human nature is the pain of a new idea. So could it just be that neither he nor Berlin was up to coping with a little pain on this one? For far from being incompatible, it is entirely possible to argue, as here I will, that 'liberty *and* equality' are simply different sides of the same coin: that is, that they are totally inseparable and mutually reinforcing.

To make my case I will also have to challenge what is often used as a second and supplementary attack on equality, but from a different source. After Bagehot but before Berlin, it was in 1900 that the Italian economist Vilfredo Pareto proposed a way of deciding between the relative desirability of differing economic systems. Called now 'Pareto Optimality', it suggests that we should not choose any system which means that another and available system capable of making everyone better off is rejected. The principle has become famous, I would say notorious, as a defence of economic inequality: it suggests that if when goods and services are distributed more unequally the poorest person is better off than he or she would be if those goods and services were distributed less unequally, it is better that we choose the more unequal over the less unequal distribution. And it is better for two reasons. First, it does not offend against the liberty of being able to have more than the next person: though only (of course) where such having can be done within the (obviously totally neutral) law. And second, it means that the poor, who (sadly) will always be with us, will be as rich as it is possible for them to be: a circumstance which in itself will also contribute to their liberty. The key idea to get hold of here, the idea which, in order to launch my counter-claims about the relationship between liberty and equality, I must deconstruct, is the idea 'better off'. Pareto Optimality suggests that we should not choose any system which means that another and available system, capable of making everyone 'better off', is rejected. But what I want to suggest is that if instead of taking it as some sort of neutral given we instead problematise that 'better off' idea, we find that a fairly nasty and for the most part totally unrecognised ideology lies coiled up in its apparent innocence and obviousness.

To expose it, try the following experiment with a group of friends. Ask each of them in turn to answer the question: 'do you think that most people want better wages?'. Encourage each of them to keep talking, prompting them with supplementary questions if necessary, until it becomes clear—it usually doesn't take long!—that they are answering a question you did not ask. While you asked them to answer the question: 'do you think that most people want *better* wages?', they are answering the question: 'do you think that most people want *more* wages'. That 'better' is not the same as 'more', that in some circumstances 'better' could actually mean 'less', will be picked up, I guarantee, by more or less nobody. Here is a problem that does not speak its name. That 'better' equals 'more', that 'better off' equals 'having more', is deeply ingrained, and for the most part totally unconsciously, in our animal-human (normal person) awarenesses. It is ingrained there, or so I suggest, by what I claim is the ideology of all ideologies: the one I am calling 'the having ideology'; the one which, under the head of my 'master *and* slave' specimen, will get its first major consideration later. For now, I want only to point out the way in which it is manifested as confusion over the way in which the two items of my current 'liberty *and* equality' specimen are to be related.

If you are hungry and somebody gives you a piece of food, are you 'better off'? My answer to that question is twofold: that without further information, there *is* an answer to the question 'is this *Homo-sapiens* now better or worse off?'. But that without further information, there is *not* an answer to the question 'is this *person* now better or worse off?' To the first question, the answer is clearly an unambiguous 'yes': for a biological entity, to live is what matters. To the second question, the answer is only a tentative 'it depends'. For an I AM self, a self for whom its capacity for loving is still at only a 'having' level, the answer remains a 'yes': but a 'yes' increasingly tentative; a 'yes' which produces an increasing level of uncomfortableness, as such a self reaches the higher stages of its development towards a non-having loving. For an I IS self, things are changed utterly. For such a self, the issue is this one: if by taking the food he or she has decreased the level of inequality (and thus increased the level of love) between herself or himself and other selves, he or she is now better off than before—as are those other selves too. But if by taking the food he or she has increased the level of inequality (and thus decreased the level of love) between herself or himself and other selves, he or she is now worse off than before—as are those other selves also. This is my position. And what it means is that that which daily haunts the embryonic I IS self—and for now, almost all I IS selves are only embryonic—is the realisation, at first only quite dimly sensed but at last only too terribly registered, that over and over and over again, ten and twenty and more times a day, he or she cannot help but function in structural relations with other persons such that most of her or his everyday actions necessarily stunt the humanity of both her or his own self and the selves of those others. This is the haunting, or so I believe, which, at some level or other, is coming to be known to ever greater extents, by ever more human persons, as we grow up. This is the haunting, or so I believe, which makes the very worst feature of the capitalist societies in which we all live—and try to love—their extreme moral ugliness.

In *Self-ownership, Freedom and Equality* (1995), Gerry Cohen refers to those who in societies of inequality are the 'have-nots' as those who inhabit 'the short end of inequality'. He is wrong. We need to learn to conceptualise inequality as having *two* short ends (and also a short middle!): a conception which can be nicely illustrated by considering the idea of a 'zero sum game'. In philosophical conversation, it is a concept which I have occasionally heard used in a way totally at odds with the way in which it is used in the 'game theory' from which it is derived. It is used as if it referred to a situation in which there are no losers: everybody wins. But in game theory it means no such thing. There it refers to situations in which all the gains of one person or institution are matched by equal and opposite losses of another person or institution. And equality, as traditionally understood, has been seen as this sort of zero sum game. The losses of those with more (those who have their surpluses removed from them) are seen as equal to, and opposite from, the gains of those with less (those who share those redistributed surpluses among them). The equality which my hypothesis posits is quite other. It is *no* sort of zero sum game. According to my hypothesis, it really *would* be the case that everybody wins. And that is because my hypothesis envisages the emergence, via the selving process, of a transformed 'everybody': an 'everybody' consisting no longer in normal persons (I AM animal-humans) but in moral persons (I IS human-humans). What it is vital to understand here is that what is gain (or loss) to an I AM animal-human is not the same as what is gain (or loss) to an I IS human-human. The prevailing image of the pursuit of equality has been the *re*distributive one: first, the economic system distributes unequally; then the political system, via taxation, *re*distributes equally, or anyway *more* equally. But my image of the pursuit of equality is distributive through and through. Mine is an image of a community which distributes equally *in the first place*. And does so because each of its now human-human members can look after her or his now fully moral self-interest in no other way: because each of its now human-human members has developed, via the selving process, the interests of the disinterested. A moral theory which worked, a scientific moral theory, would have to be one which enabled us, by its application, to test competing claims to the good in such a way as to be able to decide between them. And the test derived from my hypothesis would be this one: that if this action makes the thus newly-resulting set of relationships constituting the world a set of the *more* equal, it is good; and if this action makes the thus newly-resulting set of relationships constituting the world a set of the *less* equal, it is not yet good. But caution: see the comments under the head of the 'praise *and* blame' specimen (below) to register, yet again, both the non-moralistic meaning given to that 'good' and 'not yet good', and the asymmetry posited as characterising their relationship. For the purposes of my hypothesis, it is *this* reading of 'equality', a reading further to be developed in Chapter 11, which needs to be considered in its relation to its 'liberty' partner. So in turning now to consider that 'liberty' partner more directly, it is *this* reading which I will use to show that the way to solve the problem of the relationship between 'liberty *and* equality' is to come to realise that, in their mutuality, there isn't one.

I want to argue that the very highest freedom we can envision can be achieved only by the recognition of the necessity to love well: and that the recognition of the necessity to love well brings with it the recognition of the necessity of making equality. In one sense, that means that it *is* possible to be free *before* the attainment of such equality: one can be free before the attainment of such equality because the recognition of its necessity is enough. But in another and rather more important sense, of course this is not possible. Marx was surely right when, in the *Manifesto of the Communist Party* (1848), he insisted that 'the free development of each' is ineluctably bound up with 'the free development of all'. And Spencer was surely right too when, in like vein, he said that 'no one can be perfectly free until all are free'. In this second sense, freedom is as necessarily social as is equality. For while it is possible to make *grammatical* sense of the phrase 'freedom of the individual' but not of the phrase 'equality of the individual', it is not really possible to make *empirical* sense of either. In reality, freedom is just as ineluctably a social concept as is equality. In its full working out, it is available only to the whole human group, only to the whole human race. And this for two reasons.

First, it is relational. If somebody else is either secondary to me or primary over me (see 'master *and* slave' below), then not only he or she but also I cannot be free. Second, it is collective. Being determined by determinations very often means being determined by the determinations of others: spectacularly so, for example, by the determinations of upbringers when we are little. Thus until the determinations of all are such as to determine each towards freedom, none can be free.

But there is nothing remotely natural about equality. In the sense that we *make* ourselves free as we *make* ourselves equal, it is literally *super*natural: an argument to be further developed in Chapter 11. What I hope to be able to show there is that it will be only when we have made the thoroughly non-natural and *super*natural equality that we will be able to make, in the image of the thereby *re*made and new I IS person, the thus *re*made and new I IS God. Indeed, one way of arguing the case for the making of equality is that it is the state essential to the *re*making of a God concept. And one way of arguing the case for the *re*making of a God concept is that it is that concept which, in being immune to a relativising critique, is essential to the *re*making of truth.

(20) master and slave

In *Beyond Evolution* (1997), Anthony O'Hear suggested that 'one evolutionary advantage of self-consciousness for human beings may be that they are able to predict each other's behaviour better and so be more successful in the competition and co-operation of social life.' He went on to comment that 'once established, this type of mutual fellow feeling leads to demands for reciprocity unknown in the animal kingdom' and that 'thus is introduced a need over and above those for survival and reproduction.' This means that, 'as is suggested by Hegel in the dialectic of the master and the slave', there is introduced 'a need which will be unfulfilled so long as it is interpreted in terms of domination and submission.'

'The dialectic of the master and the slave', to which O'Hear here refers, is a much commented upon idea of Hegel's, as expressed by him in *The Phenomenology of Spirit*. There is contained in all human dyads, or so the story goes, a potentiality for a fight to the death over status, and the recognition that it brings. Most such struggles end, however, long before any such death: they end in the 'victory' of one, who thus becomes master; the 'defeat' of the other who thus becomes slave. But, says Hegel, these are only Pyrrhic victories, only apparent defeats—and for two reasons: first, because the recognition wrested from only a slave is nowhere near as satisfying as would be the recognition freely given by an equal; second, because the master, through idleness, develops dependence on the slave, and the slave, through labour, develops independence of the master. Hegel himself is quite clear that he thinks such relationships thoroughly unsatisfactory as a foundation for human sociality. But Nietzsche developed things otherwise. He has his prized 'Overman' revel in the master approach, and fight heroically for power, excellence and independence. And he takes that to be a fight against what he contemptuously calls 'the slave revolt in morals': a revolt whereby, throughout history, it is all too often mediocrity and servility, such as he thinks are to be found in the doctrines of Christianity, which lamentably win the day.

There are many ideas in twentieth and twenty-first century thought with relationships recognisably kin to Hegel's 'master *and* slave'. There is Adler arguing that to understand goal-directed behaviour we have to understand intention; to understand intention we have to be aware that the (for the most part) unconscious aim of all persons is to be superior to other persons; and that when attempts at compensation for felt inferiority fail, the result is that neurosis which he famously labelled 'an inferiority complex'. There is Paolo Freire maintaining that there is a permanent struggle between the powerful and the powerless, and that washing one's hands of the conflict is not to be neutral but to side with the powerful. There is Sartre claiming that the powerful *wield* 'the gaze' to claim their power; the powerless *yield to* 'the gaze' in acceptance of their powerlessness. On the same theme, there is Foucault adding that 'the gaze' is always an ideological (not an optical) phenomenon, and that 'there are no relations of power without resistances'. And so on, and so on: there are many more examples I could use.

Like these others, the 'primary-secondary' vocabulary of my own position is also recognisably kin to Hegel's 'master *and* slave' language. But with a difference. For where Adler sees an 'inferiority complex' as a neurosis, but a 'superiority complex' not at all, I take both secondariness and primariness to be stuntings of our capacities for full personhood. Where Paulo Freire, along with many an other of course, sees the battle as one for power, with the powerful who have it on one side and the powerless who don't on the other, I see the battle as one for equality, with nobody able to 'have' it until everyone does. Where Sartre thinks of the powerful as wielding power and the powerless as yielding to power, I think of the primary as being stunted by the power that they wield, *and the powerlessness they yield to*, and the secondary as being stunted by the power they yield to *and the powerlessness they wield*. And where Foucault makes the single claim that there are no relations of power without resistance from the powerless, I would make the four-fold claim that (1) there are no primary-secondary relations without resistance from the secondary; (2) there are no primary-secondary relations without acquiescence from the secondary; (3) there are no primary-secondary relations without acquiescence from the primary; and (4) there are no primary-secondary relations without resistance from the primary. In other words, both primary and secondary both resist and acquiesce. Or in still other words, the struggle is one *out of* lower-level (primary-secondary) normal relations *in to* higher-level (equal) moral relations. And in the course of such a struggle, the struggle that is to make equality, the powerful and the powerless are to be found on both sides of the divide. As normal I AM persons, both the powerful and the powerless are to be found, for the most part unconsciously, *resisting* the making of (equal) moral relations. As moral I IS persons, both the powerful and the powerless are to be found, increasingly with consciousnesses raised, *encouraging* the making of (equal) moral relations.

This is the reworking of the 'master *and* slave' pair which I believe to be necessary now. Where Marx famously claimed to have turned Hegel on his head, what I want to suggest is that the more appropriate thing to do is to turn him right-side-up again, but then re-work this his famous 'and' in the light of an I IS conception of a person rather than an I AM conception of a person. It seems to me that the failure of progressive thought (before such a reinterpretation) to recognise 'primariness' as a problem with an import equal to that of 'secondariness' has been a critical failure. It is a failure now very well documented, for example, as the failure of progressive thought (before feminism, and the problematising of masculinity) to recognise 'the man question' as a problem with an import equal to that of 'the woman question'. When trying to identify 'the oppressed', animal-human, normal-person revolutionaries are wont to round up the usual suspects: women, the workers, blacks, homosexuals, and so on. Human-human, moral-person revolutionaries must round up us all.

For the error of all the traditional characterisations of power dichotomies has been to see what I am calling 'primariness' as the norm, from which its Others diverge. Charles Mills for example, writing primarily on race, can use the concept 'sub-person' to get at the ways in which women, blacks, homosexuals, the secondary in general, have been labelled: a usage which illustrates very well the asymmetry which is ubiquitous in mainstream social science between the powerful (who are seen as the norm, and hence the normal) and the less-powerful (who are seen as the problem). Even the most revolutionary of reformers will talk of levelling up to the norm, as though 'up' was an entirely non-problematic idea. It is my claim that 'up' is as problematic as 'down': that the opposite of a 'sub-person' is not a person but an 'over-person'. Normal *over*-persons are, *as persons*, stunted varieties of the kind: normal *sub*-persons are, *as persons*, stunted varieties of the kind. Each are stunted, differently but stunted, by the unequal relationship between them. The traditional failure to recognise this has meant that while there has been 'a woman question' there has not been 'a man question': while there has been a 'problem' of homosexuality, of blackness, of children, there has not been a 'problem' of heterosexuality, of whiteness, of adulthood. When the issue has been 'what are *we* going to do about X, Y or Z', the often totally unconscious construction to be put upon that 'we' meant it was the 'we' who were *not* females, *not* homosexuals, *not* black people, *not* children who had to solve the problem of the 'you' who were. And for all the changes now going on in these areas, spectacularly of course in the area of gender studies where books and books and more books are now being written about 'the problem of masculinity', there remains one power dichotomy whose primary axis is hardly problematised at all. When we get to the dichotomy 'rich' and 'poor', in my vocabulary the primariness of 'having more' and the secondariness of 'having less', it is hardly ever registered that we need to conceptualise the primariness, the 'having more' or the 'being rich', also as problem rather than as norm. And also as problem *for those who have more*: also as problem *for those who are rich*. This is a particularly significant absence I think precisely because this is the power relationship which underlies and buttresses all the others. What I am suggesting here is that when the problem of inequality is aired, we need to check our almost knee-jerk response of thinking that it's the problem of poverty which is under discussion. It isn't. It's the problem of inequality. And that's different.

But coming back to my 'primary *and* secondary' formulation of Hegel's 'master *and* slave' duo, let me now turn to the way in which it plays itself out in reality. Any two specific persons, in (definitionally linguistic) interaction of course, manifest many 'master *and* slave' or 'primary *and* secondary' relationships. Take a man and a woman in Britain today. The division by sex will normally give primary status to him. If he is homosexual and she heterosexual, the division by sexual preference will normally give primary status to her. Will it be sufficient to outweigh the other? If she is older and he younger, maybe. As long as she is not *so* old as to have become invisible, of course. Unless, that is, she is very old and has acquired some of the (fast-diminishing) status of ancient wisdom. And so on. If one is black and one is white; if one is more highly educated than the other; if one is childless and one a parent; if one married and one single: in an hierarchically structured society, which is in all societies, each variation tips the primary/secondary scales differently. And it is not only that the actually- lived reality of every inter-relationship is multi-faceted. It is also that the various facets are differentially mobile. As I have already pointed out, while many of those I have used as examples above have shifted, some of them dramatically, in the recent past, the one based in more or less wealth and/or more or less income has shifted hardly at all. My claim is that the relationship *in total*, with all its relatively shifting and relatively stable elements, will have a mode around which it fluctuates. And it is that mode of which I am speaking when I speak of a relationship of inequality having a primary and a secondary participant.

Now since I have already argued (see 'individual *and* society') that the way to escape the reduction of the societal to the individual is to see society as the sum total of the dyadic *relationships* within it rather than the sum total of *individuals* within it—or better, that society is constituted by relationships *between* persons rather than *by* persons—it follows that societies also can be characterised in primary-secondary terms. They too, all of them the world has so far known, are constituted by characteristic modes of primary-secondary relationship. I am suggesting that in such societies of unequal dependence, the master on the slave and the slave on the master, the primary on the secondary and the secondary on the primary, the struggle to shift from them is indeed centred around Anthony O'Hear's previously quoted 'need which will be unfulfilled so long as it is interpreted in terms of domination and submission.' But such a need (in my terminology actually a want rather than a need: see 'needs *and* wants' below), being situated as it is pretty high up on a Maslow-like 'hierarchy of needs' table, emerges only quite late in selving development. We have to *learn* to want equality. It doesn't come naturally at all. And for as long as we have not yet learned to want it, for as long as we are normal I AM persons, our situation is both that we don't need it, and that we can't want it. Via the further development of our selving process, we have not yet learned *how* to want it. And when we have so learned, we will so want. Notice that, consistent with my stance on the 'is *and* ought' specimen, this is a formulation with not an 'ought' in sight.

In the terms of my hypothesis, it will be remembered, a self is (definitionally) selfish. And an I IS self which *has* learned to want equal relations between persons rather than primary-secondary (master-slave) relations between persons is just as selfish as its I AM precursors which have not yet so learned. It's just that when *such* a self is selfish, when *such* a self 'ishes' (looks after the interests of) its self—as, if strangely, I put it earlier—it is trying to 'ish' (look after the interests of) not just a different *sort* of self but a different *level* of self. It is trying to look after the interests of a level of self which is coming to recognise, at first in a way hazily untheorised, eventually in a way sure and compelling, that an 'I' is not something that 'AM' but something that 'IS'. And that it 'IS' equally in all persons. I am suggesting that it will be this recognition, when once it is widely available and thoroughly theorised, which will precipitate a great raising: that is, a great consciousness raising which, already underway if for now still haltingly, will literally remake the world.

In one sense, what it is that is wrong with primary-secondary ways of relating is to no small extent recognised already. But it has perhaps been rather more that such ways of relating have been *felt* to be unjust, or have been *felt* to be just plain silly and demeaning, rather than that they have been *understood* to be unjust, or have been *theorised* to be just plain silly and demeaning. Let me illustrate what I mean here with just one example of each such 'feel'.

That primary-secondary relations have been felt to be unjust can be illustrated by such examples as are provided in the work of the anthropologist Glicksman (1956). He studied the ritual of 'symbolic inversion', at that time still present in some African societies, whereby the 'king' was insulted, temporarily rejected, ridiculed, denounced, blamed. He identified the main function of such a practice as that of 'safety valve': by allowing the letting off of steam concerning injustice, it actually reinforced a social order in some sense already recognised as being unjust. And he added that perhaps the 'festival of fools' held in many European countries in the Middle Ages, and the various military traditions of days when officers serve men in armies even today, had a similar function. And that primary-secondary relations have been felt to be just plain silly and demeaning can be illustrated by what I think would have been the widely-recognised risibility of a story coming out of Oxford University some years ago. On Sunday 7 July, 1996, it announced that it was appointing one hundred and sixty two new professors. None would get any more money. None would have any new duties. And academics had to nominate themselves for the honour. Claiming that they needed the titles to compete for research funds which would otherwise go to other institutions, the result was that three hundred and sixty one became professors in that manner. And the number of readers jumped from one hundred and seven to two hundred and six. That this is surely a demeaning silliness is only compounded by the fact, rather than challenged by the fact, that the reasons given for the (of course reluctant) self-aggrandisement are highly likely to have been valid!

Faced with examples such as these, quite a lot of us *feel* the injustice, *feel* the demeaning silliness, which they involve. What I am claiming is that it's the *understanding* of the feel, the *theorising* of the feel, which is, so far, conspicuous by its absence. We have in place only half-understood recognitions that in the search for a 'view from nowhere', since all our vantage points have, to date, been points of *ad*vantage or *dis*advantage, everything we have seen has been skewed. And once again, what I am suggesting is that such lacunae are in large part due to a failure to recognise the fact that fast accumulating scientific evidence is inexorably requiring of us that we rethink our concept of 'I'.

For with 'I' posited as that which *is* experience rather than that which *has* experience, the shedding of 'the having ideology' in such a fundamental area will open the way for its more general shedding across the board. As far as 'master *and* slave' ('primary *and* secondary') relations are concerned, it will open the way to the shedding of the theoretically disabling notion that it is in 'having' (being primary) that we flourish, in 'not-having' (being secondary) that we don't. It will help us to come to understand that it is the *differential* having, the 'having-less-than' *and* the 'having-more-than', which is the problem: that it is this which, in stunting the development to full humanity of both those 'having-the-less' *and* those 'having-the-more', stunts the development of us all. And these will be understandings which will make it possible so to theorise the problems of moral philosophy that the construction of moral science becomes a possible project. Foucault believed that the search for a form of morality acceptable to everybody, one that everybody *should* accept, was catastrophic. I believe that the search for a form of morality acceptable to everybody, one that everybody *will* accept, is splendid.

(21) mind and body

An Oxford philosopher of my acquaintance once summed up for me, I think graphically, the nature of the problem presented by this specimen. While science, he said, has given us a pretty good view of how the world is, we just can't work out how to put ourselves in the picture. In addressing that problem, as now I will here, I am involved in a debate which, among current professionals, goes frequently under the heading of the 'monism *and* dualism' debate, or the 'monism *and* pluralism' debate: yet more 'and' usages, of course.

Modern dualists tend to be of two kinds. First, there are those taking the two entities mind and body to be *causally* related. Amongst them, the 'interactionists' believe that the causal relation holds in both directions (physical to mental, and mental to physical), while the 'epiphenomenalists' maintain that it holds in only one direction (physical to mental). Second, there are those taking the two entities mind and body to be only *correlationally* related. Their position is commonly labelled 'psychophysical parallelism'.

Modern 'monists', by contrast, come in three varieties. First, there are those of an idealist persuasion. They argue that there is only one entity and it is mental: physical events are reducible to mental events, often but not always events in the mind of God. Second, there are 'materialist monists'. They maintain that there is only one entity and it is physical: mental events are reducible to physical events. And third, there are those who hold some version of what is called an 'identity theory'. They take consciousness to *be* a brain process in much the same way as heat *is* mean kinetic energy or a gene *is* a length of DNA—two analogies quite commonly cited in defence of such a stance.

Very widely among the professionals, much less widely among the non-professionals engaged in philosophical conversation, there is now a pretty general agreement that dualism just won't do. It's founding father, Descartes, 'solved' the mind-body problem, and spawned the modern and derivative 'solutions', by positing two different *things*, different *stuffs*. There was brain stuff (a *body* sort of thing) and mind stuff (some *other* sort of thing). And the current and largely hostile attitude to such an approach among the professionals can be neatly summed up in the words of Tim Crane. Writing in the *THES* (30.1.98) he put it this way:

> What opponents of Cartesianism want to attack is a collection of the following ideas: the mind is a thing; the mind is not necessarily linked to the body or to the environment outside the mind; the mind is an inner theatre; the mind knows its own contents with a special authority, by means of a

> mechanism of inner sense or introspection. Many anti-Cartesians want to say instead that the mind is not a thing, but a kind of activity, it is essentially embodied and linked to its environment and the contents of the mind are not known in a special or authoritative way.

Of course I want to associate myself fully with such an approach. But I have what I think is my own particular 'take' on it. To try to illustrate that particularity, let me look in turn at each of the three claims Crane attributes to those he calls the 'opponents of Cartesianism'.

Firstly, no, 'the mind is not a thing': yes, it is 'a kind of activity'. But while thinking of it as a thing is a reification which certainly queers our pitch when we try to think clearly, it seems to me that the problem is not only reification. To invent a word where none exists, it is also 'habification'. We reify, but we also 'habify': to invent another. Our reifications involve us in a substitution *between* grammatical categories: we use a noun instead of a verb. Our 'habifications' involve us in a substitution *within* a grammatical category: we use a 'having' verb for a 'doing' verb. And when thinking about the mind-body problem, it is commonplace to do both at once: that is, to use the noun 'mind' as something we 'have' as substitute for the verb 'minding' as something we 'do'. The consequence of both the reification and the habification is an erroneous dualism. Reifying 'minding' so that it becomes 'mind' gives us *two* entities (mind and brain) instead of one entity (brain) and a process happening in it (minding). And habifying 'doing minding' so that it becomes 'having a mind' gives us *two* entities (an I AM which *has* a mind) instead of one entity (an I IS which *is* 'minding').

Secondly, yes, 'it is essentially embodied and linked to its environment'. But this is a linkage rather stronger than such somewhat inappropriate vocabulary would suggest. Mind is to body not as *pips* are to apples, but as *ripening* is to apples. Pips are *in* apples, are 'embodied', if you like. But ripening is 'an apple happening'. It is not *in* apples, it is *of* apples. To describe one 'it' (a mind) as embodied in another 'it' (a brain) doesn't quite hit the mark. And not least because it still contains a goodly chunk of that residual twoness characteristic of dualism.

And thirdly, no, 'the contents of the mind are not known in a special or authoritative way' to the person who has it. But this is not because introspection is a notoriously unreliable process, but because the person doesn't have it. The person *is* it. The person *is* the mind, or rather, *is* the minding. Translated into what I hope is my more careful vocabulary, what that means is that 'the minding' of a brain *without* language consists in that non-linguistically structured experience which does *not* constitute a person. And 'the minding' of a brain *with* language consists in that linguistically structured experience which *does* constitute a person. Such persons, extant always as one or another type of self, are then to be 'put in the picture' – in my Oxford-philosopher friend's phrase – not as free-standing

I AM viewers of the canvas who are the source of an appreciation of it which they *have*, but as under-standing I IS constituents of the canvas which are the sites (or better the moments) of an appreciation of it which they *are*. To reiterate: it's not that I AM *a viewer who* is outside the picture as observer, but that I IS *a viewing which* is inside the picture as constituent. As a constituent part not of the universe, not of the biosphere but of the world, 'viewing', along with 'believing' and 'appreciating' and 'desiring' and so on, is one of the linguistically structured experiences which a person *is*. And 'minding' just serves as appropriate shorthand for all such.

These then are my three 'takes' on Tim Crane's summary. What I want to claim is that, on the problem of the relation between 'mind *and* body', they amount to a particular version of a general anti-Cartesianism which does a better job of articulating that position than do some of the alternatives. To defend such a claim, let me consider again three example. Let me consider 'bats', 'headless females' and 'the empirical mapping of symbolic representation'.

First, bats. It has become commonplace, at least since Nagel's engagingly titled piece 'What is it like to be a bat?' (1974), to think of the central fact concerning the conscious state as being that there is something that it is like to be in it. Indeed, much of what is now called 'consciousness studies' is riddled with this 'what-is-it-like-to-be?' phenomenon: a phenomenon which, it seems to me, is in need of serious challenge. Put me into a room with my eyes open and a table in front of me. It's not that I *have* a visual experience of a table, and there's something that it's *like* for me to *have* it. It's that the current I consists in, *is*, a visual experience of a table (among other things of course). And that that is not *like* something, it just *is* something: that is, it *is* the experience of a table. The question framed with a 'like' in it, the 'what is it like to be?' question, is ineluctably framed in terms of the I AM paradigm. Its unproblematised dualism lies curled in its presupposition of (1) a that which is (tableness experience), and (2) a that which assesses that 'thatness' (an I AM which views its tableness experience) to determine what it is 'like'. The alternative question, the one framed without a 'like' in it, the 'what *is it* to be?' question, is the one that we need. But it can only be framed in terms of an I IS paradigm. Its monism is proclaimed in its presupposition of only a that which is: a that which is sometimes a *viewing* happening, sometimes a *believing* happening, sometimes an *appreciating* happening, sometimes a *desiring* happening. Or in brief: always a *minding* happening. It is not *like* something to *have* such a happening: it just *is* something to *be* such a happening. Nagel was both reifying and habifying. And neither will do.

Second, headless females. It was in 1968 that the Australian philosopher D. M. Armstrong wrote an equally engagingly titled piece 'The headless woman illusion and the defence of materialism'. He was trying to illustrate the ways in which our intuitions concerning the nature of what we call our minds can be thoroughly faulty. The argument was that when we think that the properties of experience, now quite commonly called 'qualia', are *non* physical, we are mistaking *not* being aware of our inner goings on *as* physical for *being* aware of our inner goings on as *non*-physical. The analogy was with a conjurer's illusion whereby because a woman is standing against a black background with a black cloth over her head, *not* seeing that the woman *has* a head is confused with *seeing* that the woman has *no* head. Amongst today's cognitive scientists, such faulty intuitions are well understood to be just faulty intuitions. Writing in the *Journal of Consciousness Studies* (Vol. 7, No. 4, 2000) under the brave title 'How to solve the mind-body problem', Nicholas Humphrey summarises such understanding as being that according to which it is assumed that 'the human mind and brain are, as Diderot anticipated, aspects of a single state—a single state, in fact, of the material world, which could in principle be fully described in terms of its microphysical components.' But then in the same article, he goes on to produce what seems to me to be a complete non-sequitur: he attacks Dennett's (1991) suggestion that 'sensations are nothing other than complex behavioural (even purely linguistic?) dispositions' on the grounds that 'while defensible in his own terms' it has 'proved too far removed from most people's *intuitions* [my italics] to be persuasive.' Re-enter a headless woman!

To banish her completely, it seems to me that while Dennett's insistence on the likely centrality of language is thoroughly helpful, it needs to be taken rather further than it seems to me that he takes it. We need to conceptualise the mind-body problem in such a way that bodies (things) and what goes on in them (happenings) are all that there is. And that what happens in a brain *without* language constitutes the non-linguistically structured experience which is *not* a person, and what happens in a brain *with* language constitutes the linguistically structured experience which *is* a person. Then 'not being aware of our inner goings on as physical' can no longer be confused with 'being aware of our inner goings on as non-physical' because the dualism of 'me' and 'my inner goings on' has been eliminated. Exit, but rather more finally, the headless wonder!

And third, 'the empirical mapping of symbolic representation'. This time the phrase is from E.O.Wilson. And since it seems to me to be exactly the phrase I need to describe the process by which the neural substrate of linguistically structured experience is likely to be revealed, I'll borrow it, with thanks. When the science which will give us such mapping is more securely in place than as yet, perhaps it is not too wild a speculation to suggest that what we might well learn thereby—an idea prompted for me after listening to a talk by Steven Pinker on such themes—is that just as the genetic code is 'a discrete combinatorial system' which provides the instructions for building brains, so the linguistic code will turn out to be 'a discrete combinatorial system' which provides the instructions for building minds. In such a context, if one considers Locke's famous 'blank slate' image, a slate is of course not blank. It has characteristics, such as that you can't write on it very effectively using a biro. But its non-blankness is in terms not of the symbol it is possible for it to carry, but of the instrument it is possible to use in that carrying. The same stuff as you can write on a paper using the biro you can write on a slate using chalk. Analogously only, the non-blankness of a brain incapable of mediating sound, and thus the spoken word, is also in terms of the instrument it is possible to use, not the symbol it is possible for it to carry: you can write the same stuff on it using sign language. In my terms, only a brain whose characteristics included an inability to process language *in any way* would be a brain which could not generate a person. For the rest, *brains* are much more like each other than they are different from each other. It is the *minds* which are generated from them, or the *minding* that they *do*, which results in persons ever more various.

* * *

And so ends my second 'experiment' chapter. I hope that in getting this far you are now convinced that, along with some unavoidable repetitions, there are also some new understandings of my position to be gained by going with me through my specimens right to their, albeit lengthy, completion.

Chapter 8

EXPERIMENT THREE

From 'natural and cultural' to 'time and space'

With eleven of my thirty-five specimens explored in Chapter 6, and ten covered in Chapter 7, that leaves fourteen to be investigated in this the last of my three 'experiment' chapters. But since, in the interests of reader-comfort, I have also arranged things so that the three chapters are of roughly the same length, that means that the entries in this chapter are, on average, shorter than those in the other two. Of course, I would claim that the reason for that is that it's what is appropriate to their importance. But I suppose it's just possible that a slight element of 'let's get through this now' has entered, even for me. I hope that to the extent to which it has also entered for you, you will persevere. This *is*, after all, the last of the trio: this *is*, after all, the last lap!

(22) natural and cultural

Even in the relatively uncontextualised domain of a general dictionary, definitions of the words 'natural' and 'cultural' already show them to be particularly difficult notions to pin down. Shifting up a gear to specialised dictionaries, to things such as Raymond Williams' heroic attempt to explore their meanings in his book *Keywords* (1976), the complications multiply. And shifting up yet again to books attempting a full contextualisation within debates which have been on-going for centuries, to books such as Kate Soper's *What Is Nature?* (1995), the full difficulty of getting a purchase upon all the baggage, historical and current, that such words carry becomes even clearer. As I suggested when introducing this 'and' in Chapter 4, perhaps this is why it has to be one contributed by me rather than found in my *Oxford Companion* source. Perhaps for the professionals it has acquired a baggage far more capacious than, for their purposes, promotes careful unpacking.

But for philosophical conversation, 'natural *and* cultural' just has to be here. No matter that the natural has to be held up against at least two different *specific* 'others' (the non-natural and the un-natural), and at least two different *general* 'others' (the cultural and the moral), the use of 'natural *and* cultural' language is so ubiquitous in philosophical conversation that it would be absurd to pass over this particular 'and' in silence. With an hypothesis centrally claiming that we need to distinguish much more carefully than is habitual between the organic (of the natural) and the personic (of the cultural), clearly this is a critical pairing for me. But even if it were not, in any text which was making some attempt to get at the way in which 'the big issues' are articulated in everyday philosophical conversation, it would still need to figure prominently. In the course of such articulation, we are notoriously prone to many confusions. We get 'nature' confused as we try to figure out what might be meant by the 'natural' of 'natural selection' vis-à-vis what might be meant by the 'nature' of 'human nature'. We get 'culture' confused when we fail to register the significance of what became a shift in the meaning of the word, a shift from the aesthetic to the anthropological, begun perhaps by Hegel and followed up by such as Bakhtin and Cassirer. And in particular, we get 'nature *and* culture' confused by the problem of how to characterise the relationship between them: in particular, whether to see Darwinism as a universal algorithm working out identically in both; a universal algorithm working out differently in each; or a non-universal algorithm working out only in nature, not at all in culture. But even if the articulation of the big issues in philosophical conversation is more usually bedevilled than illuminated by such confusions, as one of the fundamental 'ands' of such talk, 'nature *and* culture' must still be granted its 'specimen' status.

It is my belief that by far the biggest current confusion is caused by the many attempts, in various guises, to make the cultural part of the natural. Contained in such terms as 'second nature' and 'meme', it is an attempt to 'Darwinise' across the board. In *Consilience: The Unity of Knowledge* (1998), for example, E. O. Wilson has, as his explicit goal, a theory which will establish the fundamental unity of the natural sciences with the humanities. But as in his earlier *Sociobiology* (1975), and in most of the currently fashionable evolutionary psychology, it seems to me that the real aim is not a unity but a take-over. First, reduce persons to animals; then explain animals; and hey presto—'unity'! Nonetheless, the precise way in which my ideas on 'nature *and* culture' engage with such ideas on 'nature *and* culture' is complicated. I do *not* take the view that 'evolutionary psychology' is a lot of nonsense: just that it isn't psychology. And I do *not* take the view that what lifts us 'out' of our biology is our God-given (or anyway somehow-given) free-will or capacity for agency: rather that the 'us' which refers to humans as persons is not 'in' any biology in the first place. This is a position which, because it recognises the *different* referents of such tiny but massively significant little words as 'we' and 'us' and 'our', allows our so-called special place in nature to be characterised much more accurately. That of mine which is natural is different from that of mine which is cultural. But that is not because the natural and the cultural are two different things, of which the one 'I' has different proportions. Rather it is because the natural thing which is the 'not-I' and the cultural happening which is the 'I', the one *being* a thing and the other *being* a happening, are two different entities entirely. And it is because I think of this as a somewhat idiosyncratic position that I want now to try to explicate it further, via the careful 'picking over' of an alternative.

The alternative concerned is the one presented by the inimitable Dan Dennett. In *Darwin's Dangerous Idea* (1995), he suggests that 'in the same way that spiders make webs and beavers make dams, we make … books'. He then follows that up by adding that such cultural capacities so to make 'must emerge non-miraculously from the blind, mechanical processes of Darwinian mechanisms *of one sort or another*'. And insists that even though the process of getting from 'prokaryotic lineages to the mathematical investigations of Oxford dons' will take many 'cranes-on-top-of-cranes' to accomplish—this is the language of his to which I referred in my first chapter—that nonetheless is what has happened. He acknowledges that there is resistance to such a notion: a resistance which he characterises as coming from those who want to posit 'some discontinuities somewhere, some skyhooks, or moments of Special Creation, or some other sort of miracles, between the prokaryotes and the finest treasures in our libraries'. But he thinks that such resistance is just misplaced.

Nailing his provocative colours to the mast as usual, much of this seems to me to represent a welcome breath of the scientific, blowing through a baffling fog of the ideological. Nonetheless, I think I could argue the whole case that I want to make concerning my 'natural *and* cultural' specimen out of a critical analysis of this *different* case of his. So I will.

First: there is no 'same way' in which beavers make dams and we make books. Unpacking the 'we' into its three possible referents, means that we have to consider three different scenarios: we have to ask whether *human-animals* make books as beavers make dams; whether *animal-humans* make books as beavers make dams; and whether *human-humans* make books as beavers make dams. And the answer is 'no' in every case. In the first case, human-animals don't make books at all. And in the second and third cases, animal-humans and human-humans make books, but not in the same way as beavers make dams. Place some loud speakers near beaver territory. Camouflage them appropriately. Play through them the sound of running water. And watch them being plastered with mud by eager beavers. I do not make this up. This is an experiment that has actually been done.

Second: of course our cultural capacities to make books, as to make everything else that we make, 'must emerge non-miraculously'. But that only means that they 'must emerge non-miraculously from the blind, mechanical processes of Darwinian mechanisms' if it can be shown that there is no other and equally non-miraculous process from which they might emerge. But there is. And somewhat ironically, it's one which Dennett himself has done so much to illumine. In my terminology, it's the personic process whereby the collective accumulation of linguistically structured experience, the selving process, facilitates, via language, the emergence not of 'kinds of minds'—as he called one of his other books—but of 'kinds of minding'. For when the minding common to all the higher animals becomes, in just one of them, a minding structured linguistically, then 'our talents as designers', as he calls them, can emerge completely non-miraculously from the collective accumulation of the steadily growing personic: an accumulation which so far has been cognised (by animal-humans) as the given 'free-will' of an I who AM; but an accumulation which increasingly can be *re*cognised (by human-humans) as the achieved 'free-won't' of an I which IS. And *what* it is is linguistically structured experience.

It is because this is my suggestion that when Andy Clark suggests—in *Mindware* (2001)—that Dennett's notoriously-titled *Consciousness Explained* (1991) would have been better called *Consciousness Achieved*, I cheer. And to get some purchase on the spectacular consequences of that 'achievement', remember an experiment carried out with bonobos. A bonobo chimp with no training in the symbolic was presented with one collection of three sweets and one collection of five sweets. He was set the task of learning to choose the three in order to be given the five, but it was beyond him. However, when he was trained in such a way that he accumulated just the marginally symbolic of which chimps are capable, things were different. Presented instead with his *symbol* for three sweets and his *symbol* for five sweets, and set the task of learning to choose his *symbol* for three in order to be given what is represented by his *symbol* for five, he coped. Thus at the very margins of incipient personness is the planning-control function of the linguistic made manifest. Here is a process entirely 'non-miraculous'—if as yet little understood—which fits our bill very nicely. And whether or not the 'discrete combinatorial system' which is the linguistic works in ways analogous to those of the 'discrete combinatorial system' which is the genetic is an empirical question, and one primarily of the future. There is no pre-given (miraculous?!) reason why it should.

Third: of course 'resistance comes from those who think there must be some discontinuities somewhere, some skyhooks, or moments of Special Creation, or some other sort of miracles'. And of course it is a resistance thoroughly illegitimate. But resistance can also come, does also come—and this time legitimately—from those who recognise the discontinuities of emergence (see 'quantity *and* quality' below). Such folk—and I am one of them—reject (as miracles) the skyhooks supposedly made by a supposed person living vaporously in the sky, but accept (as the products of linguistically structured experience) the skyhooks actually made by actual persons living firmly on the ground. They are living firmly on the ground, but to use somebody's lovely image of kite-flying, they are 'fishing in the sky'. 'Between the prokaryotes and the finest treasures in our libraries' stands the transition from the organic to the personic. And that is a transition which doesn't merely make all the *difference* in the world. It makes the very world.

What I am arguing for is nature 'all the way down' at the level of the organic; culture 'all the way down' at the level of the personic; and humans, organic as organisms and personic as persons, 'all the way down' at both. And on a different page of the same book—on page 338 rather than page 135 of *Darwin's Dangerous Idea*—Dennett would at least appear to be doing the same. His comment there that 'virtually all the differences discernible between the people of, say, Plato's day and the people living today—their physical talents, proclivities, attitudes, prospects—must be due to cultural changes, since fewer than two hundred generations separate us from Plato' is surely one which suggests he is doing the same. So how come he does not see that other and non-biological source of the 'non-miraculous'?

Whether it is true or not in his case I have no idea. But in anyway much of the philosophical conversation with which I am familiar, it is often the claim of a political stance which can gum up the works when we are trying to think in this 'natural *and* cultural' territory. And to illustrate just what I mean by that, let me now consider what I think is a quite flagrant example of that very thing. Notwithstanding Peter Singer's attempts to argue for 'a Darwinian left', people of progressive political tendencies taking part in philosophical conversation tend to attack 'evolutionary psychology'. They attack it as reductionist. They argue that in reducing the mental to the physical, minds to bodies—in short, culture to nature—evolutionary psychologists are denying all that is special about people: their agency, their altruism, their morality, their freedom, their spirituality. And they attack it as politically right-wing. They argue that in presenting humans as inherently hierarchical, competitive, 'natural' animals, it quite wrongly presents the task of making egalitarian, co-operative and just societies as an impossible, pie-in-the-sky dream. From all of which it is supposed to follow that, for somebody as thoroughly committed to the making of equality as I am, theirs is the 'hooray' position, theirs the barricade to defend. That I think all this is just ideological nonsense, that I will not therefore defend it, is perhaps best illustrated by a look at one of the most-discussed of the many examples of research done in the 'evolutionary psychology' mould.

In the early 1990s, Devendra Singh, a then professor of psychology at the University of Texas at Austin, suggested that what men are biologically programmed to want in a female sex partner is a waist measurement that is 70% that of her hip measurement. Actual sizes are not what matters: what matters are ratios. And the reason why they matter is that this body shape gives out various messages about her suitability as a mate. It signals (1) that she is neither pre-pubertal nor menopausal—because in both these states the ratio is naturally higher; (2) that she is not pregnant with another man's child—because the waist-hip ratio goes up right at the beginning of pregnancy, in the very first month; and (3) that she does not have higher than average testosterone levels—which increase with waist-hip ratio. Whether or not these are findings with good-enough credentials on grounds of the scientific methods used to establish them I don't know. And although they sound sort of likely to me, I'm not really interested enough to want to find out. But what certainly *does* interest me is what I take to be a hopelessly partial reaction to such findings among my concerned-about-equality friends. They splutter. They protest. They attack. And as far as I can see, all for no good reason at all.

The cross-purposes to which discussions about the possibility of equality are prone are nothing if not engagingly spectacular. And the particular one at work here—yet again, for it is one in which I anyway quite regularly find myself hopelessly entangled—is the one whereby when I am arguing that *persons* (1) *ought to be* (2) *equal* (3), my protagonist takes me instead to be arguing that *organisms* (1a) *are* (2a) *identical* (3a). This is just daft. This is the (already-mentioned) mess which involves a three-way slippage such that 'persons' are read as 'organisms'; 'ought to be' is read as 'are'; and 'equal' is read as 'identical'. If Singh has shown such mate-choice to be 'natural', fine: he is making what can only be a valuable contribution to the natural sciences of the organic. It's just that the natural is neither what the cultural *is* (or can only be), nor what the cultural *ought to be* (or will be). As a human-animal (non-person), a male naturally chooses so for his human-animal (non-person) projects to flourish. *As development upon that,* as an animal-human (normal person), a male normally chooses part so, part otherwise, for his animal-human (normal person) *different* projects to flourish. *And as further development upon even that,* as a human-human (moral person), a male morally chooses totally otherwise, for his human-human (moral person) *different again* projects to flourish. That this is a positive process rather than a normative process means that what drives the change in male behaviour is a function of the more complex problems with which his selving process presents him. It is not what he 'ought' to do which changes him. It is what it 'is' for him to flourish—what he 'is' as flourisher—which changes him. See a man making his choices in the old way when in order now to flourish he needs a friend and equal rather than a brood-mare and a clothes horse, and see that which will increasingly disappear—*is* increasingly disappearing—as morality comes into its I IS own. (To put some flesh on the bones of my example, call him maybe Charles, imagine him maybe a prince, and see that which glaringly became just horribly tragic.)

But there is an even more important reason than this for we egalitarians not to pooh-pooh the work, as opposed to the pretensions, of the misnamed evolutionary 'psychologists'. Insisting that what they are up to is behavioural biology rather than any sort of psychology is important. But learning from them something about the implications of their biology (of the natural) for our sociology and psychology (of the cultural) is even more so. To make this one clear, let me go right into the lions den—to the work of the much-maligned Tooby and Cosmides—to illustrate the way in which we egalitarians can not only enter such a place with safety, but also leave it unscathed and unfazed.

Tooby and Cosmides have been seen as the founding co-parents of (so-called) evolutionary psychology. And in *The Language Instinct* (1994), Steven Pinker, a fellow-traveller, provides for a non-specialist audience a way into understanding some of the significance of their work. The issue which I want to highlight here concerns sex: specifically, the way in which the work of the pair, not as founding co-parents of a new evolutionary psychology but as continuing researchers in a long established behavioural biology, makes it perfectly clear that, in Pinker's words, 'differences between people must be minor quantitative variations, not qualitatively different designs'. The argument he presents goes something like this.

In sexual reproduction, 'the chromosomes are randomly snipped, spliced and shuffled during the formation of sex cells'. So if a female and a male *Homo sapiens* really were in some fundamental way different, their offspring would not be viable. He uses the analogy with cars. If such snipping, splicing and shuffling took place amongst the various parts of, say, a Ferrari and a jeep, then the 'offspring' of such an enterprise, even supposing it could be built which would be unlikely, would go nowhere. It would be only if 'the two designs were extremely similar to begin with' that what he calls 'the new pastiche' might function. And his conclusion from this is quite unambiguous: it is that the reason why genetic variation is so 'microscopic', why it consists in nothing much more than 'differences in the exact sequence of molecules in proteins whose overall shape and function are basically the same', is because if it were not sexual reproduction would not work. What this means is that 'thanks to natural selection, all normal people must be qualitatively the same.'

Could anything more powerfully illustrate the vital importance of identifying the two distinct levels of the organic and the personic, and keeping them distinct in one's arguments? What we egalitarians *actually* need is not to knock the role of sociobiology and evolutionary psychology (renamed 'evolutionary behavioural biology', perhaps) *at the level of the former*, but to develop the role of sociology and developmental psychology *at the level of the latter*. For apart from anything else—like the not insignificant fact that the work of such as Tooby, Cosmides et al can offer us some well validated hypotheses capable of considerably advancing our understanding of that organic which *in its own right* is important—the sorts of argument presented above surely just shouts at us findings which are exactly what a 'left' project needs: that while the biological (natural) base makes 'us' very similar, it is the psychological (cultural) superstructure which makes 'us' very different. It seems to me that a major part of the reason for this fact getting lost when philosophical conversation gets into a 'left-right' ding-dong on the nature of the relationship between 'nature *and* culture' is a mis-perception of the levels at which arguments are being mounted. Of course such mis-perception is fuelled by the self-aggrandising mis-labelling of 'evolutionary behavioural biology' as 'evolutionary psychology': on that, suffice it to say that because psyches don't evolve but instead develop, 'evolutionary psychology' is just an oxymoron. But fuelled thus or not, it is important to register that 'evolutionary behavioural biology' is not concerned with individual differences but with group samenesses. And that many of those samenesses, such as the one outlined above, offer absolutely no threat and frequently some buttress to that historical and developmental account of the personic of which we are so much in need.

In *Civilisation and its Discontents* (1929), Freud argued that in order to get the social harmony which is civilisation, people's natural appetites had to be constrained, and that such constraint represented an inhibition of personal freedom. It is a classic example of the hopeless confusions entailed in a mis-cognition of the relationship between the two members of this 'natural *and* cultural' pair. Personal freedom is of persons. And the I IS hypothesis is suggesting that persons, far from being *inhibited by* the constraint of the natural are actually *constructed from* the constraint of the natural: that is, from that linguistically structured experience which enables the emergence of a *making* (of the cultural) where previously there had only been a *selection* (of the natural). It follows from such a position that while it might on occasion be appropriate to condemn as 'speciesism' that attitude which privileges some non-human animals over other non-human animals—as, for example, when what have been called 'charismatic mega-fauna' are more readily conserved than other less-appealing non-human animals—it is only thoroughly *in*appropriate to condemn as 'speciesism' that attitude which privileges *Homo sapiens* above all the rest. *Homo sapiens* is to be privileged not as biological species but as the source of that which is not biological at all: as the source of personness. And what that means is that the way to resolve the 'nature *and* culture' debate is to see that there isn't one. There is the natural: a biosphere of species which compete to live. There is the cultural: a world of selves who co-operate to love. *Homo sapiens* is one of the former. And persons are all of the latter.

(23) natural and moral

When we smile spontaneously, naturally, the behaviour is mediated by a part of the old brain known as the limbic system. When we smile to order, as in all those 'say cheese' photographs, the action is mediated by the cerebral cortex. The romantic view is that the first sort of behaviour is 'hooray' behaviour because spontaneous and natural, the second sort of behaviour is 'boo' behaviour because planned and non-natural. But then be present when somebody tells a racist joke. One person smiles: spontaneously, naturally. Another person struggles to get control over the twitching mouth corners, and succeeds in frowning instead: thoroughly *non*-naturally. Which is the more fully human response? Because we do what we do 'naturally' just *cannot* be made coherently to mean that what we do is in any necessary sense 'good'. Indeed, I am arguing that 'doing what comes naturally' is *necessarily* doing that which is, at the very least, *a*moral. And in circumstances in which a better non-natural alternative is available, could be argued to have become *im*moral . How often do we anyway say, of some rather unpleasant piece of behaviour, that it is 'only *natural*'? When the parents of Jamie Bulger wanted his young killers to stay in jail for life, for example, we say that that is 'only *natural*'. What I am saying is that that is exactly what it is: '*only* natural'. And therefore thoroughly insufficient as a fully human response.

If the natural had been made 'on purpose' by God, I could only have every sympathy with Empson's depiction of such a deity as 'laboriously vile'. As somebody else observed, his only excuse could be that he didn't exist. For the naturally evolving species of the biosphere come into existence, live and then die, by a process so wasteful, so indifferent to pain and suffering, so mind-less and thoughtless, that it is very hard to see how anyone could ever have afforded it any *moral* worth, any *moral* value, at all. Housman was just right to refer to a Nature which could 'neither know nor care' as 'heartless' and 'witless'. Roger Scruton is just right to point out that things organic are the product of 'a process of random change and repeated disaster', which means that 'for every lovely horse, a million misfits have miserably perished'. But perhaps nobody more acutely captures both the full *wastefulness* and the full *hideousness* of the natural than

Richard Dawkins: throughout his writings, but perhaps especially in *River out of Eden* (1995). Consider (below) just three short extracts from that text: the first to illustrate the wastefulness; the other two the hideousness.

> Why are forest trees so tall? Simply to overtop rival trees. A 'sensible' utility function would see to it that they were all short. They would get exactly the same amount of sunlight, with far less expenditure on thick trunks and massive supporting buttresses. But if they were all short, natural selection couldn't help favouring a variant individual that grew a little taller. The ante having been upped, others would have to follow suit. Nothing can stop the whole game escalating until all trees are ludicrously and wastefully tall. It is ludicrous and wasteful only from the point of view of a rational economic planner thinking in terms of maximising efficiency. But it all makes sense once you understand the true utility function – genes are maximising their own survival.

> It is better for the genes of Darwin's ichneumon wasp that the caterpillar should be alive, and therefore fresh, when it is eaten, no matter what the cost in suffering. Genes don't care about suffering, because they don't care about anything.

> The total amount of suffering per year in the natural world is beyond all decent contemplation. During the minute it takes me to compose this sentence, thousands of animals are being eaten alive; others are running for their lives, whimpering with fear; others are being slowly devoured from within by rasping parasites; thousands of all kinds are dying of starvation, thirst and disease. It must be so. If there is ever a time of plenty, this very fact will automatically lead to an increase in population until the natural state of starvation and misery is restored.

These are the sorts of realities which make 'naturalistic reductionism' so appalling. For the slightest whiff of a suggestion that persons also are 'only capable of the natural' comes with its licence to add 'and thus less than capable of the moral' ready printed. From their celebration of the 'naturalness' of mothers being more involved than fathers in the daily care of infants, to their celebration of the 'naturalness' of market capitalism's survival of the fittest, the reducers are those who want to wallow and/or glory in our being 'part of Nature'. From their lamentations over the 'non-naturalness' of food-stuffs ceasing to be seasonal, to their lamentations over the 'non-naturalness' of humans ceasing to be subject to evolution, they are those who want to deride and/or lament our coming 'out of Nature'. And sadly some of the very best of folk can be found to be joining in. Both Hegel and Marx, for example, see alienation from Nature as 'a problem'. They both offer solutions, *different* solutions, to 'the problem' without appearing to register that the reason why neither solution is particularly compelling is because there isn't really 'a problem' in the first place. What's so wrong with alienation from the mindless and thoughtless? Sounds like a pretty sure sign of our maturity to me. Which is why it is the argument of this book that 'naturalistic reductionism' is fast becoming a seriously immoral stance. Let me illustrate that via just a couple of examples.

It was once a given—for the blind and callous watchmaker dictated that it should be so—that some proportion of foetuses would be damaged in such a way as to result in the development of a Down's Syndrome child. It was given by our ignorance of what caused it to be so. The only freedom then available to us, and thus the morally appropriate course of action, was to help to make the resulting stunted lives as good as they could be for the persons involved. But now it is a taken—for the work of loving and compassionate scientists and others have *made* it so—that the amniocentesis test will tell us whether or not a foetus is so damaged. The higher freedom now available to us, and thus the now new and different morally appropriate course of action, is to ensure via abortion that such foetuses do not develop into persons. The very meaning of care, of love, of the good, has been changed by knowledge. It now becomes morally unacceptable not to abort. On this issue, a fundamentalist Christian ideology becomes just immoral.

And again, it was once a given—for the blind and callous watchmaker again dictated that it should be so—that males should choose their female mates based on the waist-to-hip ratio discussed earlier. It was given by our ignorance of what caused it to be so. The only freedom then available to us, and thus the morally appropriate course of action, was to help to make the resulting stunted liaisons as good as they could be for the persons involved. But now it is a taken—for the work of loving and compassionate scientists and others have again *made* it so—that such choice is known to be based only on wordless signals as to whether the females were menstrual, or pregnant, or menopausal, each of which states alters the ratio. The higher freedom now available to us, and thus the now new and different morally appropriate course of action, is to ensure, via the spread of such knowledge and the process of altering the relationship between men and women consequent upon it, that such liaisons do not take place in the first place. Again, the very meaning of care, of love, of the good, has been changed by knowledge. It now becomes morally unacceptable to encourage the formation of relationships based on such criteria. On this issue, any part of a 'woman-as-body' ideology becomes just immoral.

With these two examples I am suggesting two things. First I am suggesting that when the natural is all that is available, then while delighting in it might be more than a little insensitive, accepting it (while mitigating its frequently dreadful consequences) is just an inevitable acceptance of the *a*-moral. But when the moral *becomes* available, continuing so to delight and accept becomes not passively *amoral* but (and increasingly) actively *immoral*. If this book is anything, it is a plea for an end to that knee-jerk defence of the natural which, when its moral alternative is available, becomes—let me repeat—not passively *amoral* but (and increasingly) actively *immoral*. But second I am also suggesting that in order to achieve an understanding of these realities, what we are in need of is a new theory of I: a theory of I which is not an 'only natural' I AM theory but a 'justly moral' I IS theory. And to illustrate this second point, let me now experiment with the insertion of my new theory of I into a 'worked example' of the 'natural *and* moral' duo to see what a difference it can make.

In 1996, a well-publicised case of a small child—I will call her child X—hit the headlines. She was terminally ill, and her grief-stricken parents were desperately trying to extend her life by a series of ever more costly treatments on both sides of the Atlantic. The discussion that arose at the time, and that in other contexts has arisen many times since, concerned the appropriateness of the parents actions. Was what they were doing a 'good' thing to do? If, experimentally, we now analyse the case using not an I AM theory of the person but an I IS theory of the person, then the question 'are the three humans X, father-of-X and mother-of-X better off as the result of X having her life extended via her treatments?' has to be seen as dependent upon a prior answer to a prior question. We have first to ask: 'better off *as what*?'. We have first to ask: 'better off *as which kind of human*?'

For a human-animal (natural non-person), life is what matters: and life is a good *which exists independently of its distribution*. For animal A to have life at the expense of animal B is a good for animal A. Which means that if I am human-animal A and you are human-animal B, it's just good that I have life, regardless of whether you do or not. The answer to the human-animal version of the question therefore is clear. If the question is 'are the three human animals X, father-of-X and mother of X better off as the result of X having her life extended via her treatments?', the answer is an unambiguous 'yes'. For an animal-human (normal person) the position changes. Life matters, certainly. But other things start to matter as well. Goodness, and how to interpret what it means, matters. Love, and how to interpret what it means, matters. To the extent to which a normal person rests in its natural animalness, to that extent it also experiences life as a good independently of its distribution. But to the extent to which a normal person stretches after her or his moral humanness—and it is precisely because we are all at some stage of our development across normal-personness towards moral-personness that the issues arose for debate in the first place—something different comes into play. Because he or she begins to search for that theory which will require of her or him that he or she takes account of other I-persons in making decisions about the good, in my terms because he or she begins to search for an I IS theory of the self, the answer to the animal-human version of the question becomes ambiguous. There is an uncomfortableness.

For a human-human (moral person) things are changed completely. When person A is not 'natural person A' but 'moral person A', then for her or him to have life at the expense of person B is not a good. Differently from non-persons and from normal persons, for a moral person there is something it is worse to be than dead. If I may be excused, for the moment, the nonsense of that oxymoronic 'being dead' phrase—when you're dead, you don't 'be'—that 'something' is being alive *at the expense of another*. For the moral person, it is not life that matters but love. And love, unlike life, is a good which far from existing independently of its distribution *is actually created by, actually made by, its distribution.*

For animals, life is all. It just is the case that 100% of X's genes, 50% of which are also those of her father, 50% of which are also those of her mother, struggle that she should live. It is only *natural*. But it is also *only* natural. And to moral persons, it is not the '*only* natural' or the '*merely* natural' that matters. If X's treatments meant, for example, that ten, twenty, thirty other persons did not get their more-likely-to-succeed treatments as a result, or if she herself suffered more than appropriately as the consequence of being treated, then while she and her father and her mother *as individual members of the species Homo sapiens* are better off because she has life, she and her father and her mother *as persons* are worse off because they have not love. I repeat my oxymoron: as a person, there is a worse thing to be than dead.

It is to get a purchase on realities such as these that we so urgently need a theory which will separate the natural and the moral for us. And not in some vague and woolly and spiritual vocabulary: for growing numbers of us, concepts such as 'the holy' and 'the sacred' will no longer cut the mustard, if ever they did. We need a theory which will effect that separation *scientifically*. And to get one, there are things we need to register, and things we need to understand.

When in *Man's Responsibility for Nature* (1980), Passmore observes that 'natural processes are entirely indifferent to our existence and welfare—not *positively* indifferent, of course, but *incapable* of caring about us'—we need to register: 'yes!'. When in *The African Queen*, Katharine Hepburn tells Humphrey Bogart that 'Nature, Mr Allnut, is what we are put in this world to rise above', we need to register: 'yes!'. When the sociologist Mauss insists that we see everyday concepts such as small and large, inanimate and animate, left and right, as in no sense 'natural' but instead as 'priceless instruments of thought which human groups have laboriously forged through the centuries', we need to register: 'yes!'. We need to understand that it is at least ironic, but better maybe just sad, that when first presented with my distinction between a natural person and a moral person, many folk, because they will wrongly read 'good' for 'natural', because they will wrongly read 'moralistic' for 'moral', will almost certainly want to claim a preference for being the first sort of person rather than the second. We need to understand the ways in which all those 'natural' metaphors, products of our I AM pasts, do nothing much more than mislead. We need to understand that no child was ever 'born' (or worse, 'unborn'), but that those that are born are foetuses, on the way to becoming neonates. Analogously, we need to understand that no nation was ever 'born' either. We need to understand that neither the *making* of childhood nor the *making* of nationhood are in any sense *natural* processes at all. And I IS theory allows us to register and understand these things: to see coming out of the natural not negatively but positively.

By helping us to see the ways in which we are moving *out of the range where mindless evolution, via careless natural selection, will be able to shape our animalness,* the geneticist Steve Jones is already playing a helpful role on this last. When young babies cease to die for genetic reasons, such as genetic variation in full-term birth weight, nature cannot select on the basis of *that* variable any more. When isolated gene pools cease to exist and cousin marriage continues to decline, situations where both parents are carriers of damaging genes causing inherited disease in their offspring are becoming fewer, making nature less able to select on the basis of *that* variable. When the mapping of the human genome makes us more able first to spot 'spelling mistakes' in genes, then to correct 'spelling mistakes' in genes, nature will become less able to select on *that* basis. And when economic development markedly diminishes the range of family size, most families are of similar size, and so no one person becomes markedly more successful than any other one person in passing on her or his genes, nature will become less able to select on *that* basis either. Splendid! But what we are now in need of is that 'social' scientists should start to help us to see the corollaries rather more clearly: that is, the ways in which we are moving *into the range where mindful development, via careful moral making, will be able to shape our humanness*. And it is my claim that to be able to come so to see, one of the most important things we need to do is to transform our attitudes towards making.

Interviewed in 1996 on Radio 3, Ernst Gombrich insisted that the tonal system in music was *discovered* not *invented*. This is a position held by many people on many things. The mathematician Roger Penrose, for example, argues that mathematics is *discovered* not *invented*. All and sundry seem to think that genius is *found* not *made*. It seems to me that all such talk represents at best a failure of nerve before the best of the human, at worst a positive alienation from the best of the human. It is as if such folk feel that by saying something is discovered (*found* by us) rather than invented (*made* by us) they somehow enhance its significance. My argument is that that is nonsense. My argument is that it is precisely because things *made* by us are capable of the moral (being *good*) while things *found* by us are capable of only the natural (just *being*), that it is exactly the opposite which is the case. For humans, what naturally there is *is* as the result of evolution: what morally there will be *becomes* as the result of development.

One way of expressing the difference is to say that where the natural process which is *evolution* proceeds via the exhibition of variation, followed by the submergence (by natural selection) of all but the fit, the moral process which is *development* proceeds via the inhibition of variation, followed by the emergence (by moral creation) of only the good. But for now, let me present that case more simply by suggesting that the whole development from tradition to modernity can be centrally captured by thinking it in terms of this shift from discovering (or *finding*) to inventing (or *making*).

Tradition stresses that which we cannot make; that which has made us; that which has been given. It stresses the natural, and its conservation. Traditional attitudes to knowledge discover it in truths sacred and essential. The sources of such truths are fundamentally mysterious. The revelation which makes for us the sacred text is a mystery: the genius which makes for us the essential canon is a mystery. Mysteries cannot and must not be understood. We can only stand before them in wonder: awe-struck and worshipful. We cannot make revelation: we cannot make genius. We can only try to ensure that, when they occur, they continue to be religiously preserved. So that that which is not known can remain what it should be: mysterious. Ignorance can be bliss.

Modernity, by contrast, stresses that which we can make: that which we are making; that which can be taken. It stresses the moral, and its development. Modern attitudes to knowledge invent it in validities mutable and provisional. The sources of such validities are fundamentally transparent. The formulation with which we make the mutable hypothesis is transparent: the construction with which we make the provisional theory is transparent. Transparencies can and must be understood. We must ever stand before them in enquiry: curious and interrogative. We make formulations: we make constructions. We must try to ensure that, when we make them, they continue to be scientifically questioned. So that that which is not known can become what it can be: transparent. Ignorance can only ever be ignorance.

And it is fear of this making, a fear which is essentially Fromm's 'fear of freedom', which means that as the I AM God has died, all manner of nonsenses have sprung up to fill the 'God-shaped hole': nonsenses such as the ideology of environmentalism (as opposed to the science of ecology), which at precisely the moment when what we most need to understand is the urgent need to *make* that which is *fully moral*, as opposed to *rest in* that which is *merely natural*, comes along as a spectacularly dangerous and reactionary 'religion of the natural' to try to hinder our progress. For when it is a confident *making* which is needed, ultimately as I will argue a confident making of a new idea, immune to a relativizing critique, it is both dangerous and reactionary to hinder that process by appeals to the merely natural. In the transition from the merely natural to the justly moral—and it is as such a developmental sequence that the 'natural *and* moral' pair needs to be understood—the greater freedom of greater knowledge first *makes* possible, and eventually *makes* mandatory, the *making* of the good.

(24) natural and social

To get at that which, under this head, is important to my hypothesis, I must first—and not before time—attempt some terminological sortings. As I have already pointed out, 'natural' has several candidate alternatives: among them, 'non-natural', 'unnatural', 'cultural', 'moral'—and now 'social'. In coming to consider this last, to consider 'natural' as it relates to 'social', I think it would be appropriate that, as a kind of summary of my approach to all of them, I should try to be as clear as possible about the way I am using what is potentially weasel vocabulary. Such clarity is vital not only to these three 'natural *and* something' entries, but also to much of the rest of my defence of the I IS hypothesis. And I think I can best try to achieve it this way: first, by presenting a table summary of the language I am using to think these things; second, by commenting upon what I take to be its significance to my thesis. So first: the language.

level	*stage*	*category*	*collectives*	*singletons*	*orientation*	*type of person*
2	4	natural	group	creature	living	not-I non-person
3	5	normal	society	individual	having	I AM first person
3	6	moral	community	soul	loving	I IS just person

Table 4: Summary of the vocabulary of my hypothesis as it pertains to 'the social'

By way of explication of the above, you will perhaps remember that *level* refers to the three chronologically emergent levels of reality (universe, biosphere and world): though in this table, only levels 2 and 3 (those at which the two *levels* of humans—as animals and as persons—are respectively to be found) are represented. And *stage* refers to the six stages of emergence across the three levels of reality: with in this table, only stages 4, 5 and 6 (those at which the three *stages* of humans—as human-animal non-persons, as animal-human normal persons, and as human-human moral persons—respectively emerge) being shown. Both *levels* and *stages* are necessary to my thinking because fully to capture the development of humankind I have found it necessary to have a classificatory system which respects both a 'twoness' about that development (from animals to persons) *and* a 'threeness' about that development (from human-animal non-persons, to animal-human normal persons, to human-human moral persons).

About one of the words I am using, I have to say more. Although I have used it a couple of times already, each time without comment, I have been extremely nervous about using the word 'soul' to delineate a singleton at level 3 stage 6. I have wondered about whether it perhaps now carries just too much religious baggage to be rescued for what are my totally secular purposes. But it is a good word, even a beautiful word. It fits well with my designation of persons as immaterial: that is, as (completely non-mysterious) 'happenings' in time rather than 'material things' in space. It fits well with my designation of persons as immortal: that is, as (completely non-mysterious) 'experience' passed on linguistically from generation to generation, and never lost. And it fits well with my designation of the moral as that which, in superseding the natural, is the supernatural. In short: it just fits well. And so I'll try to 'reclaim' it. In the same spirit as that of the Women's Movement campaign to encourage women to 'reclaim' the night; in the same spirit as that of the Gay Movement's choice to reclaim' pink triangles—alongside the yellow stars of the Jews, they had been what the Nazis had required them to wear—as their badges of pride: I'll risk it. When I want to refer to singletons who have made it to I IS understandings—as yet of course, anyway completely, none of us: already of course, anyway to some extent, many of us—I will see if I can get away with 'reclaiming' the word 'soul'.

Returning then to the table, what I hope will be immediately obvious is that 'the social' is to be found everywhere: that is, at each of the two levels, and at each of the three stages. At level 2, stage 4, there are non-person not-I creatures living in natural groups. And such groups are social in the way that the dominance hierarchies and pecking orders of ants and hens and monkeys are social: they are *naturally* social. At level 3, stage 5, there are first-person I AM individuals, having selves in normal societies. And such societies are social in the way that the castes and classes and meritocracies and nations and cricket clubs and world banks are social: they are *normally* social. And at level 3, stage 6, there are just-person I IS souls, loving selves in moral communities. Such communities are social in the way that friendship is social: they are *morally* social.

What is of course different is what is *involved* in 'the social' at each level and stage. The natural sociality of level 2 stage 4, that found in many animals including *Homo sapiens*, is a totally biological phenomenon: a gene-driven evolutionary device structuring totally natural hierarchies. The normal sociality of level 3 stage 5, that found in human persons in the earlier stages of their selving process, is one whereby social norms are traced timidly over natural models to give nature-validated values of various types of inequality. And the moral sociality of level 3, stage 6, that found in human persons in the final stage of their selving process, is one whereby the totally non-natural moral value of equality comes to structure human-human moral relations. What this means is that, with some type of 'the social' present at both levels and at all three stages, a 'natural *and* social' pairing clearly will not do. It is incoherent. Reality as I hypothesise it cannot be carved at any obvious joints by such an 'and', and so it must go—as redundant. As with it, of course, must go the notion that there will be something called 'the social sciences' as some sort of partner to the natural variety. Given that the social is present in both natural and non-natural places, a categorisation of our knowledges which counterposes 'natural sciences' and 'social sciences' produces nothing more than confusion. And as I now hope to show, a damaging confusion to boot.

Consider, as example, the London School of Economics and Political Science. When William Beveridge, as director of the school, created a chair of 'social biology', and appointed Lancelot Hogben, friend of the school's founders Beatrice and Sydney Webb, to be its first occupant, his was a thoroughly *political* act. Of course. For his was a thoroughly *political* institution. 'Social biology' was supposed to link the social (as in 'social sciences', meaning 'human sciences') with the biological (as in 'natural science'). And by that appointment, the social scientific institution was committing itself to the appropriateness of the idea of searching for social science observations, hypotheses, experiments and theories, in a natural science source. With the Centre for the Philosophy of the Natural and Social Sciences at the current LSE being one of the more important British sources of the recently fashionable 'evolutionary psychology', it is a tradition still very much alive. The trouble is that that which was a thoroughly political act did not take its politics half-way seriously enough. Dominance hierarchies and pecking orders are found (when they *are* found) in the naturally social groups of animals: it is appropriate to look for them (or their absence) in, for example, ant colonies and elephant herds. But democracy and equality are found (when they *are* found) in the normally social societies or the morally social communities of persons: it is appropriate to look for *them* (or their absence) in, for example, Mondragon co-operatives and National Health Services. In short: social groups, social societies and social communities are different entities: they are social in different ways. And it will be when one of the world's leading social science institutions becomes a place to study not groups of the naturally social but societies and communities of the normally and morally social that it will have become a home for the moral sciences of the personic. To do that, it is my claim that it will need to have been permeated by a new understanding of what 'I' designates, from which a new understanding of what 'we' and 'us' and 'them' designate will follow. Only then will it be able to become a truly scientific institution.

(25) needs and wants

Philosophical conversation concerning 'needs *and* wants' is usually conducted in terms of what I *need* being necessary to my very survival and what I *want* being in some sense rather less vital. As has been the case with each of the other 'ands' I have considered, it seems to me that such conversation would do better to begin by establishing just what sort of an 'I' we are talking about. For fairly obviously, my hypothesis would contend that since there are three different types of I – the non-I of a human-animal, the I AM of an animal-human and the I IS of a human-human – there are going to be three different types of requirement. As examples only: a human-animal requires food and water; an animal-human requires recognition and status; and a human-human requires equality and community. The real question then becomes that of determining whether those different sorts of requirement are best understood as 'needs' or as 'wants': a task which would for the most part come down to a consideration of how best to make the two concepts operational, largely perhaps for the purposes of research. As I have now repeatedly spelled out, in my terms the human has both a twoness and a threeness about it: a twoness of levels (animals and persons); and a threeness of stages (human-animals, animal-humans and human-humans). How then are 'needs *and* wants' to be referred to these levels and stages? Do they make (complete) reference to the twoness, such that needs are of one level and wants of another? Do they make (incomplete) reference to the threeness, such that needs are of one stage, wants are of another, and something-as-yet-unnamed is of the third? Or do they readily impinge on my categories hardly at all, undermining perhaps thus, if only somewhat, their very legitimacy?

I want to suggest that it is the first which is the case: that living species which *evolve* do so to acquire what they *need*; and that having and loving selves which *develop* do so to acquire what they *want*. This is a differentiation which can usefully be illustrated by considering an oft-quoted way of distinguishing 'training' from 'education': while persons could be 'trained' to use a gas chamber to exterminate concentration camp inmates, it is said, they could not be 'educated' so to do. And the difference which I am proposing between the 'needs' and the 'wants' of the 'needs *and* wants' pair is that which postulates the 'wants' as being able to be educated, but the 'needs' as being able only to be trained. Thus where the needs of a human-animal non-person for, say, nutrients can be trained so as to be satisfied in different ways, the wants of an animal-human normal person, as it selves its way towards becoming instead a human-human moral person, can increasingly be educated, indeed *are* increasingly educated, from being 'having wants' to being 'loving wants'. It is Habermas who suggests that a rational desire is one we would continue to feel even if the limitations of knowledge which we experienced when forming it were to be removed. If I think of his 'rational desire' as corresponding with my 'loving want', his 'removal of the limitations of knowledge' as corresponding with my 'selving from I AM to I IS', then what I am saying is something similar. I am suggesting that while the more limited understandings of I AM persons are manifested in a preponderance amongst them of the *having* wants, the more complete understandings of I IS persons are manifested in a preponderance amongst them of the *loving* wants. And this means that while, in terms of my hypothesis, the 'needs *and* wants' pair are indeed to be viewed developmentally, needs *are* prior to wants, they are in no way to be conceptualised in such a way that we can talk of needs being for essentials and wants being for optional extras. Such a conceptualisation is often grounded in the Aristotelian notion of 'flourishing': needing something is to be distinguished from wanting something by how necessary-to-flourishing it is that that 'something' be acquired. If it is essential, that is a need. If it is not essential, that is a want. But in the terms of my hypothesis, this is not the case. In my terms, the loving wants of human-humans are as necessary to the flourishing *of those human-humans* as the having wants of animal-humans are necessary to the flourishing *of those animal-humans* as the needs of human-animals are necessary to the flourishing *of those human-animals*. In other words, needs are to be distinguished from wants not by how necessary-to-flourishing is their satisfaction but by *whose* flourishing it is to which they are necessary.

The great advantage of this way of thinking is that it avoids a hopelessly incoherent objective-subjective distinction, further explored under the head of my next specimen, whereby we are to think in terms of *objective* human needs and *subjective* human wants. This is a distinction which regularly produces and re-produces the hopelessly incoherent notion of there being a 'human nature' or even a 'naturalistic ethics'. And recognising the human as having its three different references—references which mean that not-I natural not-persons are characterised by their *natural needs*, I AM normal persons by their *having wants*, and I IS moral persons by their *loving wants*—saves us from such nonsenses. *All* needing and *all* wanting is definitionally subjective: in the sense that it emanates from a needing and/or wanting 'subject', how could it be other than 'subjective'? But so is *all* needing and *all* wanting definitionally objective: in the sense that it is directed to a needed and/or wanted 'object', how again could it be otherwise? Of course it is true that the extent to which what is needed or wanted is *appropriately* needed or wanted—that is, the extent to which the current 'subject' is a good or a bad hypothesis as to the nature of the biosphere or world in which it finds itself—will determine the extent to which said current 'subject' flourishes. But it is also true that 'said current subject' is not *a* human but one or other of three *different* humans, for each of which 'flourishing' has a *different* meaning. This is a conception of the 'needs *and* wants' pair which, not before time, allows us to stop trying to make sense of concepts such as 'human nature' and 'naturalistic ethics'. And that is a contribution not at all inconsiderable to the project of making the moral sciences.

(26) objectivism and subjectivism

This is a truly troublesome duo, if only because it is so difficult to pin down just exactly which of the several debates to which it *can* refer is the one to which, in any particular philosophical conversation, it actually *does* refer. Indeed, it is sometimes in the course of a single sentence that one of the distinctions it tries to label slips and slides into another of the distinctions it tries to label without anybody noticing that it's happened. As we talk about this 'and', we talk about many and various things: about how 'objective' biological facts about brains can produce 'subjective' experience; about 'objective' judgements as those made about objects (and thus verifiable?), and 'subjective' judgements as those made by subjects (and thus not?); about the relativism which claims that everything is subjective, nothing objective; about what might be the significance of adding the 'ism' to the two words; and sometimes about all these things (and more) at once.

So let me begin by making it quite clear what of all this will be concerning me here. I am going to claim that, as far as philosophical conversation is concerned, the 'isms' are more or less irrelevant, and can be safely ignored. For such conversation, it is how we are to characterise the duo 'objective *and* subjective' which matters. And on that one, by far and away the most common reading of their relationship is to think of 'the objective' as that which just *is* the case, whether any particular knower knows that it is or not: and 'the subjective' as that which is only *thought to be* the case, by any one such knower.

I want to offer a different reading. It is a reading which involves the rejection of the idea that the emergence of the human mind, or better of human minding, was the emergence of the human 'subjective' into an 'objective' already-there world. In other words, it involves the rejection of the formulation whereby there was what there was (the objective), and then there emerged that which could *know* what there was—if only 'subjectively'. Instead, I want to suggest a reading whereby mind(ing) emerged from non-mind(ing) not as subjectivity emerging from objectivity, but as subjectivity and objectivity *both* emerging from the prior existence of *neither*.

It is a reading which is grounded in the differences between the three levels of reality with which I am working. There was a reality which was a universe 'already there' before there was human minding. And later, there was a reality which was a universe and a biosphere 'already there' before there was human minding. It's just that there was not a reality which was a world 'already there' before there was human minding. And both subjectivity and objectivity, each of them aspects not of the universe, not of the biosphere but of the world, come into that world as co-emergent twins, both the offspring of human minding. This is a formulation which suggests that just as all the knowing done by persons is 'subjective' (knowing by a subject) so all the knowing done by persons is objective'(knowing of an object). But what I am adding, and crucially, is that our protection against the relativism that usually shouts from such a formulation is to be found in examining which type of person, and thus which type of human minding, we are talking about.

As a consequence of being persons in the earlier stages of the selving process, those who thereby are I AM persons have been defended against relativism by the making of one great all *knowing* all *powerful* and all *good* monotheistic God: that is, by the making of I AM THAT I AM. Such a great I AM, sitting as it does in its 'great tautology', made the world from outside the world, and provided thus that view from nowhere which could ground, for believers, their definitions of the good, the true, the beautiful—and any other absolutes that might be needed. It is when (as now) such a creation is inexorably becoming more and more incredible, that then (as now) the fog which is relativism creeps over the intellectual landscape, and the view from nowhere is steadily removed. But hold. There is a way out of this enveloping gloom.

For as a consequence of becoming persons in the last stage of the selving process, those who thereby have become I IS persons can be differently defended against relativism. They can be defended by making one great *all* knowing *all* powerful *all* good I IS THAT I IS: that is, by the envisioning and then the making of a world consisting not of an I AM first person plus others who are second and third, but of an I IS just person plus another I IS just person plus another I IS just person—up to that very 'all'. Such a just I IS makes the world from inside the world, and provides *in the guise of the view from everywhere* a new view from nowhere. It is a view which will ground, but so much more securely, new definitions of the good, the true and the beautiful—and any other absolutes that may be needed. And it is a view which can be brought into focus, which will be brought into focus, only as a consequence of the making of equality. That crucial transition from one 'all *knowing*' to '*all* knowing', from one 'all *powerful*' to '*all* powerful', from one 'all *good*' to '*all* good' can be one, will be one, which because it shifts us from the interests of the interested to the interests of the disinterested allows us to make a qualitatively different level of 'the objective'.

Fully to understand the emergence of this new level of 'the objective', it is first important appropriately to have characterised the old one. In particular, it is first important to realise that we have always *made* the objective. It was never given (data): it was always taken (capta). The reality of 'the universe' and 'the biosphere' was given. But 'the objective' has always been made by us. What this means is that it has always been not *what* there is, but what some 'we' *knows* about what there is. In other words, it has always been the *inter*-subjective. So before we can understand the new (intersubjective) 'objective', we must first be clear about the old (intersubjective) 'objective'. And it's the full significance of that 'inter' that we need to get hold of.

The 'inter' of the old 'intersubjective objective' was a 'between' linking unequal subjects: linking I AM persons in various stages of the selving process operating in myriad master-slave, primary-secondary relationships of inequality. It resulted thus, and necessarily, in an 'intersubjective objective' which was partial (as opposed both to whole and impartial) and interested (as opposed to disinterested). An 'intersubjective objective', thus limited, has nonetheless been of some value. It has been good enough to get us *considerable* purchase on a scientific understanding of 'the universe', in the study of which the interfering partialities and interests of normal persons are of *least* significance. It has even been good enough to get us *some* purchase on a scientific understanding of 'the biosphere', in the study of which such interfering partialities are of *rather more* significance. But it has been hardly good enough at all to get us *any* purchase on a scientific understanding of 'the world', in the study of which they are *maximally* significant and *critically* disabling.

There is what can perhaps be a useful analogy available here. What I am saying is that just as for more limited purposes (up to and including getting a rocket launched) Newton's more partial theories will for the most part do, so for more limited purposes (up to and including getting the inorganic and the organic sciences launched) I AM theories of the person will for the most part do as well. But when it comes to the bigger picture, specifically when it comes to getting our very 'selves' actually *into* the picture, then just as Newton is valid only 'locally' (at a particular place) and has to give way to Einstein to explain more generally, so I AM conceptions of the person are valid only 'momentarily' (at a particular time) and will have to give way to I IS ones for the more all-embracing explanations of the nature of persons.

Which explanations just possibly start to appear over the horizon. Their appearance is just possible because the more sophisticated I AM knower is now becoming thoroughly baffled, and flays about looking for enlightenment. The difficulty such a knower is in is that by positing itself as in some way 'over against' and 'outside' that which it knows, it creates for itself the totally insoluble (but happily totally imaginary) problem of how it can possibly know 'the thing in itself', expressed often as how it can possibly be 'objective'. After all, it can only know through *its* history, *its* sense organs, *its* biases, *its* capacities (or lack of them). But an I IS knower—and my hypothesis would suggest that, though still few, the number of I IS knowers will definitely be on the increase—is one who avoids these mistakes by coming to appreciate that the 'it' which is 'knowing' in the above sentences is the same sort of 'it' as is contained in the sentence 'it is raining'. Just as the one 'it' means 'raining is happening', so the other 'it' means 'knowing is happening'. It's not that an I AM *has* its history, its sensations, its biases, its capacities, or lack of them—and can only know 'through' them. It's that an I IS *is* its history, its sensations, its biases, its capacities or lack of them—and there isn't any 'through' to worry about at all.

And with no 'through' to worry about, the fog can lift. For as long as the interests involved are those of the disinterested—as will be the case in conditions of equality—then a qualitatively new level of 'the objective' will be able to be intersubjectively made. Remembering of course that *all* levels of 'the objective', including the spectacularly successful one involving the I AM God with his view from nowhere, have been inter-subjectively made—invented, that is, not discovered—of course it will not be the *fact* of its intersubjectivity which will make it different. Rather, it will be the *type* of its intersubjectivity which will make it different. For the first time ever, the intersubjectively made 'objective' will have been made by IS equal subjects rather than I AM unequal subjects. And given that they will be those who have the interests of the disinterested, *their* creation will be able to be 'objective'—and about the world too—in a qualitatively different way. It has been Habermas, of course, who has devoted a working lifetime to developing the case for modernity as an unfinished project: an unfinished project whose main task has been, and is, to find some way of grounding our claims about what is to be considered true—or rather, valid—in the *ob*jective inter-*sub*jective. Here, it seems to me, is the way in which it can be done. And its doing will mean that, for the first time ever, a *science* of the good (moral science) as opposed to various *ideologies* of the good (moral philosophies) will become available.

I suggested above that the I IS hypothesis would predict that, though few, the number of I IS knowers will be growing. To end my consideration of this 'objective *and* subjective' specimen, let me illustrate that point by turning to the work of just one psychologist.

In a brief piece which he wrote for the *Journal of Consciousness Studies* (Vol. 6, No. 2-3, 1999), Max Velmans, of the Department of Psychology at Goldsmiths College London, began by suggesting that 'the study of consciousness in modern science is hampered by deeply ingrained dualist presuppositions about the nature of consciousness'. He went on to insist that while 'phenomena can be objective in the sense of inter-subjective, investigators can be objective in the sense of truthful or dispassionate, and procedures can be objective in the sense of being well-specified', nonetheless 'observed phenomena cannot be objective in the sense of being observer-free'. Of course not: observation is that which is done by an observer. So in what sense then *can* 'observed phenomena' be objective? They can be objective, he holds, 'to the extent that observed entities and events are subject to similar perceptual and cognitive processing in different human beings': a 'commonality', as he calls it, which means that 'while each experience remains private', it may be that it is also 'a private experience that others share'.

'A private experience that others share': that's a nice image. And even when I take things further, when I suggest that his 'similar perceptual and cognitive processing' is going to need to be similar not only to the extent to which the organic entities called *Homo sapienses* are similar but also to the extent to which the personic entities called *persons* are similar too, it would seem that Velmans thinks so as well. He writes:

> *How* different observers establish intersubjectivity through negotiating agreed descriptions of shared experiences is a complex process that we do not need to examine here. Suffice it to say that it involves far more than shared experience. One also needs a shared language, shared cognitive structures, a shared world-view or scientific paradigm, shared training and expertise and so on. To the extent that an experience or observation can be *generally* shared (by a community of observers), it can form part of the data base of a communal science.

Behold, the emerging of I IS consciousness! For my money, all that needs adding to the above is that the absolutely fundamental dissimilarity between persons, the dissimilarity which more than any other vitiates not the 'search for' objectivity (for it is not something which can be discovered) but the 'making of ' objectivity (for it is instead something which must be invented), is the dissimilarity expressed in inequality. *This* is the dissimilarity which most hinders the making of objectivity: even in the physical sciences; considerably in the organic sciences; and almost completely in the therefore almost non-existent personic sciences.

But things change. For one thing that changes, the 'shared experience' which Velmans speaks of is rapidly on the increase, and world-wide. As only a tiny example of that, it was just such an increase that struck me as I watched the eclipse of the sun which took place in August, 1999. As a consequence of new technologies, almost all of which are vastly increasing the very possibility of 'shared experience', millions of people along the path of totality in Europe, the Middle East and Asia experienced at the same time, *and were aware that they experienced at the same time*, a single 'global' event. In addition, millions more, away from the path but aware of the event, experienced it at the same time via television. These are the sorts of experiences, experiences unknown before the advent of such technology, which can promote in us an absolutely vital understanding. By allowing us to experience together, maybe particularly by allowing us to experience *that which is other*, together with *those who, in experiencing it with us are not*, they can only make us more aware of the ways in which we are kin. The example, to repeat, is tiny. But by hastening the emergence of what will become a steadily increasing dissatisfaction with the disgusting inequalities which stunt the development of our full humanity, all such developments can only help to weld that consensus which, in time, will end them.

To some not inconsiderable extent, the physical sciences of the inorganic and the biological sciences of the organic know already something of that sort of consensus. Perhaps it is even already true, for example, that a scientific practice thoroughly based on consensus within a scientific community makes any modern equivalent of the anyway *simple* and *gross* mistakes of the past—phlogiston maybe, or phrenology—just no longer possible. The physical sciences (as opposed to the use to which some of their understandings are put) seem certainly quite remarkably resilient to the encroachment of 'interest'. And if I am right in putting this down to the fact that we all already stand so much more equal in the face of their subject matter than we do in the face of the potential subject matter of the potential moral sciences, then let them to that extent stand as role models for a new form of knowing. In making for us a qualitatively new level of 'the intersubjective objective', the 'negotiated co-ordination'—I borrow the phrase from the Manchester economist Pat Devine—of the ideas of equals will allow us so to re-conceptualise things that an additive pairing of 'objective(ism) *and* subjective(ism)' will become redundant because incoherent.

(27) praise and blame

Because praising and blaming involves taking attitudes towards the good and the bad—praise being given for the former, blame for the latter—my consideration of this pair must be very much of a piece with my earlier consideration of the 'good *and* evil' duo. And just as my approach there was to shed both 'evil' and its 'bad' alternative in favour of 'not yet good', so my approach here will be to shed 'blame' and replace it with 'not yet praise'. The *form* of the argument for doing so in this case is identical to the *form* of the argument for doing so in the previous case. Its *content*, nonetheless, needs its own and separate rehearsal.

For me it is still the work of B. F. Skinner which has made the biggest single contribution to my understanding of these issues. As I have earlier made clear, it seems to me that his experimental work, together with just masses and masses of everyday evidence from child-rearing practices and any other 'behavioural shaping' one might want to consider, demonstrates quite unambiguously that while being reinforced ('praised') for what we do *will* make it *more* likely that we will do it again, being punished ('blamed') for what we do will *not* make it *less* likely that we will do it again. It therefore seems to me that the challenge facing my hypothesis here is to see if it can provide a sound theoretical underpinning to what I take to have been already clearly demonstrated to be the case empirically. In other words, given that the essential asymmetry of 'praise *and* blame', coterminous with the essential asymmetry of 'good *and* evil (alias 'good *and* bad') which I have already considered, is now well understood as a reality, my experiment here must be to see if my I AM and I IS characterisations can validate that reality theoretically.

Let us consider – by way of introduction – some of those we are encouraged to blame and punish. Bob Johnson worked as a psychiatrist in Parkhurst, a top security prison on the Isle of Wight. In July 1991 he began to work with the 12 inmates of C Wing: all of them violent; many of them multiple killers; all of them diagnosed mentally ill. Who were these people? As a small boy, Des had grown up in a regime of ceaseless, almost casual violence, and was still terrified of the mother who had inflicted it. Geoff had a father who beat him and a mother who abandoned him. A young schizophrenic was terrified by visions of his father appearing in his cell: the father who, when he was eleven, had tied him to the staircase and thrashed him with a belt because he thought, wrongly, that he had taken the car keys. There was a father who, when walking across a field with his son, told the boy to keep walking while he disappeared into the trees to take pot shots at him with his air rifle. There was a mother who was not there when a small boy woke up in the morning, and when he knocked on a neighbour's door to ask if they had seen her, heard her say from inside the house, 'tell him you don't know where I am': and he never saw her again. There was a father who tied a rope round his son's ankles and dangled him upside down over the side of a bridge. These were their sons: this is who they were. These are the people our system is out to 'deter': a notion about which Johnson, in an article in *The Guardian* (5 March 1994) is positively scathing:

> Deterrence makes the assumption that people are behaving rationally, and they're not. These people may be offenders, but they are also victims – victims all the way through. They are desperate. Children are not born evil. You hit them and reject them and terrify them and make sure they know that the world is a terrifying place. You train them to be anti-social. – These people have been battered and abused all through their childhood, and then the prison system comes along and carries on battering and abusing them. I want to stop punishing them and help them.

And he did. By talking to them about the parents they so feared; by treating them as equals; by reducing and reducing and reducing their medication whenever he could; by understanding that an important source of much mental illness is to be found in childhood terror of abusive and violent parents: he did.

Now one of the main issues in the field of child abuse—physical, psychological or sexual—is about how to regard the abuser. Is 'abuse' a specific syndrome, making 'abusers' specific sufferers from 'the condition'? Or is there just a continuum along which is ranged the population at large: from the never-abusive-in-any-way to the frequently-abusive-in-several-ways? Not surprisingly, the 'syndrome' position has tended to be that of psychiatrists and paediatricians: the 'continuum' position that of psychologists. And it seems to me that it has been primarily the work of the former, working within 'the disease model', which has fed the image of such people which is favoured by the tabloid press: that they are 'monsters of depravity'. Psychologists, by contrast, have tended to see different levels of abuse, across all social classes and types of personality, with greatly various consequences. They have been the ones from whom most frequently has come the recommendation of education for parenthood, particularly for boys. My point here is that by working with the 'syndrome' model, the model in which abusers are one sort of person and 'we' are another, we actually encourage blame. It is so easy for us, even as we feel not quite right about blaming these dreadfully abused men for the terrible acts of violence they have perpetrated, to fall instead into the position of blaming the parents who first abused them. And to do so sometimes with increased vehemence precisely *because* we feel so uneasy. Working by contrast with the continuum model in which, as Maslow put it, our own so-called 'normal' behaviour is so often 'the psychopathology of the average, so undramatic and so widely spread that we do not even notice it ordinarily', maybe we can more easily be helped to understand that blame is not just ridiculous but also immoral.

The making of that which is praiseworthy (good) is essentially an inhibitory process. What has to be inhibited is not that which is blameworthy (bad) but that which is not yet praiseworthy (not yet good). The total inappropriateness of punishment seems to me to rest in the total inappropriateness of a blaming stance: a stance revealed to *be* inappropriate just as soon as we understand an I not as an I AM which *has* experience but as an I IS which *is* experience. Geoff doesn't *have* the experience which put him in Parkhurst, Geoff *is* (was) the experience which was put in Parkhurst: just as his father *is* (was) the experience which beat his son; and his mother *is* (was) the experience which abandoned her child. The past tense in brackets is vital to the sense in which what any of them were then *cannot be* what any of them are now. As Bob Johnson knew full well, what they are at any particular 'now', distanced as it has to be, either more or less, from any particular 'then', will depend on the experience sedimented down as their person-ness in the interim.

By the I AM dualist we are invited both to praise and to blame. The agent free to take both good and bad decisions should be praised for the former and blamed for the latter. Thus the agent who, by choosing the non-violent form of resistance to aggression defuses that aggression, perhaps by refusing to the aggressor some apparent justification for her or his action, is to be praised for avoiding the 'violence' which might otherwise have occurred. And the agent who, like all those Parkhurst men and the parents who abused them as children, chooses instead to interfere with the body of another against the will of that other is to be blamed for perpetrating the 'violence' of which such is the standard definition. Because they see good and bad as equal and opposite, I AM selves can both praise and blame. But it is my case that in doing the second, they undermine, albeit often unwittingly, the very good being done by the first.

I IS selves, by contrast, because they see not 'good *and* bad' but 'pre-good *and* good', can praise the emergence of the one from out the other simply as achievement. But just as surely as they understand that such praising of the good which is a *reinforcing* attention helps that burgeoning good to become established, so they also understand that an alternative blaming of the bad which is a *punishing* attention helps that burgeoning bad to become established too. They are both, after all, examples of attention. One might expect them to function equivalently. And they do.

It is frequently suggested that determinism, in removing the concept of responsibility, invalidates all systems of punishment. It does. It is also frequently suggested that determinism, in removing the concept of responsibility, invalidates all concepts of morality It doesn't. Indeed, by removing the blight of blame and replacing it with the struggle first to understand, then to change, far from *invalidating* all concepts of morality it makes available for the first time a sophisticated understanding of what it is.

For blame is a blight, and difficult indeed to eradicate. I first became fully aware of this when I registered that I had frequently encountered a 'move' in argument which had me baffled. I would have been suggesting that one course of action is better than another—that under torture it is better not to betray beliefs than to betray them, for example—and I would have inexplicably found my opponent accusing me of blaming the person who so betrays. One form of this challenge was to suggest that since we none of us knew whether we would betray under such circumstances (true) it therefore followed that we none of us had the right to label non-betrayal as better than betrayal (a total non-sequitur). Strictly, such an argument just isn't one. But it appeared so frequently that I had to begin to wonder why. And it took me quite some time to realise that although, as often as not, my partner in argument 'knew' what my attitude to blame was, 'knew' that I thought it always inappropriate, this was nonetheless a stance so very far away from what he or she took to be just 'obvious' that he or she just could not remember to take it into account when assessing what I said. It was as though he or she had *listened* to what I said but could not *hear* it: something I came later to think was often a sign of there being a (fish-not-seeing-water) ideology at work.

For of course I was not arguing that the person who betrays should be *blamed* for so doing. How could I be? When I am also arguing that it is never appropriate to blame anybody for anything, how could I be? Instead, what I *am* arguing for is its being better—that is, its being a further emergence of the good (which can be praised) from out the pre-good (which cannot be blamed)—not to betray than to betray. Given that the only notion of free-will which it seems to me morally acceptable to defend is the previously discussed and slowly *emergent* free-wont version, then praising first the successful inhibition of a 'natural' (human-animal) behaviour in favour of a more highly developed 'normal' (animal-human) one, and then the further inhibition of that 'normal' (animal-human) behaviour in favour of a yet more highly developed 'moral' (human-human) one, is a recognition that as persons *we can only do what we can do*. At bottom, it is because persons are entities existing in irreversible time, because once an I *is* what an I *is* it can't re-become what it *was*, that blaming is so irrational. Concerning this 'praise *and* blame' pair then, my claim is not just the empirical one that, as a way of changing behaviour, blaming and punishing *doesn't* work, but also the theoretical one that, as a way of changing behaviour, blaming and punishing *couldn't* work. It couldn't work because, yet again, it is based in a faulty I AM conception of what a person actually is.

(28) public and private

Although there can fairly easily be shown to be family resemblances between them, the various debates within philosophy to which this pair might be thought to refer are nonetheless somewhat disparate. They include the question concerning the merits or otherwise of Wittgenstein's thoughts on the impossibility of a private language; the question concerning the extent to which it is appropriate, if at all, for the state to interfere in such areas, among others, as who may have sex with whom; the question concerning whether or not, in human development, the public precedes the private (as for example Vygotsky argues that it does in the acquisition of language) or the private precedes the public (as for example all Social Contract theories of state formation have to assume); and the question concerning the right of the rich to buy for themselves or their children the private provision of those things which are otherwise publicly funded such as health and education. There is only a fairly limited extent to which such a mixed bag of concerns *can* be made to cohere into just one question: into the 'public *and* private' question, as it might be called. But given that my interest is primarily in philosophical conversation, I am here going to concentrate more or less exclusively on the last issue. For it is this one, the one concerning the ownership and control of resources, the one which asks which of those resources should be publicly (that is, collectively) owned and/or controlled, and which of those resources are better privately (that is individually) owned and/or controlled, which in my experience most frequently exercises the interlocutors in such exchanges.

I think it is reasonably uncontroversial to maintain that human societies almost certainly developed their concepts of ownership at the transition from a nomadic to a settled existence. It was then, when production as opposed to just hunting and gathering became a possibility, that land came to be conceived of as able to be owned. And the seeds sowing the beginnings of agriculture some 10,000 or so years ago became also the seeds sowing what later we would call civilisation. It was the possibility of accumulating surpluses that mattered. Production of more than was immediately needed, and by fewer than everybody, enabled some members of society to be carried as non-producers. And that meant that such folk could become instead those who controlled production in some way: as professional rain-dancers perhaps; as god-propitiators or witchdoctors. Thus was the question of 'ownership and control' put onto the human agenda for the first time—where, of course, it remains to this day.

Fundamentally, there are only two ways in which things *can* be owned. They can be mine or thine, hers or his: that is, they can be individually owned. Or they can be ours, theirs or everybody's: that is, they can be collectively owned. As ownership by one person, *individual* ownership is a relatively unproblematic concept both philosophically and legally. But ranging as it does across everything from a joint mortgage, via myriad forms of corporate ownership, group ownership, co-operative ownership and the like, to state nationalisation, *collective* ownership is anything but. (And all this without the no-way innocent little 'and control' phrase which some are ready to claim complicates things considerably.) I hope it will already be obvious, however, that this is not the place to do anything other than acknowledge the existence of such complexity, and move on. For what I want to think about is not the detail of the economic debate on ownership and control but the very general question of what might be the appropriate balance between collective (public) ownership and individual (private) ownership in a polity. In other words, I want to think about it where, by collective (public) ownership, I mean ownership not by named persons who individually come together as a couple (the joint-mortgage case), but by un-named persons who collectively constitute such a polity.

The first though no doubt obvious point to make is that this is a political question: in the language of today's political philosophy, a question about 'distributive justice'. If I start from somebody's definition of politics as a social process determining who gets what, when and how, then it is a 'how' political question. For any particular scarce resource—and on supply-side definitions of abundance, *all* resources are deemed to be scarce—do we more readily achieve distributive justice in the sharing of that resource by organising its provision collectively, that is publicly, or individually, that is privately? Logically, of course, there are two prior questions: some might want to question whether this is a distributive justice question at all; and if it is, some might want to question whether that distributive justice is something we should even want to achieve in the first place. I make no apology for taking the answers to both of these as read.

Totally unproblematically, or so it seems to me, the right balance between public and private ownership of resources is that which emerges as the result of the choices made by all the equal (and thus free) persons of a polity: as the biggest potential polity of the lot, ultimately by all the equal (and thus free) persons of the world. What will then be the outcome of such choices can in no way now be second-guessed. But for us, for now, that is a fact of absolutely no consequence. For our task now, as opposed to what will be their task then, is to help to make the equality (and thus the freedom) which they will need. It is to the theoretical understanding of that task, the task of making the equality they will need, that my hypothesis can I think make an important contribution.

For both sorts of ownership, 'public *and* private', are most frequently defended, of course, in terms of the good: the good of individuals, and 'the common good'. And I am arguing that short of development to the I IS stage of the selving process, the stage at which I IS conceptions of the self finally emerge, the good, whether common or otherwise, is a category unavailable to coherent conceptual formulation. To make it so available, to enable the science of the good (which they will have then) to emerge from the ideologies of the good (which are all we have now), the necessary selving development is one which I envisage as being first quantitative then qualitative (see also the next entry).

That which takes place across the I AM stages is quantitative. Perhaps somewhat in the way that Kohlberg outlines it, our understanding of 'the good' deepens (or doesn't) as we do (or don't) progress through his levels and stages. Perhaps somewhat in the way that Erikson outlines it, our understanding of 'the common' whose good it is appropriate that we be concerned with expands (or doesn't) as we progress through his ages. What I am wanting to add to such formulations—and there are several other and similar ones lurking at the softer end of the social sciences and among the various therapies—is this: that it is I AM conceptions of the self which, in the Kohlberg case, tend to halt development—so much so that in his later theorising he abandoned his final stage as being beyond the capabilities of almost all persons; and that it is I AM conceptions of the self which, in the Erikson case, tend to steer development inexorably towards the negative rather than the positive poles of his continua—so much so that he sees 'despair' as a much more common destination than 'ego-integrity'. The quantitative changes accumulating across such development, while they pose ever more clearly the problem of what is to be conceived of as the good, are in no way sufficient to offer us its solution: that is, to furnish us with a thoroughly coherent conception of what that good actually is. Such a conception comes only when the quantitative shifts across the various I AM stages give way to the qualitative shift to the single I IS stage. For the realisation that we live not in an 'I *and* thou' world but in an 'I *and* I' world will involve the biggest single development of our understandings so far seen. In giving us, at last, a coherent conception of the good as the loving of each of those Ies *equally*, via the active pursuit of the well being of each of those Ies *equally*, it gives us the basis for the development of moral science. So with this time the 'public *and* private' duo as my illustrative text, let me spell out once again why I think it is that this transition will be so important.

The fundamental limitation of the thinking of I AM selves is that contained in what I am calling 'the having ideology': the false belief that an I is an entity which stands over-against its experience which it then 'has'. And from that self-having error, source of the self-ownership error of academic philosophy, flow all the other having errors which characterise all other ownerships. I AM is a having person: and since what I have you can't have, each person is embedded in relations antagonistic to rather than co-operative with each other person. I AM is a first person: and since we can't all be first, each person is locked into an unending struggle for status with each other person. Which means that when it is such selves, such I AM selves, which take political decisions as to the appropriate balance between public and private ownership and control, or for that matter any other political decisions, theirs will be the decisions of the interested: the selfish decisions of those who, when they 'ish' their I AM selves do so in ways that seek to maintain rather than challenge 'master *and* slave' (primary *and* secondary) relations. On this 'public *and* private issue, by way of illustration, they were the sorts of decisions taken at the time of the Atlee round of nationalisation in post-war Britain: that to provide the necessary infrastructure for private profit, public ownership should be only of 'failing' (not making a profit) industries; that it should be accompanied by super-generous compensation, the payment of which was more or less bound to cripple the resulting nationalised concerns; and that it should always carry the possibility that the industry concerned might well go back into private hands if and when it ceased to 'fail'.

But when such a fundamental limitation of thinking is no more; when I IS persons have shed the having ideology and thus changed the world; when what is best publicly owned and what is best privately owned is a question concerning 'the administration of things' rather than 'the government of some people by other people' (to slightly misquote Engels), who knows what public-private balance will then emerge? In what will then be a world of equals, who from now can know what its public-private balance will turn out to be? A world where, because what is demanded is no more than (nor no less than) an equal share with others, abundance is defined from the demand side not the supply side; a world where resources are understood to be (morally) abundant when, however much there is (or isn't) of them, is shared equally; a world where something perhaps called GWP (Gross World Product) is distributed via a democratically controlled political process such that some is consumed collectively in the form of public goods, some is consumed individually in the form of private goods: this is a world we have not yet made—to understate somewhat dramatically! From now, we just cannot know what then will turn out to be the appropriate balance between the public and the private. We cannot even know whether, in circumstances where exploitation in both its technical economic sense and in its more general moral sense is no more, the two terms 'the public' and 'the private' will have, or even could have, the same meanings as they have now.

In *Conditions of Freedom* (1950), John Macmurray observes that 'it is enough to remind ourselves that our freedom, as individuals, depends upon the co-operation of others'; that we are 'fed and clothed by our fellows'; that 'the whole apparatus of our life is provided by others'; that the fact that 'the system of co-operation is impersonal and indirect makes it no whit less real'; that it is not only 'the material resources which we use at every moment that are the gift of others', but also 'the language we speak, the thoughts we think, the ideals we cherish and pursue' which are all 'only partially our own'; and that having them 'from those who went before us', it is 'the forms they take in our private minds and mouths' which 'bear witness that they are symbols of a life that is shared'. In *The Language Instinct* (1994), Steven Pinker points out that 'a common language connects the members of a community into an information-sharing network with formidable collective powers' which mean that 'anyone can benefit from the strokes of genius, lucky accidents, and trial and error wisdom accumulated by anyone else, present or past', and that 'people can work in teams, their efforts coordinated by negotiated agreements'. In *Kinds of Minds* (1996), Daniel Dennett registers that far from our individual experience being subjectively private, the thing that is so special about persons (vis-à-vis animals) is precisely that, via language, we '*share* (my emphasis) a subjective world—and know that we do—in a way that is entirely beyond the capacities of any other creatures on the planet'. What I am saying is that it will be when we come to be able to take such observations as these thoroughly seriously that we will know about the morality of equality: about having *together with* other just persons the equally shared public; about having *alongside* other just persons the equally distributed private. And in proportions the one to the other, the public to the private, to be democratically determined in the course of that previously mentioned 'negotiated co-ordination'.

(29) quantity and quality

The very widespread significance of the 'quantity *and* quality' issue, and thus the very considerable importance of what is often called 'emergence', was most recently given emphasis in Marxist scholarship. So that now, when according to some wit 'PMT' has come to stand not for 'pre-menstrual-tension' but for 'post-Marxist-trauma', there is perhaps some reluctance still to acknowledge its centrality. To the extent to which this is the case, it is I think unfortunate. For it seems to me that a formulation of the nature of change is central to all scientific thinking. And that contained in any such formulation must be, first, an understanding of its ubiquity; second, an understanding of its dual 'quantitative *and* qualitative' properties.

To understand, first, its ubiquity, it is necessary to be very clear about essences. As they appeared in the philosophy of Plato, as 'Ideas', they were abstract entities. Called by him 'the Forms', and with 'the Form of the Good' the most important of them, they existed perfectly, timelessly and unchangingly with God. By contrast, all the actual entities of the actual world existed only as less-than-perfect copies of 'the Forms'. And as such, were subject to all aspects of time-based change. It seems to me that all the many varieties of what has come to be called 'essentialism' in philosophy are stamped with this origin. An essence, like an Idea or a Form, is something definitive, out-of-time, unchanging: and thus, in what is to many of us now a simple and obvious sense, non-existent. I have argued that when we take as universally valid, general, given that which is actually historically situated, specific, taken, we are unable fully to understand that which we mis-perceive so. We do not see its development. We do not appreciate that because it is process rather than entity, there is a serious sense in which whatever it is it really isn't. But one of the many strengths of the perhaps not so widely discarded Marxism was that while it was thoroughly anti-essentialist, there was a sense in which it could be seen to have posited a single exception: that is, a sense in which 'change' was, in effect, a Marxian essence. And while that is still not a good-enough formulation—as I have already discussed, change after all also changes—it does give some sort of half-valid expression to its simple ubiquity.

To understand, second, its 'quantitative *and* qualitative' properties, it is vital to be clear about preformationism. This nineteenth century biological doctrine, the idea that a tiny human being sat curled-up and ready-made inside each sperm (of course!) and only had to grow for nine months to become a full-sized baby, might be dead. But that which by analogy might be called 'intellectual preformationism', the idea that if we need to explain something complex we need either to find that it arose from something equally complex which preceded it, or that it was always-already there 'in the first place' (wherever *that* is), is still very much with us. The ultimate in intellectual preformationism is of course the fundamentalist idea of God: the idea that only an always-already all powerful entity could possibly be responsible for the existence of something as complex as a universe, a biosphere and a world. To think about complexity scientifically, by contrast, it is important to get hold of a vital idea: the idea that in the beginning, complex things were not—they did not *then* exist; and that later, because they evolved or developed from simpler things, they were—they do *now* exist. In other words, to *get* the more complex you don't have to *start out with* the more complex. Perhaps the single most important example of this reality, one to which I will later return (see Chapter 12), is that whereby you don't need to start out with a soul, put there by some prior soul-creator, for a soul to emerge. As I will there suggest, there is a perfectly respectable sense in which Keats was right to see the world as 'a vale of soul *making* [my emphasis]'.

Another way of labelling this preformationist tendency is to call it 'originism', understood as the ideology which asserts that explanations in terms of *origination* will serve also as explanations in terms of *potentiality*. It is made manifest in statements such as 'the *ultimate* function of sex between human beings *is* reproduction', as opposed to 'the *original* function of sex between human beings *was* reproduction'. Or in philosophical positions such as that which holds that because our intellectual capacities were *originally* evolved to deal with survival and reproduction, they must be *ultimately* incapable of understanding the universe. But however it is labelled—preformationism, originism, whatever—the idea that to *get* the more complex you have to *start out with* the more complex is just wrong. As with David Chalmers' notion that to *get* consciousness you need to *start with* consciousness—that is, that consciousness has to be thought of as an always-already phenomenon on a par with matter and energy—it just flies in the face of all the evidence concerning the emergence of the qualitatively new from the quantitatively accumulating.

For it seems to me that the way to handle the disputes that occur under the 'quantitative *and* qualitative' head is twofold: first to *resist* the idea that they indicate two different *types* of change; but second to *embrace* the idea that they indicate two different *consequences* of change. It is this approach that I take, for example, when I posit three levels of reality. And I can perhaps best illustrate what I mean by this stance by considering some examples relevant to just that three-fold classification.

At the level of the inorganic, consider 'superconductivity' and 'superfluidity'. These are phenomena which emerge from a material in a very particular circumstance: that is, when its individual atoms reach a high level of coherence. Such coherence is produced by the steady reduction of thermal oscillation in the material as its temperature is lowered to near absolute zero. The *type* of change is incremental throughout: a degree- by-degree reduction in temperature giving an atom-by-atom more orderly alignment. It is the *consequence* of the change which is first quantitative (more and more alignment) then qualitative (at a critical degree of coherence, the emergence of the totally new phenomena 'superconductivity' and 'superfluidity').

Other inorganic examples abound, from amongst which I choose only three. One: a degree-by-degree increase in the temperature of ice results first in water then in steam. Two: an iron magnet heated to above 770 degrees centigrade loses its magnetic power because increasing thermal oscillation, thermal noise, disrupts the alignment of all the microscopic magnets which are its tiny constituents. Three: as discovered only in 1997 by chemists working at University College London, the fluidity of water emerges not at one, two, three, four or five molecules, but at six molecules, precisely. In each case, the *type* of change is incremental: the *consequence* of change is first quantitative then qualitative.

At the level of the organic, consider 'learning'. This is a phenomenon which probably emerges from a central nervous system in a very particular circumstance: that is, when some of its cell assemblies reach a high level of synaptic smoothing. And such smoothing is produced by the steady reduction of synaptic resistance in the assembly as its firing thresholds are lowered by repeated activation. The *type* of change is incremental throughout: a firing by firing lowering of synaptic threshold giving a synapse-by-synapse, less-resistant, smoother impulse flow. It is the *consequence* of the change which is first quantitative (lower and lower firing thresholds) then qualitative (at a critical level of smoothing, the emergence of learning).

Other organic examples abound too, from amongst which again I take just three. One: the common ancestor living six million years ago in Africa had offspring that went on to evolve into modern apes, *and* offspring that went on to evolve into human beings, although at no point along those lines of evolution did a parent animal reject an offspring animal as any different from itself. Two: the bare simplicity of light sensitive cells on the surface of the mollusc nautilus becomes the lidded complexity of a human eye via a very long series of tiny and discrete changes. Three: (as discovered only in 1996 by scientists working at the Research Institute of Electrical Communications in Sendai, Japan), in groups of between 10 and 100 individual animals, if the ratio of the body size of the animal to the product of its flying speed and the time taken for each wing beat is more than 1, as in bees, the animals swarm; if the ratio of the body size of the animal to the product of its flying speed and the time taken for each wing beat is less than 1, as in birds, then the animals flock, meaning that while changes in ratios are incrementally continuous across species, changes in behaviours are not. In each case again, the *type* of change is incremental: the *consequence* of change is first quantitative then qualitative.

And finally, at the level of the personic, consider 'the sentence'. This is a phenomenon which emerges within a discrete combinatorial system called language in a very particular circumstance: that is, when its phonemes and morphemes are combined in a particular way. Such sentences are produced by the steady accumulation of linguistic capacity in speakers as their language acquisition capabilities develop. The *type* of change is incremental throughout: a phoneme by phoneme, morpheme by morpheme, holophrase by holophrase accumulation, of linguistic competence. It is the *consequence* of the change which is first quantitative (what in *Language and Human Behaviour* [1996] Derek Bickerton calls the 'grammar-less chains' of early hominids) then qualitative (the arrival in the eighteen month old child of Roger Brown's eight types of two-word sentences).

Other personic examples finish my case. One: fewer than twenty-four successive stills a second, and we see—a succession of stills; more than twenty-four successive stills a second, and we see—'a movie'. Two: as illustrated in the work of Robin Dunbar, fewer than one hundred and fifty people in a human group, no gossip; more than one hundred and fifty people in a human group, gossip. Three: the so-called 'epidemiological transition' whereby a country moves from predominantly infectious causes of death (still common in poor countries) to predominantly degenerative causes of death (now dominant in richer ones) as the consequence of an incremental growth of Gross National Product (GNP). It is critical to register that in all these examples—inorganic, organic, personic—it's not that *simpler* versions of the emergent properties existed already at the lower level: it's that *no* versions of the emergent properties existed at the lower level. That which emerged just did not exist before. It is a *brand new* reality. And this is because the essence of the concept of 'emergence' is that while the lower-level underlying substrate must *permit* the higher level over-lying 'superstrate' characteristics to form, it does not itself *share* those characteristics: neither H_2 nor O is wet. The properties of emergent systems are a function of the *organisation* of their component parts, not of the *characteristics* of their component parts. In *The Symbolic Species: The*

Co-evolution of Languages and the Human Brain (1997), Terrence W. Deacon illustrates this point with a memorable anecdote to the effect that it was the question of an eight year old schoolchild which led him to think about a paradox. The child asked why animals don't have *simple* languages rather than *no* languages. And the paradox presented itself as the mis-match between what Deacon called 'the unbroken continuity between human and non-human brains' and 'a singular discontinuity between human and non-human minds'. Consistent with my position on the 'quantitative *and* qualitative' pair, my hypothesis would resolve any such paradox in levels-of-reality terms. With the emergence of language characterised as facilitating conscious co-operation (a newly emergent phenomenon) rather than sentient communication (an already well established phenomenon), it is the development of that 'knowing-with' which is the source of the third type of reality called the world. And just as there is a qualitative shift between the first type and the second (between inorganic reality and organic reality), so there is a qualitative shift between the second type and the third (between organic reality and personic reality). Animals don't have 'simple' languages in the same way that rocks don't have 'simple' life. Animals don't have 'simple' ways of *making* love in the same way that rocks don't have 'simple' ways of *evolving* irritability, directed movement and sentience.

If nothing else, such ways of thinking about 'quantitative *and* qualitative' change would save endless amounts of scholarly time. From the Cambrian explosion to the industrial developments of the eighteenth and nineteenth centuries in Europe, *various* forms of the phenomenon 'revolution' have taxed those of the learned—and they have been many—for whom *any* form of the beast was anathema. Thus Rupert A. Hall devoted many (thoroughly ideological) person-hours trying to iron out 'the Scientific Revolution' of the sixteenth and seventeenth centuries, while Dorothy George did the same for 'the Industrial Revolution' of the eighteenth and nineteenth centuries. But all such effort could have been avoided if they had been able to see the processes they were describing as ones in which the 'inevitability of gradualism' refers to their incremental (quantitative) *causes*, while the 'behold I make all things new' refers to their emergent (qualitative) *consequences*. As if preferring it could make it so, it has been a commonplace in the thinking of the traditional conservative that he or she is in favour of evolution rather than revolution: that is, of gradual change rather than sudden change. But the thinking that I am proposing enables us to see this as a totally false dichotomy. As the man said, gradualism is indeed inevitable: those changes which seem sudden and revolutionary in fact take place only at that point when a final small shift is added to an accumulation of already accumulated small shifts. It is the *effects* of a final straw that are indeed sometimes dramatic, even revolutionary. But it can only *be* a final straw because all the others are already in place. In *The End of Theology* (1999), George Pattison puts the idea this way:

> Even if it is only all of a sudden that we notice that it has indeed become a new picture, the likelihood is that the final paradigm-shift will have been long-prepared by a history of gradual transformations, invisible fermentations and the accumulation of unrecognised pressures.

And it is an approach to an understanding of the nature of change which can be recognised as the important philosophical stuff it undoubtedly is when we see what can happen to our arguments when it has not been appreciated.

In *Darwin's Dangerous Idea* (1995), Dennett writes of what he calls 'the most general form of the schema of Darwinian explanation'. He stresses that 'getting from the early time when there wasn't any X to the later time when there is lots of X is completed by a finite series of steps'. And as they are taken, it becomes steadily less clear whether there is any X there or whether there isn't, until eventually all debate is stilled and, *without ever having drawn any lines*, we acknowledge the quite obvious existence of 'lots of X'. Now there are aspects of this description which I find somewhat disingenuous. What is described is surely not in any sense specifically Darwinian: to be a specifically Darwinian formulation it would have to contain (at least) the mindless production of more of something than can be coped with, followed by the mindless culling of all of the worst fit of it, leading to the mindless survival of only the best fit of it. In addition, and as Darwin himself insisted, given that little 'mindless' word, the 'something' would need to be of the organic. No: that which here is surely not in any sense specifically Darwinian is instead in some sense specifically dialectical. It is a general description of change as exemplified in 'quantitative *and* qualitative' formulations. And when a little later in his book Dennett argues that 'at *some* level of magnification, any evolutionary ramp must look like a staircase', I want to demur. Thinking of the real thing from which his metaphor is taken, then as every wheelchair user knows, the emergence of a ramp from a staircase is the emergence of a totally new phenomenon: one whereby flowing and smoothly continuous wheeled transportation can replace the broken and discretely step-by-step variety. All of which means that the way to avoid the additive (and thus primitive) nature of the formulation contained in this particular 'and' pair is to register that it is a *quantitative* accumulation of discrete changes as cause (the inevitability of gradualism) which leads to the emergence of a *qualitatively* new phenomenon as consequence (the equal inevitability of revolution). In the case with which this book is most concerned, it is a *quantitative* accumulation of discrete changes across developing I AM selves as cause, which leads to the emergence of the *qualitatively* new phenomenon of an I IS self as consequence.

(30) realism and anti-realism

Einstein somewhere commented that the real mystery about the world—translating into my terms, I take him to have meant the universe, the biosphere and the world—was that it should be comprehensible. Another way of putting this, and a way adopted by many engaged in philosophical conversation, is to ask why our nervous systems, evolved to promote the survival of our genes, should be able also to fathom, say, the second law of thermodynamics. It is via an exploration of this issue that I want to conduct the investigation of this 'realism *and* anti-realism' pair.

In making his comment, Einstein was clearly in I AM mode: there is an I that am and (separately) a world that is. And what is mysterious is that the first should come to be able to know the second. But this is not the only possible paradigm. What I want to do here is to point up the different way in which the 'realism *and* anti-realism' pair can be understood from within the alternative I IS paradigm. And what I hope to show is that when we do that, any mystery fast disappears: that as with all the others, when we do that we are left with a much more coherent account of this 'and' also.

I need to begin with a brief comment on what I am taking each of the two members of this specimen 'and' to mean. The entry on it in my *Oxford Companion* source begins by suggesting the following: that when we say that something is mind-independent we are moving in a realist direction; and when we say that something is mind-dependent we are moving in an anti-realist direction. Of course, a vocabulary which requires us to associate only the mind-independent with the real cannot be my vocabulary. Given the way in which I characterise reality as just 'all that there is', in my conceptual system minds (as temporal states of spatial entities) and brains (as the spatial entities of which they are the states) are clearly both thoroughly *different*, the one from the other, and equally *real*, the one with the other. But leaving that aside for the moment—of course it will figure later—the way in which this duo figures in philosophical conversation leads me to make a terminological shift here.

For in philosophical conversation, although it is true that mind-independence is associated with realism and mind-dependence with anti-realism, this is a debate which is more often than not mediated through claims about 'relativism'. In such debate, 'realism' is (more often than not) understood as the position which claims that there is a real world, independent of us, and we can come to know what it is (really) like. And 'relativism' is (more often than not) understood as the position which claims that, since 'we' consists in about six and a half thousand million individual persons, each geographically, socially, genetically, intellectually, politically *differentially* situated, it follows that anything that any one of them 'knows', being relative to such situation, cannot possibly give us certainty about what is (really) the case. In extreme versions of the idea, in thoroughly sceptical versions of the idea, not only can it not possibly give us certainty about what is (really) the case: it cannot even give us certainty about whether there is anything we can be only uncertain about! It is for *this* reason, because I want to get at *this* argument, that henceforth I will refer to this duo not as the 'realism *and* anti-realism' pair but as the 'realism *and* relativism' pair. Such a usage does not accommodate well the fact that there are many different types of relativism. But it accommodates very well the fact that, in philosophical conversation, they are rarely distinguished. So: I will adopt it.

On 'realism', therefore, I want to suggest that when realists claim that truth is knowledge of an *autonomous* universe, biosphere and world, they are wrong. But that when they claim that truth is knowledge of a *real* universe, biosphere and world, they are right. And on 'relativism', I want to suggest that when Pilate put his 'what is truth?' question, he was certainly posing a real corker. But that the fact that it is difficult to take seriously any sort of 'that-which-is-true-for-me' response is rather less because of problems in being clear about 'true' than it is because of problems in being clear about 'me'. In other words, it is rather less an answer to the 'what is truth?' question, rather more an answer to the 'what is me?' question, which will help us to make subordinative as opposed to additive sense of this now 'realism *and* relativism' specimen.

In *Philosophy* (Vol. 72, No. 279), the philosopher Hans-Johann Glock wrote a piece under the heading 'Truth without People?'. Writing on realism, he suggested that there is 'a venerable tradition according to which the concept of truth is totally independent of human beings, their actions and beliefs, because truth consists in the correspondence of mind-independent propositions to a mind independent reality', and added that 'for want of a better general term, I shall refer to this position as realism.' So defined, realism is surely just untenable. It is only necessary to ask 'who proposed the propositions?' for it to be shown to be untenable. Indeed, faced with such ideas as to its nature, it is perhaps easier to see why many should reject realism and take refuge instead in Nietzsche's comment to the effect that 'there are no truths, only interpretations'. If nothing else, the current round of relativism has performed a useful function in pointing up the considerable weaknesses of some formulations of realism. But if we see the task as being not to so define realism that it becomes nonsensical, but to so define realism that it becomes tenable, there are other traditions just as venerable which can help.

Having been initiated by Lorenz, the approach to the theory of knowledge sometimes known as 'evolutionary epistemology' has been developed for the most part in German scholarship. In defence of his realist position, Lorenz suggested that it is precisely because our eyes and ears have evolved in response to light waves and sound waves that they will tell us something true about the real nature of the same. Indeed, how could it be otherwise? Any living things whose adaptations were to a world not real would not survive. As hypotheses about their environments, they would be not just null but pretty soon void: they would become extinct. In *Philosophical Darwinism: On the Origin of Knowledge by Means of Natural Selection* (1993), Peter Munz calls his version of this position 'hypothetical realism'. And translating it now into the terminology of my hypothesis, I think it can suggest something interesting.

In my terms, what theorists such as Lorenz and Munz are saying is that if the genetically structured entities which are human-animals are aggregates of hypotheses as to how the universe and the biosphere are, *which hypotheses are increasingly wrong because the universe and the biosphere are changing,* then they will become extinct: fewer and fewer such human-animals will come into existence; they will last less and less successfully for shorter and shorter times; and eventually there will be none of them at all – as now there are no Neanderthals. It is by analogy with this that what I am saying is that if the linguistically structured entities which are human-persons are hypotheses as to how the world is *which hypotheses are increasingly wrong because the world is changing,* so too will they: fewer and fewer such human-persons will come into existence; they will last less and less successfully for shorter and shorter times; and eventually there will be none of them at all – as now there are no persons with theories of medicine premised in the four-humours, and as then (a prediction of my hypothesis!) there will be no persons with theories of the inter-personal premised in inequality. This is the type of realism I seek to defend. The relativism which therefore I want to challenge is that which posits three entities where I would posit just two. Such relativism suggests that there is [1] an I that am; [2] a my-theory-about-reality that is; and [3] a reality that is (or may be). And that these three entities are related to one another in such a way that since [2] must always come between [1] and [3], [1] cannot know [3] directly. By contrast with this, the realism that I am defending suggests that there is [A] an I that is (the linguistically structured experience of a human nervous system) and [B] a not-I that is (everything else). Because it is actually a constituent part of [A], the second term of the relativists, their 'my-theory-about-reality that is' term, has gone. And what that means is that the only now two entities [A] and [B], there being no third to come between them, can be related the one to the other directly. In Kantian terms, if you like, [A] as an I 'thing in itself' can know [B] as a not-I 'thing in itself' in an unmediated and direct way.

What this amounts to is that we solve the 'realism *and* relativism' problem of the veil which hangs between me and 'things in themselves' by *dis*solving. But we *dis*solve not the veil but rather the distinction between the veil and the me. In my language, I AM persons are those for whom there is an homuncular 'I' having their experience, including having their theories, and I IS persons are those for whom there is only a non-homuncular 'I' which is their experience, which is their theories. This means that I can know what a 'thing in itself' *is* only when I know that an I *is* also. Such a thorough-going realism, a realism about what I is along with a realism about what not-I is, is available only to the I IS consciousness of an I IS person. In other words, it is not that knowing stands over-against reality, which reality is to be known by an I that AM. It is that knowing has emerged within reality, within which reality an I that IS is the *site* of the knowing. Or rather, to once again avoid one of those pesky space words, an I that IS is the *moment* of the knowing.

Such is the realism which my hypothesis involves. But it is a realism which challenges not only more or less all of current relativism, but also at least some of (other) current realism. Both realists and relativists are concerned with truths. And one way to get at the difference between their approaches is to say that while for realists truths are universally 'given' (there to be 'discovered' by anyone), for relativists they are individually 'taken' (not there until 'invented' by someone). Within such a formulation, and acknowledging that of course there are others, I would be for claiming that, when offered a choice, it's often better to take both: in this case, of course to take that aspect of realism which posits *something* as being universally given, but see it as reality not truth; but also to take that aspect of relativism which posits *something* as being individually taken, but see *it* as validity not truth.

What this formulation does is to point up the way in which both some stripes of realist and some stripes of relativist are apt to undervalue that 'making' on which I am putting so much emphasis. Realists can seem to undervalue it by wanting some truths to be independent of it. Relativists can seem to undervalue it by arguing that because such independence is not possible, truth itself is therefore not possible. Such realists want to insist that truth is *not* merely human-made: and seem to think that that *boosts* its status. And such relativists want to insist that truth *is* merely human-made: and seem to think that that *diminishes* its status. What I want to claim, in contrast to both of them, is that that which is made by humans, as opposed to that which merely is, can be the strongest, surest, most certain stuff there is *precisely because it is so made.* And that that which determines whether or not it will be is the stage of development that the selving process has reached in its makers: make within an I AM paradigm, and make that which will only be able to be ideologically crooked because of the interests of its unequal makers; make from within an I IS paradigm, and make that which will instead be able to be scientifically straight because of the disinterest of its equal makers.

It is sometimes the optimism of such an approach which is questioned. It is sometimes suggested, for example, that there may be some truths of which human brains just cannot become aware. I think this is just a tautological nonsense. Where *are* such truths? Where *could* they be? *Whose* truths are they? For truths are like that: they adhere to beings; they are of beings; they are made by beings for beings. Until such time as somebody shows it to be otherwise, we just have got to get used to the idea that the only entity in the universe capable of making truths is us. We are the truth makers. Which means that the idea of a truth which our brains cannot deal with is just daft.

So let me conclude this section by way of reference to a philosophically celebrated example of making in the 'realism *and* relativism' stakes. The philosopher Nelson Goodman famously if not infamously once claimed that because representations of stars are human creations, so stars are, in an important sense, 'made' by us. A collection of papers from him and three colleagues which came out in 1996 was thus titled *Starmaking*, and sub-titled *Realism, Anti-Realism and Irrealism*. Since I am dealing here in part with similar issues, and since I am writing under the title *Making God*, it is perhaps important that I orient myself with respect to such a claim somewhat carefully. On stars, I would want to argue that we are not star makers in two senses. But that we are star makers in a third.

The *inorganic* reality which is the hydrogen-helium reality which we call 'star' was *not* made by us. It was not made at all. Of course we experience it in ways that are a function of the capacities and limitations of human nervous systems, rather than in ways which are a function of the capacities and limitations of, for example, frog nervous systems. We are humans: we are not frogs. But the inorganic reality differentially configured by the two species, a reality which is of course different from the hydrogen-helium truth configured by only the persons which emerged from just one of them, existed prior to either. This is the first sense in which we are not star makers.

Likewise, the *organic* reality which is the species-specific characteristics of the nervous systems of the two configuring species, frog and human, was also not made by either of them. It also was not made at all. It was not (a) designed (b) by them (c) with intention. It was (a) evolved (b) as part of them (c) without intention. This is the second sense in which we are not star makers.

But the *personic* reality which is the state of a nervous system structured by language, the state of a human nervous system, is made. It is made in that 350 millisecond gap between the mindless and natural of the organic and the mindful and moral of the personic. But it is made there not as the result of free-will: not as the result of individual and singular action by an agent. It is made there as the result of free-won't: as the result of communal and collective structuring by a language. And the folk way in which, up to now, we have expressed this difference, the difference which means that we are not star makers in the first two senses but we are star makers in the third, is to call it 'meaning'. The sense in which we are star makers lies in the fact that we are the makers of meaning. And stars, along with much else, mediate our meanings. When we reach, we reach not for the gravel but for the stars. We do not bless our lucky pebbles. We do not catch a falling aeroplane. To stars we hitch our wagons: with stars our loves are crossed. These are the ways in which we are the makers of stars.

(31) reason and emotion

Perhaps the single most important difference that the application of my hypothesis makes to the study of emotion, and thus to the study of the way in which 'reason *and* emotion' are to be understood as related, is that it lends support to one particular current approach over what are several other current approaches. To see how this is so, some basics.

With 'basic' meaning (1) physiologically distinct, (2) expressively distinct, and (3) biologically based, it was Darwin who suggested that fear, anger, surprise, disgust, happiness and sadness were the six basic emotions. And much work since, for example the work of Ekman, Levinson and Freisen (1985), suggests that all six certainly occur very widely across many if not all cultures. Such apparent universality, together with the possibility that they can also be found in some of our primate relatives, leads many to suppose that they will have been naturally selected. At present, four of the six—fear, anger, happiness and sadness—are quite firmly in the generally agreed canon of the basics, with disgust being probably the most frequently added fifth possibility. Surprise, while certainly still widely thought of as a sixth, is perhaps less frequently cited. And other contenders—hatred, contempt, interest—each have their 'these-are-basic' defenders.

The theoretical problems here are considerable. It is very difficult to know just how to conceive of '*an* emotion'. For one thing, what if anything might be the significance of our old friend reification being at work here again? For another, how has it become possible for perfectly respectable philosophers, Janet Radcliffe Richards for one, to speak of 'the fact' that women show emotion far more readily than do men? If it makes sense for anything to be labelled '*an* emotion', it surely has to make as much sense for the 'hard' emotions to be so labelled as for the 'soft' emotions to be so labelled: the man kicking hell out of the rival football fan or road-raging a fellow driver is being 'emotional'? And for another, isn't it just as questionable to say that a person is 'showing no emotion' as it is to say that a person 'has no personality'? In each case, isn't it the apparent absence which *is* the emotion, which *is* the personality? Until there is a generally agreed theory of the emotions, which as of now there isn't, these are the sorts of issues which remain difficult to think about.

In *Best Laid Schemes: the Psychology of Emotions* (1992), Keith Oatley suggests that there are currently discernible three main theoretical approaches to the study of the emotions, each with some respectable credentials. First, there is the approach exemplified in the work of William James, whereby the terms used to identify the different emotions 'arise from folk theories in ordinary language' but 'have no coherent psychological status'. Second, there is the approach which is still perhaps the most common among those taking part in philosophical conversation, whereby emotions are the leftovers from an earlier and more primitive state, the state of an animal ancestor, an infant, or an as-yet-immature adult, which essentially distort adult rational life. Third, there is the approach which is perhaps the most popular among those whose specific research area is emotion, and the one which seems to me to make the most sense, whereby emotions are evaluations. Since it is this last which I think sits most comfortably with my hypothesis, it is this last which I want to consider further.

Oatley describes this third approach as one whereby each goal and plan has 'a monitoring mechanism that evaluates events relative to it'. He argues that it is when 'a substantial change of probability occurs of achieving an important goal or subgoal' that 'the monitoring mechanism broadcasts to the whole cognitive system a signal that can set it into readiness to respond to this change'. And he concludes that it is 'these signals' and 'the states of readiness they induce' which we humans 'experience as emotions.' It seems to me that this is an approach which carves reality at joints significantly different from, indeed hardly related to, the essentially 'folk' joints embedded in the 'reason *and* emotion' duo. It is an approach which chimes much better, for example, with Nietzsche's comment to the effect that he had at all times written his writings with all his heart and soul, and did not know what purely intellectual problems were. It is an approach which insists upon there being not two systems of cognition but one, not two ways of knowing the world but one. And is thereby an approach whose consequence is that when cognition is faulty, that is just as likely to be the result of 'faulty emotion' as of 'faulty reason'. As I anyway experienced it, coming to appreciate this way of posing things was a two-stage affair.

For the understandings of the first stage, I have again to thank the splendid John Macmurray. It was primarily from reading his *Reason and Emotion* (1935) that I first came to understand something vital: that is, that the often-heard phrase 'I can't help it, it's just how I feel' was no different from the nonetheless hardly-ever-heard phrase 'I can't help it, it's just how I think'. His presentation of what he called 'emotional rationality' was thoroughly compelling. He argued that 'the capacity for reason belongs to our emotional nature, just as much as to the intellect', and went on to insist that 'our feelings may be illusory, just as our intellectual ideas often are'. In thus insisting that 'our emotions can be real or unreal', he explained that 'to say that a feeling is unreal does not mean that we do not feel it, any more than to say that an idea is false means that we do not think it'. And concluded that 'an unreal or illusory emotion may be very strongly felt, and it may influence our conduct profoundly.'

This chimed with everything which I half knew from much else that I had read: with Spinoza's insistence that when we experience a negative emotion, it just means that we are working with an incomplete, inaccurate hypothesis as to the nature of the reality we are dealing with, and have got the issues wrongly posed; with Brecht's notion that 'our feelings compel us towards the maximum effort of reasoning, and our reason purifies our feelings'; with the opposition of such as the neuroscientist Antonio Damasio to the idea that reason is somehow to be understood as 'over against' emotion. It began to seem just obvious that emotions are just attitudes *towards* things, evaluative ideas *about* things. And that the somatic representation of such attitudes and evaluations (in the autonomic nervous system and the endocrine system) would be what constituted the 'feelings' of emotion just as surely as the somatic representation of such attitudes and evaluations (in the central nervous system) was what constituted the 'thoughts' of reason. From all of which it would follow that because emotion is just as much able to be in error as is reason, it is therefore just as much in need of education as is reason. And Macmurray, of course, had argued just that when he had suggested that 'we should recognise that our emotional life does need educating; that our feelings, emotions and passions are often nonsensical; that they do not correspond to anything real'. So important to my understanding was the gaining of this insight that for some long time I had pinned up over my desk an extract from *Reason and Emotion* (1935). And although it is rather long, indeed by far the longest quotation in this book, I want to include it in full. He wrote:

> We must really believe that a great many of our feelings ought not to be felt, and that our ordinary apprehension of values is very poor and often false. We must admit, at least, that our feelings are always liable to be wrong. We feel that something is beautiful when it is not, and ugly when it is not. We are liable to be shocked by things that are not really shocking, or to be terrified when there is nothing to be afraid of. On the other hand, we are often emotionally blind. We fail, very often, to feel anything at all when we ought to be profoundly moved. We notice this when somebody points it out to us.
>
> Besides this, we are full of vague and indefinite feelings which we fail to recognise for what they are, and as a result we often think we are feeling one thing when we are really feeling something else. We misrepresent our real feelings to ourselves. We have to distinguish three different ways in which our emotional life may be unreal. We may have clear and definite feelings which are illusory. We may have feelings which we misrepresent

> and mistake; and we may be too insensitive to have feelings that are called for by the nature of the situation. I must therefore repeat what I said in my first lecture, that we are very apt to take the view that one feels what one feels, and that's all there is about it. If one is jealous, for instance, one is just jealous, and it can't be helped. One must just make the best one can of the situation.
>
> Now that isn't so. There is in the emotional life a capacity for growth and development. What is apt to blind us to this is simply that it is impossible to be in error and to know that we are in error at the same time. That is true in the intellectual field. We cannot believe something that we know to be false. It is equally true in the emotional field. If, for example, we feel very angry or ashamed, and our anger or shame is an unreal emotion, then we cannot know that it is unreal. If something shocks us which is in fact not shocking, it is precisely because we do feel shocked that our feeling is false. There is no way of revealing the falseness of our emotions except in the process of changing them.

Splendid stuff! In teaching psychology, many was the time I introduced my students to these ideas.

But slowly I came to register a problem. I registered that Macmurray, for all his wonderfulness, presents the 'I-can't-help-it' concerning emotion as equally as *false* as the 'I-can't-help-it' concerning reason. The second stage of my understanding came when I was able to stand at least half of this argument on its head: that is, to maintain the insight of their being two equivalent statements; but to junk the error of their being two equivalently *false* statements. Instead, I came to think of them as two equivalently *true* statements. And to do *that*, I needed to ask some fundamental questions about that 'I': about the 'I' which could not 'help it' either intellectually or emotionally. They were the sorts of fundamental question which came eventually to enable me to formulate my I AM to I IS hypothesis. And when I asked them in the context of this 'reason *and* emotion' duo, the picture that emerged looked something like this.

In the lower mammals, it is information which flows *raw* to the limbic system which results in an emotional response mediated by the hormones and drives specific to the particular type of (lower) species concerned, and which constitutes the non-linguistically structured experience of such (simpler) creatures. In even our nearest mammalian cousins, it is information which flows *cortically processed but still not linguistically processed* to the limbic system which results in an emotional response mediated by the goals and plans specific to the particular type of (higher) species concerned, and which constitutes the (still) non-linguistically structured experience of such (more complex) creatures. But in humans, anyway in those humans whose linguistically structured experience constitutes them as persons, things change. It is information which flows *both cortically and linguistically processed* to the limbic system, which results in an emotional response mediated by the attitudes and evaluations specific to the particular type of self concerned, and which constitutes the (now) linguistically structured experience of what are (now) persons.

This is how I understand the 'emotion' member of our 'reason *and* emotion' pair. And in order to fit it together with its 'reason' partner, it seems to me that it is particularly important that I get hold of Macmurray's point about error: the idea that it is equally impossible to be in emotional error *and know that we are,* as it is to be in intellectual error *and know that we are.* The full force of Schopenhauer's quip to the effect that life has to be lived forwards but can only be understood backwards comes into play here: we know that we were in error—and on both counts—only when we have ceased to be so. And this means that the claims contained in such phrases as 'she feels very strongly' or 'he feels very intensely' must be scrutinised no less closely than the claims contained in such phrases as 'he believes very strongly' or 'she believes very intensely'. What Macmurray's point about error is making clear is that neither the beliefs of our reason nor the feelings of our emotion are to be validated by the strength or the intensity with which they are believed or felt. Instead, as twin parts of the same cognitive process, both 'reason *and* emotion' are to be validated by using that method which has shown itself to be by far and away the most reliable we have of escaping subjectivity, partiality and interest: that is, by using the scientific method. And when we do that, then 'reason *and* emotion' can maintain their additive mode, but only if we can remember to subsume both under the more general head of 'cognition'.

(32) reasons and causes

The essence of the problem presented by this 'and' can be summed up in the question 'are reasons causes?' And taking that question at face value, my simple answer would be 'no'. To fill that 'no' out a little, I would want to say that reasons are the explanations given by reasoners, that is, by persons, for what happens in the universe, the biosphere and the world. But they are not the causes of those happenings. Of course this does not mean that such reasons cannot represent those causes. They do: sometimes rather well; and sometimes even perfectly. And when they do, what we have is science. But it is my contention that the cause remains the cause, and *other than* the reason that represents it, whether that representation is accurate or not. In other words, the cause is independent of the reasoner and the reason is not. In still other words, while *all* reasons are caused only *some* causes are reasoned.

What's more, it seems to me that this is *as* true when the reason on offer is 'I did not park illegally because I did not want to get into trouble' as it is when the reason on offer is 'the crystal spheres cannot exist because if they did the moons of Jupiter could not circle the planet without crashing through them'. The not-parking of the car and the non-crashing of the moons both have causes. But those causes, whatever they are, remain those causes whether the reasons offered by the reasoner accurately represent them or not. In a nutshell, while everything is caused, some causes are represented (by reasoners) *as* causes when they *are* causes; some causes are misrepresented (by reasoners) *as* causes when they are *not* causes; and some causes are neither represented nor misrepresented (by reasoners) *as* causes—but *are* causes just the same. These are the propositions required by my hypothesis. And I want to begin the process of defending them by outlining what I think is a three-stage process in our thinking about such things.

For stage one reasoners, I AM thinkers at earlier stages of the selving process, *all* their representations of causes are as reasons. This is a style of thinking both animistic and immanent. In explaining the rolling of the thunder as the anger of the gods (rather than the rushing back of air into the vacuum created by lightening) it is animistic: and in explaining the origins of the universe using a creator God (rather than a big bang) it is animistic. Likewise, in explaining the child's fall from the swing as her getting her comeuppance for stealing the apple (rather than as her getting her centre of gravity in the wrong place) it is immanent: and in explaining the death of the gay man from AIDS as his being punished for 'slopping about in a cess-pit of his own making' (rather than as the result of his being infected by the HIV virus) it is immanent. The key to all such thinking, primitive thinking which when it is given voice by modern adults is frequently obnoxious, is intention. Things happen because somebody, a god-person or a human-person, intends them to happen. And that intention is causative. The god-person or the human-person has a purpose; the purpose supplies a reason for action; and the result of the action is that the happening happens.

For stage two reasoners, I AM thinkers at later stages of the selving process, things have changed. Some of their representations of causes, particularly of those causes that are operative at the inorganic and the organic level, are no longer as reasons. For things that happen in the universe and the biosphere, there is no longer a *reason as to why* they happen but a *cause as to how* they happen. It is recognised that to require a reason of the inorganic or the organic is to make a category mistake: though all such realities are thoroughly caused, stones don't have reasons for falling, any more than daffodils have reasons for being yellow or moths have reasons for flying into flames. Only at the level of the personic, they maintain—or would if ever they used the word!—are there 'reasons for actions': elsewhere there are just 'causes of happenings'. This is the considerable conceptual progress consequent upon the development of the physical and biological sciences. Equipped with such understandings, mature I AM thinkers can make considerable sense of their universe and their biosphere.

But not yet of their world. For as long as they remain wedded to the notion that while of all else that exists it is appropriate to say that it *is*, of only I it is appropriate to say that it *am*, then what might be called their 'personic exceptionalism' will not let them make coherent sense of the human. Such thinkers escape animism and immanence in their thinking about the universe and the biosphere alright. But limited by their homuncular conception of I as an entity separate from it and over-against it, their thinking about the world remains thoroughly enmeshed in both. In explaining the I as an entity which has experience rather than an entity which is experience, their thinking remains animistic. And in explaining the I as an entity which reasons its way to free choices rather than an entity whose choices are deterministically caused, their thinking remains immanent. I AM stories concerning persons maintain, of course, that animistic and immanent thinking about them is thoroughly appropriate. And that morality would be inconceivable on any other basis. But stories are not explanations (see below). The explanation of persons, or anyway the one which is offered by my hypothesis, makes it clear that, far from its being appropriate and facilitative of morality, animistic and immanent thinking about persons is that which legitimises a thoroughly inappropriate, thoroughly immoral, attitude towards them. An attitude couched in terms of blame and punishment, it is one which exemplifies nothing so much as a failure of love.

It is stage three reasoners, I IS thinkers, who can at last sort out what are the vital differences between 'how' and 'why' questions. For them, every happening—in the universe, the biosphere and the world—is caused: which means that, concerning every happening, there is a 'how-did-this-come-about?' question to be asked. But when the happening to be investigated is one in the world, what they understand is that some of the most revealing versions of that question are versions such as 'how did this *why* question come to be asked?', or 'how did this *reason* come to be given?', or 'how did this *story* come to be accepted?'. In a philosophical culture in which the Socratic position (that rational behaviour is behaviour for reasons rather than behaviour as the result of causes) is still perhaps mainstream, all this will take some swallowing. But if, in the spirit of my hypothesis, you were to ask (of the Socratic position) the 'how-did-this-story-come-to-be-accepted' question, the answer, I would suggest, would be the following. It came to be accepted because the thinkers concerned were working from within the I AM rather than the I IS paradigm.

In the terms of my proposed paradigm shift, the shift from I AM to I IS thinking, the way to understand the 'reasons *and* causes' duo would be as a pairing characteristic of the more advanced level of the I AM stage. At that stage, things inorganic and organic are understood to take place as the result of causes alright. But things personic are still thought to take place as the result of reasons: hence 'reasons *and* causes'. It will be when I AM thinking finally collapses before the advance of its I IS successor that this particular 'reasons *and* causes' pairing will be able to be given up to the history books as just 'a way we used to think'. And that will be good.

But there is a not insignificant caveat here. In the current context, the context of an exploration of the 'reasons *and* causes' duo in the light of an I AM to I IS shift, it has been both appropriate and necessary to register the primacy of causation: indeed, to see reason itself not as some sort of super-duper first cause, but rather as itself caused. But this of course is not the only context in which the idea of reason can be situated. There are other contexts, several of which have already been considered in earlier parts of this book. And in many of those cases, the last thing we need is a downgrading of reason's significance. Its superior rationality vis-à-vis logic; its continuing centrality to what is still the unfinished project of the Enlightenment; its role in the development of the scientific method: all these and more are such contexts. Certainly, reasons are caused: but then, so is everything else. Certainly, 'reasons *and* causes' is a misleading and inappropriate pairing: but just as the idea of 'cause' as decoupled from 'reason' is important, so also is the idea of 'reason' as decoupled from 'cause'. In *Philosophical Enquiries into the Nature of Human Freedom* (1809), Schelling, as only to be expected from a Romantic thinker, tried to downgrade its significance with his concept of an 'irreducible remainder': that which, at the end of all debate, stays unable to be reached by reason. And it seems to me that, the misleading 'reasons *and* causes' vocabulary notwithstanding, it remains as important as ever to challenge such woeful mysterianism wherever it is to be found. For it seems to me that the staking out of some territories as those in which reason is to be declared 'off limits', it is nothing much more than one of the favourite ideologies of those afflicted with a fear of freedom.

(33) right and wrong

As I pointed out when introducing this 'right *and* wrong' pair in Chapter 4, the distinction sometimes made by the professionals between it and its 'good *and* evil' companion is one which, anyway in my experience, hardly ever figures in philosophical conversation. And as already defended at some length under my earlier 'good *and* evil' entry, the position which my hypothesis requires on this one as well is that we interpret it positively rather than normatively: that is, that we recognise that the notorious problem of deriving an 'ought' from an 'is' does not arise when, as is the case with my hypothesis, no 'oughts' are posited in the first place. In other words, it is my claim that that which is good (right) is not that which all persons at all times *should* do, but that which some persons at a particular time—or rather at a particular stage of development—actually *do* do. The stage of development concerned is that at which I AM thinking has become I IS thinking. And given that there is not really much that needs adding to such formulations here, I want to make just a single point before passing on. The single point I want to make is this. When I put moral content into my positive I IS developmental category, when I for example somewhat contentiously suggest that I IS moral persons are those who would never lie but who might sometimes kill (see Chapter 11), of course I am making a moral judgement. But the case I am arguing is that whether or not such a judgement is right (good) or wrong (bad) is an empirical question. We will only know whether I am right or not when (and if) persons emerge who *on other counts* behave in ways that suggest they may appropriately be described as I IS thinkers, and who *on this count also* are to be found defending a lying-is-always-wrong but killing-is-sometimes-right position. Each time such a person is shown to exist, so my hypothesis is strengthened. It will be when (and if) they come to exist in some numbers that we will be able to say that my hypothesis is well validated. Let such judgements become the norm, and we would have a scientific theory. Let them become universal, and we would have a scientific law. Meantime, what we have, what I have put forward, is an hypothesis.

(34) stories and explanations

Perhaps the most common thing to want to know when listening to a story is 'what happens next'. And of the various 'whats' that could happen next, some will be those traditionally classified as events (the sun goes down, the wind gets up), and some will be those traditionally classified as human actions (the woman starts to prepare her speech, the man starts to prepare their dinner). These two sorts of things, events and human actions, have been absolutely central to debates concerning both the general status of the 'stories *and* explanations' duo, and the particular nature of its 'explanation' component.

Historically, by far and away the most common approach to such debate has been to argue that events and human actions are fundamentally different: briefly, that events are effects, and determined; that human actions are (first) causes, and free. And as I know that by now will be tediously clear, this is not the position I take. In my conceptual scheme of things, the 'human action' happening is as determined as the 'event' happening. And the only thing that really needs adding here is some comment on the way in which 'stories *and* explanations' terminology, common among the professionals, offers another way of presenting my position. It does so by permitting a formulation whereby the idea that human actions are (first) causes and free is a *story* offered by I AM persons, while the idea that human actions are effects and determined is an *explanation* offered by I IS persons. Let me comment, if briefly, on such a formulation.

In introducing this specimen in Chapter 4, I suggested that the difference between stories and explanations is in the nature of the patterns created in them. Stories create narrative patterns made out of intentions, motives, purposes: explanations create non-narrative patterns devoid of intentions, motives, purposes. Stories are works of art, relatively simple to make, which use local and specific materials: explanations are works of science, rather more difficult to make, which use a universal methodology. Stories come first: explanations come second. To give an example which makes it clear that stories are the product of I AM thinking, explanations the product of I IS thinking, consider the boy who has stolen the apple.

I AM stories would say he stole the apple because he was dishonest; because he was led astray by feckless friends; because he was badly brought up: or for other such local, specific and simple *reasons*. I IS explanations, by contrast, would need to deal in such realities as that all societies built on the institution of property, and with unequal ownership of the same, exhibit the phenomenon 'stealing'; that some statistically known proportion of young boys pass through a 'stealing' stage in their adolescence; that which boys they will be will be a function of upbringing procedures, with authoritarian styles encouraging such 'deviance': or other such universal, general and complex *causes*. And then just to compound the differences, and at first flush somewhat ironically, those telling the local and specific and simple *story* would generalise wildly to 'he is a thief', while those offering the universal and general and complex *explanation* would particularise cautiously to 'he stole an apple'. In what then, besides this relative modesty, does the superiority of the I IS explanation consist?

In *The Fabric of Reality* (1996), David Deutsch suggests that an explanation is about 'the inner workings of things'; about 'what must be so' as opposed to 'what just happens to be so'. It is better when it 'leaves fewer loose ends', 'requires fewer and simpler postulates', and 'meshes more easily with good explanations in other fields'. With such a characterisation in mind, I feel reasonably confident in claiming, for the I IS hypothesis, that in coping so well with so many of the 'and' problems of philosophy, it certainly 'leaves fewer loose ends'; that in postulating only 'events' rather than 'events and human actions', it certainly 'requires fewer and simpler postulates'; and that in removing 'free-will exceptionalism', it certainly 'meshes more easily with good explanations in other fields'. But I would claim also yet more.

In *Three Rival Versions of Moral Enquiry* (1991), Alasdair MacIntyre asks, of his three versions, whether there is 'any way that one of these rivals might prevail over the others?' He then suggests that a possible answer to his question is to be found in the writings of Dante. As MacIntyre puts it, the poet argued that 'that moral narrative prevails—which is able to include its rivals within itself, not only to re-tell their stories as episodes within its own story, but to tell the story of the telling of their stories as such episodes'. It seems to me that Dante's answer provides an excellent basis not so much for judging between different stories (narratives), but for showing the superiority of explanations (non-narratives) over stories of any kind. For when it at last emerges—as I have said, explanations are always later on the scene than stories—a good explanation is one that precisely fulfils both of Dante's criteria: it re-tells the stories as aspects of its own account of things; and it enables us to understand how such stories came to be told in the first place.

This is exactly the kind of claim that I am prepared to make for the I IS hypothesis. It seems to me that, via its concept of a three-way classification of the human and a two-way classification of the person, it offers that which enables us to solve (or dissolve) many of the erstwhile problems of philosophy. And that, via its concept of a 'having' ideology, it also offers that which enables us to understand how it came about that the I AM stories it replaces should have been created in the first place.

(35) time and space

It is my experience that discussion of the 'time *and* space' pair, in philosophical conversation but also among the professionals, can very quickly degenerate into what for me is a somewhat exasperating 'triumphant resignationism'. Particularly attractive to the temperamental mysterian, for whom too much understanding is a threat to decent ignorance, this is a pair which, because the questions under discussion are so engagingly huge, and the answers (where there are any answers) so reassuringly incomprehensible, permits that urbane splaying of uncomprehending hands, that poised shrugging of punctilious shoulders, which in the face of the unintelligible is allowed to pass for wisdom. Steven Hawking's *A Brief History of Time* (1987) was translated into forty different languages, remained in the Sunday Times best-seller list for eight months, and sold in excess of one copy for every seven hundred and fifty people on the planet. It would be nice to think that, as a result, at least the 'time' member of this 'time *and* space' duo would have ceased so to bewilder. But I doubt it. And I doubt it not so much because the concept of time is one intrinsically difficult to comprehend—though it is!—but because it's also a concept with thoroughly ideological ramifications. And that means that, together with its 'space' mate or alone, it's not going to be readily amenable to rational exploration.

In most philosophical conversation on the issue, this is something which becomes very rapidly obvious. Discussion is likely to swing all over the place. Someone recalls that, for Whitehead, our understanding of time is derived from the various passings of things natural: seems likely. But then someone else remembers that for St. Augustine, sounding remarkably like Paul Davies—or should it be the other way around?!—the world was made not *in* time but simultaneously *with* time: and that seems monumentally mysterious. Except that maybe some combination of such as Durkheim, Saussure and Wittgenstein ('social constructionists' speaking up now!) can get us off the hook by persuading us that 'time' and 'space' —like 'reason' and 'nature' and other such troublesome things—are socially constructed, and thus endemically unstable in meaning anyway. In the absence of much scientific light, it is my experience that, more often than not, it is ideological heat which wins the day.

These are confusions yet further compounded when, as is often the case, discussion of the 'time *and* space' pair goes on in the context of a discussion about mind. In *Mind, Brain and the Quantum* (1989), for one example, Michael Lockwood argues—if I have understood him aright—that since Special Relativity makes space and time into space-time, then mental states, given that they are in time must also be in space: an error, or so it seems to me, which arises from failing to separate out the 'time' which is a constituent part of the 'space-time' concept from the 'time' that makes me late for my train. And on a March 1998 edition of the radio programme 'On giant's shoulders', for another, the former editor of *Nature* John Maddox, refers to the 'taxing question of where the mind is' as one which 'all the neurophysiology of the past century hasn't answered': again an error, or so it seems to me, which arises from failing to separate out the extensional from the durational. The mind, like an hour, is a concept concerning the durational in time, not a concept concerning the extensional in space. And that means that since Maddox's question is not really a question, it can't have an answer. As Bergson among others was wont to insist, there are only three dimensions of extension: the fourth is a dimension of duration. The mind, like an hour, isn't anywhere: it's 'somewhen'.

Among academic philosophers, there does seem to be some sort of professional consensus which traditionally *does* separate out different types of time. Many of them would probably recognise a threefold division: a cosmological perspective, concerned with the time of Nature; a subjective perspective, concerned with the time of individual consciousness experienced subjectively as duration; and a social perspective, concerned with the time of human history. But first, these are distinctions not always acknowledged by all of the professionals. And second, these are distinctions hardly ever acknowledged by any of those taking part in philosophical conversation. Indeed in such conversation, or such is anyway my experience, the most common position adopted is one of apparent bamboozlement before cosmology-speak. 'Cosmo-babble'—as the psychologist in me is tempted to call it!—is apt to claim such things as that time is 'just another dimension' like the (curved, of course) three dimensions of space; that it might be possible to travel in it, backwards and forwards as we do in space; that Einstein's theory of relativity has conclusively shown that there are no absolutes any more concerning the stuff. And so on. For the purposes of understanding anything at all, such talk seems to me to be monumentally unhelpful.

As I understand it, all dimensions above three are just mathematical formalisms. Developed for the most part by nineteenth century mathematicians, a 'space' of nine dimensions is just an ordered set of nine numbers which can be manipulated in the same sort of way as can an ordered set of two numbers in a grid. The existence of such a totally formal 'space' is just a matter of whether or not a consistent set of rules for talking about it can be formulated. They can. So it exists. Fine. But the supplementary question as to what it exists *as*—is it a totally formal set of rules invented by humans, a characteristic of physical reality, or both?—seems frequently to go unasked. And perhaps also unrecognised.

This is also true about questions concerning the famous curvature of space. In much of the philosophical conversation I know about, it does not readily emerge that whether we say that a body moving in space deviates from a straight line because a force is acting on it, or instead that it continues in its straight line but the very space it moves in is curved, is just a matter of convention, just a formalism.

And it is also true, notoriously true, about questions concerning that 'relativity', latched on to as the (surely thoroughly inappropriate) name for Einstein's 'time *and* space' ideas. Once we are paddling around in notions of relativity, it's but a short step to the 'everything's *relative* you know' of relativ*ism*. And trying to suggest that all we need to acknowledge is that, for some purposes, it's not three-dimensional Newtonian 'space' but four-dimensional Minkowskian 'space-time' which is the relevant *absolute* to think with does not go down well.

When I am trying to think coherently about this 'time *and* space' specimen, I find the idea of 'betweennesses' useful. Axiomatically, all four Minkowskian dimensions, the three of space the one of time, measure 'betweenness': that's what a geometry does. But it surely in no way follows that, taken separately, the mathematical reality of the betweennesses of space and the mathematical reality of the betweennesses of time are the same as each other. Just as it surely in no way follows that, taken together, the mathematical reality of the betweennesses of space and the mathematical reality of the betweennesses of time are the same as the physical reality of either. In the case of the latter difference, the one between mathematical reality and physical reality, isn't this what accounts for the fact that when Zeno of Elea divided space and time into infinitesimal proportions thereby ensuring that Achilles came second, athletes just merrily continued to outrun tortoises? Isn't whether or not a particular mathematics has meaning for a particular physical reality a matter for empirical investigation rather than a priori dictat? And in the case of the former difference, the one between the mathematical reality of the 'betweennesses' of space and the mathematical reality of the 'betweennesses' of time, I would suggest that there are at least four crucial differences between the 'betweennesses' of the one and the 'betweennesses' of the other which such a mathematics would need to accommodate.

First, we cannot move in time as we move in space. Instead, and willy-nilly, we are carried along in it: not moving towards the millennium, but waiting for it to arrive; not going back into childhood, but remembering it (we think!) from now. Second, we do not experience time passing as (say) Oxford passes when our train rushes through its station. Instead we experience its betweennesses as differences: differences between our present understanding of present events (perceptions), and our present understanding of past events (memories). Third, time is unidirectional. While moving back and forth in space is a commonplace, moving back and forth in time is an impossibility: the thermodynamic aspects of the problem would just ensure that a time-reversible world could contain no life at all, and thus nobody travelling anywhere, or 'anywhen'. And fourth, while no two things can occupy the same space, an infinity of events can occupy the same time. It seems to me that these, and especially this last, are the 'time *and* space' phenomena towards which it would really be useful to turn our philosophical attentions. And that when we do so, when we manage to screen out the ubiquitously obfuscating 'cosmology-speak', vital understandings can emerge.

They emerge, it seems to me, out of the deconstruction of a particularly damaging ideology: an ideology which persistently and relentlessly uses the (inappropriate) imagery of space to stand in for the (appropriate) imagery of time. In *Philosophy Now* (1996, issue 16), Richard Mason, Staff Tutor in Philosophy for the University of Cambridge Board of Continuing Education, wrote a piece on the metaphors of relativism. And he pointed out that almost all the imagery used in such metaphor is spatial. There are 'points of view' and 'positions'; there is 'where one stands' and there are 'fields of enquiry'; there are 'areas of discussion' and 'intellectual domains': and this is to name but a few. And he suggested that such metaphors could be misleading for three reasons. Firstly, because while the proposition 'you must be somewhere' or 'you can't be nowhere' is a truism about space, it is not true about what he called 'the intellectual world', where the idea 'that there are intellectual *terrains*' and that 'if you are not *within* one then you must be *within* another' can lead to 'a rather simplified intellectual world which contains only two possibilities: agree, or not'. Secondly, because while 'in literal terms, *finding out where you are* does not affect your location', when it comes to 'holding opinions' this is 'the reverse of the truth', because '*understanding* where you are can *alter* your intellectual location in the most radical way'. And thirdly because while '*you can't be in two places at the same time* is a truism applied to space' and 'looks beguilingly similar to the logical truism that you can't, or shouldn't, believe both A and not-A', this does not follow because 'intellectual *positions* are just too complicated for that'. He finished his piece by pointing out that 'you can't *move* from knowledge to ignorance like *stepping* from one *area* to another', and that 'the relativists idea that we can *survey* different *areas* of thought—different schemes or frameworks—from a different, maybe *higher* vantage point is completely misleading' once we understand that 'to realise that you have a choice ... is an irreversible step, much more like moving from the present into the future than like moving from one place to another.'

Now reasonably obviously, there are aspects of this with which I would want to take issue. But it seems to me that he is not only bang-on right about the misleadingness of using space imagery for time phenomena, but that it involves a misleadingness considerably more misleading than he was perhaps suggesting. My first (and less satisfactory) way of developing his argument was to say that it is not that we *stand* in space but that we *understand* in time. My second (and more satisfactory) formulation became possible as I developed my ideas about 'I', and thus 'we'. In short, it came when I saw that the entity which stands in space and the entity which understands in time are different entities: the one an organic entity existing as part of the biosphere; the other a personic entity existing as part of the world. It seems to me that the absolutely vital difference between the two sorts of metaphor is the extent to which they differentially accommodate notions of development. For while the space (place) in which an organic entity stands does *not* exist in a developmental relationship with other spaces (places) in which it has stood or will stand, the time (moment) from which a personic entity understands *does* exist in a developmental relationship with other moments from which it has understood or will understand. Thus do the space-metaphors characteristic of relativism, as opposed to the time-metaphors appropriate to realism, so often allow us to miss the developments which constitute progress.

For they are indeed misleading. They mislead, for example, when they encourage us to think erroneously about the past and the future. Using them, for example, we 'peer into' the future to try and 'see' it. And it's not only that it's not possible to see duration—though it isn't—but that when the more appropriate time imagery *is* substituted, it is then that we discover (see 'freedom *and* determinism') that it is not the future into which we need to peer if we are to see our way forward, but the past which we need to change if we are to understand our way forward. And again, such thinking about the past, perhaps counter-intuitive but I think absolutely essential, is thinking which is discouraged by an essentially spatial dating system which tends to present it as fixed and unchanging. When zero is (arbitrarily) fixed at some 'point' in history—in the most commonly used calendar of today, at the birth of Christ—other dates are (also arbitrarily) fixed at other 'points' in history: and seem to stay there. That famous battle was *in* 1066 in much the same way as it was *in* Hastings. And will always be in both 'places'. Contrast this with the way of dating that we use for those 'very long ago' significant events: the way in which we say that the big bang was some thirteen thousand million *years ago*; that the earth was formed some four thousand six hundred million *years ago*; that eukaryotic cells came into existence one and a half thousand million *years ago*; that *Homo sapiens* emerged perhaps one hundred thousand *years ago*; and that zero was invented some two thousand four hundred *years ago*. When by contrast with this system, 'years since' (how far forward from a then which is an arbitrary zero) take over from 'years ago' (how far back from a now which is a non-arbitrary zero), a vital truth can be masked. We can fail to see that the Battle of Hastings is not fixed at some arbitrary 1066 *place-like* date, but moves from 10 to 100 to 1000 years ago *in time*.

And they also mislead, obfuscate in fact, when what we are trying to think about is progress. This is an idea more usually thought about in connection with the collective historical than with the individual developmental. But I am wanting to apply it to this latter also: to apply it, that is, to the process I have called selving. If I am arguing, as I am, that as persons we are definitionally 'self-ish', and that across an individual lifetime the selving process is one which 'progresses' towards the production of that moral self which when it 'ishes' *its* self, 'ishes' all *other* selves equally well, then to such a conception the space metaphors of relativism are thoroughly inappropriate. They can seem innocent enough: we 'pass away' (space imagery) rather than 'die' (time imagery); we spend our final hours 'on this earth' (space) rather than 'in this world' (time); we 'head towards the next millennium' (space) rather than 'age' (time). With perhaps the ultimate in the 'substitute-space-words-for-time-words' phenomenon contained in the phrase 'the space of time', these phrases—and countless others of the same ilk—can seem to be just some of the myriad ways in which we exhibit our fear of, even our refusal of, death. But because I think that while they are such fears and refusals they are by no means *only* such fears and refusals, let me deal with the whole issue rather more fully.

The fundamental cause of the conservative and conservationary fear of change, that fear (sometimes called nostalgia) which in so many people grows apace as they get older, is I think fear of death. It is a displacement behaviour whereby what is *in general* ceasing to be (the past) is to be maintained over what is *in general* coming to be (the future) as a forlorn and hopeless way of trying to ensure that what is *in particular* ceasing to be (my life) be maintained over what is *in particular* coming to be (my death). It is the cry, 'please God let what *has been* and what *is* (which includes my life) be preserved over what *will be*' (which includes my death). And all those substitutions of space words for time words help us to hide from the reality that death is not 'passing' (from one *place* to another *place*) but 'ceasing to be'.

But this is hardly a new observation. What is new, or anyway so I claim, is the way in which my hypothesis deals with such fears. It is my case that the fundamental way in which fear of death, and thus fear of change, will be alleviated is by the paradigm shift which I have identified. For with an I AM consciousness, the I is an entity in space: if it is here it is not there; if it is there it is not here; *but it is always somewhere*. So the question as to where such an I might be after death is real enough, and has urgently needed a real space/place answer: in heaven, in hell, with my father, in eternity, and so on. But it is difficult to know which of two possible thoughts about this 'place' is potentially the more terrifying. If all that I know of is *this* place that I *am* in, and yet I have to imagine *that* place that I *will be* in, a place about which I can know nothing and from whose shores no traveller returns, that is potentially very frightening indeed. Think only of all those dreams where one is in one place, one is trying to get to another place, and one's efforts—ever more desperate—are constantly and constantly and constantly frustrated. But the alternative—all that I know of is *this* place that I *am* in, and yet I have to imagine *no* place that I *will be* in—is maybe even worse: if not impossible. But suppose I AM is not. Suppose that *the place where* I AM is supposed to exist in space is actually *the moment when* I IS does exist in time. An I IS does not exchange *this* place for either *some other* place or *no other* place, because the mode of existence of an I IS is not a spatial mode at all but a temporal one. Certainly we need such things as Spinoza's insistence that the wise person thinks less of death than of anything, or Wittgenstein's understanding that death is not an experience of life, to help us contemplate dying without fear. But what I am suggesting is that there is a way *theoretically* much more substantial which will do the same job, but better. For our fears and refusals are not only our fears and refusals. They are also, and I think much more fundamentally, some of the myriad ways in which we exhibit a more or less total misunderstanding of just what it is that 'we' are, and thus a considerable philosophical confusion concerning 'I'.

These are the ways in which philosophical conversation about the 'time *and* space' pair underlie what I write. All the problems presented by 'relativism' as opposed to 'realism', for example, can be situated here. With the 'space' imagery of the post-modern, there is no progress, no development. In space, one goes round and round in circles (or anyway ellipses), forever returning to where one started from. Which is why, if we are to understand development, it is a re-worked 'time' imagery of the yet incomplete Enlightenment project that we need. Development is through a continuity of experiences, and experiences are durational not extensional. It is a gradual and graduated process, lived incrementally. It occurs as a series of stages, the successive emphases of which are to make it ever more inclusive. As I hope I have already made abundantly clear, it is my belief that one of the biggest hindrances to the development of moral science is the wholesale determination to try to derive theories about the world from theories about the universe and the biosphere: a determination which is misplaced first because it conflates different realities (the universe, the biosphere and the world are not the same), and second because it attempts to derive theories from theories, where theories can only be derived from observations, hypotheses and experiments. In the particular case of our 'time *and* space' theories, it seems to me that while those appropriate to our study of the universe have been radically transformed by Einstein and all that has flowed from him, it is totally unproductive to try to derive those which would be appropriate to our study of the biosphere, still less of the world, from such sources. Theories about the way in which the 'time *and* space' duo functions *in the world*, the ones we need for the creation of the moral sciences, have to be derived from observations of, hypotheses about, experiments on—*the world*. And the place to find such observations, hypotheses, experiments and theories—and they do exist—is in psychology texts and journals not physics texts and journals.

To conclude what I have to say about this my last specimen, let me turn to the poet Horace. As he long ago put it, 'caelum, non animum, mutant qui trans mare currunt' (the skies not the souls are changed of those who fly across the seas). Because they are not spatial but temporal entities, persons change not by moving from place to place but by selving from time to time. And that means that the most important thing to register about this 'time *and* space' pair *as it relates to the I IS hypothesis* is the fact that, differently from inorganic and organic entities, personic entities exist not as material things in space, but as happenings in time. As later we will see, this will be a fact which will play a considerable part in my thinking about equality. For historically, the only equalities easily conceded as appropriate by I AM consciousnesses, and pretty advanced I AM consciousnesses at that, have been the equalities of humans as organisms: organisms both female and male, both black-skinned and white-skinned, both able-bodied and disabled, for examples. The fundamental equality which I IS consciousness requires, by contrast, is one which recognises that since persons are time entities not space entities, then 'lives of equal value', as the philosophical jargon has it, has to translate into 'times of equal value'. And this means that, whatever the rhetoric to the contrary, social systems which permit payments of £5 an hour to one, £50 an hour to another, and £500 an hour to a third, do not respect that 'lives of equal value' criterion. But this is for later chapters!

* * *

And so, not before time, I come to the end of these three lengthy 'experiment' chapters. Difficult though it might be to contemplate with equanimity, there could have been four—or maybe even more. I could have extended my specimens list with others: with 'rights *and* duties', 'form *and* substance', 'form *and* content', 'extrinsic *and* intrinsic', 'rationalism *and* empiricism', 'Enlightenment *and* Romantic Movement'—and so it goes on. And I think I would want to claim that, had I done so, then in those cases too I could have shown that what keeps them primitively additive rather than more sophisticatedly subordinative is the essentially dualist conception of I which lies embedded in I AM usage.

The considerable advantage that this alphabetical encyclopaedic format has afforded me is that I have been able repeatedly to make the same points but in different ways and in different contexts. And although that obviously risks the 'oh not again!' reaction, I think that when ideas are new, not to say strange, then such modified repetition can be invaluable. Its considerable disadvantage, however, is that with its 'this and that' *and* 'this and that' *and* 'this and that' construction, it lays itself wide open to what might seem a particularly telling and wry criticism. For doesn't it stand, in itself, as a veritable monument to that very 'and-ism' which, I am suggesting, is some sort of indication of the relative immaturity of thought?

Indeed it does. The three 'experiment' chapters just cry out to be seriously 'un-anded'. And in the next chapter, the 'theory' chapter, I will try to do just that: that is, to sift and sort the findings of my 'and' specimens so that they can start to be presented a little less additively, start to be structured a little more subordinatively. Of course I shall fail. And the fact that I shall fail is precisely a vindication of the very point I am trying to make: that new thinking frequently starts in the (easier) 'additive' mode, and has slowly to progress to the (more difficult) subordinative one. I know that the structuring of the next chapter will not make all of that progress. But I hope that the structuring of the next chapter will make at least some of that progress.

Chapter Nine

THEORY

Emergence

A theory is an hypothesis validated. And that means that, to the extent to which I think that the experiments of the last three chapters validate the hypothesis of the previous one, to that extent I can start to think of what I have made as a theory. I am only too aware, however, that that would be ridiculous. For a theory is an hypothesis *well* validated. And that means validated not by one person's first attempt at what is anyway a very general validation, but by many people's many (and independent) attempts at what would need to be several very specific validations. Theories are precious things. And their significance should in no way be diminished by laying too premature a claim, on behalf of one's ideas, to their status. Contrary to popular opinion, for example, a theory is much more precious than a fact. Explanatory of many facts, validated by the work of many scientists, it is a vital tool for use in the making of sense. And a good theory, one such as the theory of gravity or the theory of evolution, because it is explanatory of many facts, and validated by the work of many scientists, is quintessentially so. It is therefore about only a candidate theory, only an hypothesis, that I sit down to write this 'theory' chapter.

But even as only that, my boldness (or is it foolhardiness?) gives me pause. It's not that my candidate theory has not been proved. On that one [see 'knowledge *and* science'], I have the comforting reassurance that no theory is ever proved, and thus not accepting a theory until it is proved does not mean staying neutral, whatever that could be made to mean, but staying in some other theory which is not proved either. And it's not that, at a time when grand narratives are supposed to arouse only incredulity, mine could be thought to be a very grand narrative indeed. For on that one [see 'realism *and* antirealism'], I have the comforting reassurance that in the debate about such matters not (quite!) everybody is some sort of postmodernist nowadays. It's rather that what is implied by putting forward even only a candidate theory is in fact much more audacious than anything towards which the 'not-proven' or 'grand narrative' criticism might be directed. For I am suggesting that what I am offering is not a narrative of any sort, grand or otherwise, but an explanation. And on that one [see 'stories *and* explanations'], I am making a claim which could not be more huge. So how does she have the nerve to put forward a theory, even if only a candidate theory, of such outrageously enormous proportions? Who on earth does she think she is? Well, *that* at least is a question an answer to which can get me launched. As I sit down to write my thoroughly audacious chapter headed 'theory', who on earth do I think I am? Or rather—and with each word of the different formulation being critically important—what in the world do I think I is?

I have been arguing that I is the linguistically structured experience of a *Homo sapiens*: in my case, of the *Homo sapiens* Ann Long. And on the critical question as to whether or not such an entity is free—can I decide *not* to sit at my desk and *not* to begin my chapter?—I have also been arguing that it is not: that I can't. I can't because what I shall call 'the unity of consciousness' provides only one slot, a time slot not a space slot [see 'time *and* space'], in which the decision to sit or not to sit can be taken. Within such a slot, I can't decide both to sit down and write and to get up and go: that is, I can't do the two things simultaneously. Indeed, it is not totally outrageous to say that I can't even do one thing simultaneously! For if somewhat over-simply, then nonetheless graphically, it's not only that one I becomes, in sequence, several other Ies *between* the consideration of one option and the other, but that one I becomes, in sequence, several other Ies *during* the consideration of *each* of the options. The linguistically structured experience of the *Homo sapiens* Ann Long—that is, the person Ann Long—changes on that millisecond to millisecond basis. And such changes, summed, mean that, as one I, it would *with no choice* have got up and gone: as a subsequent I, it would *with no choice* have sat down and written. I could maybe capture this point with an illustrative contrast: Heraclites is supposed to have argued that, because it is continuously becoming a different river, we cannot step into the same river twice; I am arguing that, because I am continuously becoming a different I, the same I cannot sit at (or turn from) the desk even once. The linguistically structured experience which constitutes turning away, and the linguistically structured experience which later (by milliseconds) constitutes sitting down are two different Ies. Neither has choice. Both are determined. But they act differently—the one to turn away, the other to sit down and write—because they are two different instantiations of I. Their no doubt myriad differences at a micro level are going to be those which, for the most part, we cannot (yet) spell out. But at least one of their differences at a more macro level is (already) one which we can. We can already understand—or so I would claim—that while the earlier I is not in part constituted by the later decision to sit down, the later I is in part constituted by the earlier decision to turn away. Because—as Henri Bergson put it—'the following moment always contains, over and above the preceding one, the memory the latter has left it', the later I is not only more experienced than, but also differently experienced from, the earlier. And because each such I—a durational not an extensional entity, remember—is that which *is* linguistically structured experience rather than that which *has* linguistically structured experience, each such momentarily existent I is determined.

I have also been arguing that far from this determinism being some sort of threat to our being able to have our freedom now, it is actually some sort of guarantee of our being able to make our freedom then: where 'then' is the time at which the recognition of the necessity of equality to that freedom has sufficiently accumulated. I am suggesting that Ies conceived of as *having* linguistically structured experience, and as therefore being endowed with the 'free-will-plus-will-power' that goes with such a conception, are endowed with that which is notoriously and lamentably weak. But that Ies conceived of as *being* linguistically structured experience, and as therefore being free from the constraints of an illusory free will and a puny will-power, will make together that freedom whose only possible source is determinism [see 'free will *and* determinism'].

So in answer to that initial question, to the who-on-earth-do-I-think-I-am question, reformulated as the what-in-the-world-do-I-think-I-is question, it is an I conceiving of its self in an I IS rather than an I AM way that sits down to write its 'theory' chapter. It's not that this is who I think I am. It's that this is what I think I is.

As I said at the end of the previous chapter, my main aim in this one will be to try to make my thirty-five 'ands' rather less 'andish': to try to link them not in the arbitrary way represented by their encyclopaedia-like alphabetical order, but in the rather more structured way represented by their being situated in the development of a single argument. To this end, and as already begun in my opening paragraphs here, one of the ways in which I will do this is to make square-bracketed reference to them at that point in my unfolding text at which such reference seems most appropriate. But a full shift from the additive to the subordinative mode will require, of course, much more than that. A full shift, or so it seems to me, will require a structuring concept around which I can develop my case. It is such a structuring concept, therefore, which first I must try to identify.

When presenting my hypothesis in Chapter 5, my emphasis both *between* each of the three levels of the inorganic, the organic and the personic, and *within* each level, was on change across time: was on growth, maturation, development, progress. And I presented that change, again both between and within levels, in a thoroughly stage-structured way. In presenting, firstly, the inorganic universe, I said that at 15000 million years back, from a big bang singularity, an energy-matter universe *emerged*; and that at 4600 million years back, from the cooling and aggregation of some of that that matter, this solar system *emerged*. In presenting, secondly, the organic biosphere as *emergent from* that inorganic universe, I said that at 2000 million years back (or given some recent discoveries, maybe earlier), the first living things *emerged*; and that at what might turn out to be no more than 50,000 years back, from amongst the other members of the *Homo* genus, the species *Homo sapiens emerged*. And in presenting, thirdly, the personic world as *emergent from* that organic biosphere, I said that at the (disputed) point at which language appeared normal persons *emerged*; and that as we speak (so to speak!), at the point at which I AM conceptions of a person give way to I IS conceptions of a person, then moral persons are *emerging*.

That's a lot of emergences. And the joints at which I have carved reality to make such emergences emerge—the earlier ones pretty widely accepted, the later ones more peculiar to me—are all of course chosen joints. Where other ways are obviously possible, I have chosen to carve up a sequence of events in one way rather than another. And all such choices of course are grounded in the particular theories being used. In some cases, particularly perhaps when it is something of the universe or the biosphere which is under study, the use of one theory rather than another will be sufficient to put us either inside or outside a united and mainstream scientific community. In other cases, particularly perhaps when it is something of the world which is under study, the use of one theory rather than another will not do that. Instead, it will put us either inside or outside one or other of a series of different and competing camps within a community still only struggling towards the scientific. But in either case, the decision to carve at all is highly contentious. The decision to present change in terms of some sort of stage theory of development, and particularly to do so using some sort of quantitative-qualitative distinction [see 'quantity *and* quality'] is by no means an uncontested procedure. And that being so, the structuring concept around which I will try to mount my case in this chapter is the one most important to any such decision: the concept of emergence.

Mine is an emergence theory. And as has often been pointed out, emergence theories are fighting on two fronts. We are fighting the hard-nosed reductionist monists, who sooner or later (and usually sooner) want to tell us that what has emerged is really 'nothing but' that from which it has emerged—that humans are nothing but animals, for example—and that ultimately everything is physics. And we are fighting the non-reductionist dualists, who want to claim that 'higher' things—things like minds and souls—are just something completely other than could be dreamed of by (or anyway conceptualised by) any sort of science. In this situation, the task of an emergence theory, it seems to me, is to construct a theoretical position from which firmly to resist the reductionism *but not the monism* of the first, and the dualism *but not the non-reductionism* of the second. To borrow yet again Hegel's wonderful word for a concept which would appear to encompass 'annulling' and 'preserving' and 'lifting to a higher level' all at once, what an emergence theory needs is a bit of 'aufhebung'. I don't think I can offer a word: the English language doesn't seem to have one. But in defence of my emergence theory, I will certainly try to offer something of the idea.

In one very simple and obvious sense, that things emerge is startlingly obvious. However difficult it might be to envisage that singularity which, at the big bang, constituted all that there was, it's pretty easy to imagine that it was nothing much like a bus ticket or the biosphere. So given that bus tickets and the biosphere now exist, they have to have emerged: don't they? So far so obvious. But as is usual with the big ideas—and along with such as 'reality' and 'experience' and 'existence', 'emergence' is amongst the biggest of the big—give them more than a moment's thought, and they cease to look so simple. For academic philosophers, a common way of delineating what often they call 'an emergent property', understood as a property of a more complex system, is as one which, although it arises out of the properties of its less complex parts, can be neither reduced to, nor predicted from, the properties of those parts. Thus the liquidity which is a property of water is a property of neither its hydrogen nor its oxygen constituents. Such an example seems simple enough. But for what I will suggest are thoroughly ideological reasons [see 'ideology *and* science'], whether or not emergence can be made a philosophically coherent concept remains a controversial issue.

In *Mindware: An Introduction to the Philosophy of Cognitive Science* (2001), Andy Clark suggests that the relevant literature contains four main contenders as possible approaches to an understanding of emergence. He labels them 'collective self-organisation', 'unprogrammed functionality', 'interactive complexity' and 'uncompressible unfolding'. And while by far the best approach at this point would be for readers to consult the Clark text—or an appropriate alternative—for themselves, I will nonetheless attempt a brief summary of each approach here.

On a 'collective self-organisation' account, an emergent phenomenon is (1) produced as the result of positive feedback, (2) by the many interactions of essentially identical elements obeying simple rules, and (3) in a system constituted by such elements. Examples would be the convection rolls which emerge when liquids are heated, or the flocking which emerges when birds fly in groups. They are effects which arise without external instructions: that is, in the example of the liquids, the temperature gradient does not contain the instructions for the emergence of the rolling pattern.

On the 'unprogrammed functionality' account, an emergent phenomenon is one arising as a side-effect of a sequence of entity-environment interactions, where once again the instructions as to how to produce the effect exist in neither. Clark gives as an example the wall-following behaviour of a robot which has 'an in-built bias to move to the right, and a right-side sensor, contact activated, that causes it to veer slightly to the left'. As long as the two biases are well calibrated, then the robot will, he says, 'follow the wall'. And will do so by what he calls a 'veer and bounce' routine. The point here is that an instruction 'follow the wall' is not explicitly programmed into the robot. Instead, it is a behaviour which emerges as a consequence of the interaction between the entity (robot) and its environment (wall).

On the 'interactive complexity' account, an emergent phenomenon is one produced by 'a certain class of complex interactions between systemic components'. Clark is offering this account as one under which he thinks each of the others so far detailed could be subsumed. It involves the depiction of emergence as 'the process by which complex, cyclic interactions give rise to stable and salient patterns of systemic behaviour'. And he thinks that what becomes clear with this way of thinking, is that emergence obtains in different degrees. Because it depends on 'repeated linear interactions with only simple kinds of feedback loop', the wall-following behaviour in the robot should be classed as only 'weakly emergent'. But because they depend on 'multiple, non-linear, temporally asynchronous, positive feedback', the convection rolls in heated liquids should be classed as 'strongly emergent'.

On the 'uncompressible unfolding' account, one of a totally different kind from the other three, 'emergent phenomena are those for which *prediction* requires *simulation*'. Since this is a way of thinking about the problem which those who use it are apt to stress involves the simulation of all the interactions which give rise to the emergent phenomenon, it is one which, for Clark, is unnecessarily restrictive.

It is within the domain of what in the physical sciences is called 'dynamic systems theory' that these approaches have been developed. And in that light, I am trying to conceive of 'emergence' as that which takes place in what by such theorists are called 'complex dynamic systems'. Dynamic systems are those which involve change over time: a definition wide enough to mean that more or less every system to be found in operation in the universe, the biosphere and the world is dynamic. But it is really only *complex* dynamic systems, those identified by features such as Clark was categorising under his 'collective self-organisation', 'unprogrammed functionality', 'interactive complexity' and/or 'uncompressible unfolding' heads, which are usefully analysed by 'dynamic systems theory'. And while the details of that theory are way beyond both the needs of the current explication, and the competences of its author, some of its more general ideas may be useful. In particular, it may be useful to think about systems where the observable outcome is a sequence of ever more complex states, but where such complexity can often disguise what is actually an underlying simplicity.

Consider, for example, a neural network. At time 1 and in state 1 it receives inputs which determine its transition to State 2; at time 2 and in state 2 it receives inputs which determine its transition to State 3—and so on. It 'learns' because 'behaviours' that initially work are given more weight next time around; behaviours that don't initially work are given less weight next time around; and behaviours that don't work over and over again are eventually extinguished, allowing new behaviours to emerge. The observable outcome is a sequence of ever more complex states. But this is a complexity which disguises an underlying simplicity. For although the presence of a very large number of variables can give each state a very complex description, the rules governing the transition from one state to the next are incredibly simple.

Suppose then that we think of a person—that is, of the linguistically structured experience of a *Homo sapiens*—as a complex dynamic system. If we think, for example, about what students of such systems call a mechanism—a set of possible initial states, a set of possible inputs, and the rules that determine the state-to-state transitions—it does not take much imagination to realise that if the simplicity is real, so also is the incredible complexity. Think just of the 'possible initial states' criterion. Even leaving aside the problem of what, this side of the big bang, 'initial' might be taken to mean, if the problem of getting accurate measurements of the initial state of almost anything is considerable, the problem of getting accurate measurements of the initial state of a person, so construed, is almost unimaginably so. We would need to specify the initial state of a nervous system before its structuring by the linguistic—whenever 'before' might be taken to have been. Such a specification would have to include, amongst very much else besides, the 'initial' (?) state of every neuron, every cell-assembly, every drop of cerebrospinal fluid, every microtubule, every axon, every dendrite, every synapse, every glial cell, every potassium ion, every sodium ion—and so on, and so on, and so on. And that's all before being able to give some sort of operational account of just what it is about the symbol system which is a natural language which enables it to act not as *a possible input* but as *the actual input* which, via yet-to-be-formulated rules which determine the state-to-state transitions, structures the brains of language-users in such a way that the persons which are the resulting linguistically structured experience thereby emerge.

There is not anybody, as yet, who can do much, if any, of this. What's more, there are many who think that such a conception of a person is fundamentally wrong-headed in the first place, and thus unlikely to help us find our ways forward: for readers interested in an explication of just who they are, and just where they alternatively stand, in his book *Philosophy of Psychology* (2005), Luis Bermúdez provides both. So doesn't that mean that an hypothesis based in only the *possibility* of doing perhaps only *some* of it, and perhaps only *sometime*, is, anyway for now, somewhat vacuous?

I think not. For given that they are the levels of precision which would be needed for prediction rather than for explanation, I think that 'initial conditions' specifications such as those outlined above would fall well within Clark's 'overly restrictive' category. The aim of science [see 'knowledge *and* science'] is not prediction: it is explanation.

Indeed, to the extent to which we do see prediction as the aim, and not just of an enquiry into the nature of persons but of an enquiry into the nature of anything, then to that extent we actually hinder rather than help the development of that science. Way before we even get to those 'possible-input and 'state-to-state-transition-rules' criteria, the tightly-defined predictive power of any theory about anything, even the simplest of things, is almost always severely constrained by the difficulty of precisely defining initial conditions. So to be in that situation when trying to construct an emergence theory of persons is thoroughly unexceptional. And my attempt so to construct, to construct not that which will enable prediction (except, of course, in principle) but that which will offer explanation, can proceed.

In attempting then a defence of an emergence theory presented in both levels and stages, I will need to shift from a consideration of emergence in general to a consideration of emergences in particular. But first I must draw attention once more to the all important subject of reification: that is, to the tendency to turn processes (in this case the process called *emerging*) into things (in this case the thing called *emergence*). I want to suggest that it is perhaps the biggest single mistake in all thinking to operate with logical categories which take no account of the developmental characteristics of their referents: that is, which turn processes into things. As just a tiny example, taken from some sorts of academic philosophy, there are those who can merrily compare the philosophy of, say, Aristotle with the philosophy of, say, Wittgenstein as if the several centuries between them were of no consequence whatsoever. Since their persons both belong to the *logical* category 'person', those persons are seen as unproblematically comparable: since their ideas both belong to the *logical* category 'ideas', those ideas are seen as unproblematically comparable. As things called persons (rather than as processes called selving), the thing 'Aristotle' and the thing 'Wittgenstein' can be compared: as things called ideas (rather than processes called thinking), the things 'ideas of Aristotle' and the things 'ideas of Wittgenstein' can be compared.

Such reification—or so it seems to me—is a major obstacle to understanding. And is sadly also ubiquitous. There are funny examples: examples such as that provided by Lord Salisbury defining the ether as the nominative of the verb to undulate. There are frustrating examples: examples such as that provided by me being for years unable to solve the puzzle involving Achilles and the tortoise because I did not grasp that it *dis*solves as soon as one gives up the reification of fractions. There are reifications of verbs: of course. But there are also reifications of adjectives. The adjective 'true' is reified to the noun 'truth', tempting us to forget that there can be 'true statements' without it necessarily being useful to posit something over and above the summed totalities of such statements called truth 'itself'. The adjective 'rational' is reified to the noun 'rationality', tempting us to forget that there can be 'rational statements' without it necessarily being useful to posit something over and above the summed totalities of such statements called rationality 'itself'. The trouble is that, as shorthand, such reifications are both incredibly useful and thoroughly legitimate. This book, like most books, is full of them. After all, the 'naming-of-parts' approach to reality, which begins at the age of about twelve months when the child starts to produce its first morphemes, begins precisely because reality does come *part*icular. But this does not mean that all reification is literally regressive: that it literally involves a regression from the later-acquired verbs and adjectives to the earlier-acquired nouns. The trick surely has to be to use such nouns, particularly the great big abstract ones which litter philosophical conversation, very carefully indeed: that is, to benefit from their shorthand usefulness without slipping back into thinking like a one-year-old child. And if reification is a major obstacle to trying to understand anything, then it is quintessentially such an obstacle when trying to understand emergence —that is, the verb 'emerging' reified to the noun 'emergence'—whose very characteristics are those which belie non-process thinking. So with a health-warning firmly attached to a nonetheless useful noun, let me proceed with my exploration of its plural form.

All the 'emergences' relevant to my hypothesis have been registered already in Chapter 5. To those applicable to the categories which there I labelled the inorganic and the organic, that is to the emergence of matter from energy, of life from non-life and of sentience from non-sentience, I will make no further reference here. Here I am concerned only with the emergences of the personic category. An essentially additive folk psychology lists such emergences as those of 'consciousness' *and* 'language' *and* 'minds' *and* 'persons'. But what I want to do here is attempt a rather more subordinative analysis of the way in which such emergences might fit together: in short, to defend the claim that it is the *consciousness* emerging from *language* use which results in the *minding* characteristic of *persons*.

There has recently been a veritable surge of interest in what have come to be called 'consciousness studies'. The *Journal of Consciousness Studies* was founded in 1994. At Tucson, Arizona, there have been biennial conferences on the subject ever since. And in collaboration with researchers at Oxford, Cambridge and University College London, Warwick University received a substantial sum from the Humanities Research Board in 1996 to investigate 'awareness'. In the case of the journal and the conferences, it is interesting that many of the founding parents of both have been people out of some kind of religious background. And in the case of the grant, it is interesting that it is the Humanities Research Board, not the Science Research Council, which is putting up the funds. But to whatever is the extent to which it might still be the case that consciousness is not a scientifically respectable object of study, it is almost certainly true to say that it is steadily becoming rather more so, not rather less so. Such study, of course, faces huge conceptual problems, not the least of them being that any sort of scientific investigation is bedevilled at the outset by definitional dilemmas. A precise definition of a concept does not precede a good theory: it is the consequence of a good theory. But attempts to get to scientific grips with the consciousness problem are considerably hampered by there being not even an imprecise definition, not even a working definition, which is anyway widely accepted. So let me begin by trying to make as clear as possible the way in which I will be trying to carve my way through what is often the very fluid language used in this area. (And if carving fluid is a mixed metaphor, it's meant to be!)

Among what are many and varied mutually exclusive and competing conceptions of consciousness, there is I think just one consistency. It is that almost everybody uses the word to refer to two different states: to the state of an animal, non-human or human, which is awake, sensing and feeling; and to the state of a person who is *aware* of being awake, sensing and feeling. When it is both recognised to be possible to do so, and thought to be appropriate to do so, the standard way of distinguishing between the two is to call the first 'consciousness' and the second 'self- consciousness'. But as I have already explained (see Chapter 5), this is not my usage. To avoid the interpretation whereby one can *have* a self of which one is *not* conscious, and to save the very word 'conscious' to go with its stable-mate 'conscience' as applying only to the human, my preferred language is different. That which others call 'consciousness' I prefer to call 'sentience'. That which others call 'self-consciousness' I prefer to call 'consciousness'. But I know already, and from exasperating experience, that this is a usage which can be confusing. So I am going to take an 'emergence' approach to its introduction. Instead of trying to establish such usage immediately, I am going to begin by using the portmanteau word 'consciousness' in much the same—I think muddled—way as almost everybody else. And then, hopefully as some sort of resolution of the muddle, allow my alternative vocabulary slowly to 'emerge' across my argument. In this way, what I hope to make clear is the way in which I conceive of sentience as being an organic phenomenon appropriately studied by the (already flourishing) biological sciences, and consciousness as being a personic phenomenon appropriately studied by the (not yet flourishing) moral sciences. For there is, of course, a considerably bigger difference between my approach and that of what I think could be called the mainstream than might at first be realised by registering only a change of terminology.

It seems to me highly likely that the hardly better than intuitive concept of consciousness which we have now will turn out to be thoroughly inadequate to the development of a theory. In *Neurophilosophy: Towards a Unified Understanding of the Mind-Brain* (1986), Patricia Churchland suggests that, like such concepts as impetus, gems or dirt, consciousness is probably not a natural kind: that is, it probably does not carve nature at her joints. For while the brain 'undoubtedly has a number of mechanisms for monitoring brain processes', what she calls the 'folk psychological categories' of awareness and consciousness probably 'indifferently lump together an assortment of mechanisms'. She then invites a comparison with the concept of fire, arguing that *it* was used, historically, 'to classify not only burning wood, but also the activity on the sun and various stars (actually fusion), lightning (actually electrically induced incandescence), the Northern Lights (actually spectral emission), and fire flies (actually phosphorescence).' And goes on to point out that 'as we now understand matters, only some of these things involve oxidation, and some processes which do involve oxidation, namely rusting, tarnishing, and metabolism, are not on the "fire" list.'

This is all true. Indeed, because it is all true, some even go so far as to deny the very existence of 'consciousness' as anything other than a 'folk-psychological' term. 'Quining'—as in honour of the philosopher W.V.O. Quine, who did a lot of it, it is called—means denying the existence of something which, on the face of it, seems obviously to exist. And quining consciousness, suggesting that it is a concept which deserves to go the way of the ether or phlogiston, of vital fluids or karma, is a perfectly respectable philosophical position. It nonetheless seems to me that because there are some things that we already know about consciousness, particularly about what it is not, it is probably worth hanging on to the concept, at least for a while, to see what theoretical work it is capable of doing. So let me do that. In particular, let me begin by detailing just seven of the things I think we already know about what consciousness is not.

First, we know that it is not experienced. I somewhere heard it quipped that for an entity that experiences consciousness, that consciousness is the only thing which would still be experienced as real after it had been shown to be an illusion. But the thing is, nobody does experience consciousness. We experience not consciousness *per se*, but only consciousness *of* things—like buses and trees and arguments and the like. It is a familiar theme in philosophy—as here expressed by Paul Grice (1962)—that 'if we were asked to pay close attention, on a given occasion, to our seeing or feeling as distinct from what was being seen or felt, we should not know how to proceed; and the attempt to describe the differences between seeing and feeling seems to dissolve into a description of what we see and what we feel.' Of course, says my candidate theory: for as describers, we *are* the seeings and the feelings. The relation of an I to *its* awareness, *its* experience, *its* consciousness is constitutive: an I *is* these things. Of course it seems as though things are somehow offered up to a spectator called 'consciousness', 'self' or 'me'—the 'too-many-words-for-ourselves' phenomenon is very clearly at work here—but it isn't really so. What is really so is that consciousness, self-ness, me-ness is actually created or made. First it doesn't exist (in non-human animals, earlier *Homo* species and human infants): then, as the product of the linguistic structuring of experience, it emerges (as persons). 'What was I before I came to self-consciousness?' asks Fichte. 'I did not exist at all', he replies to himself. And if he drops the 'self', I am with him.

Second, we know that it is not plural. Even in the much-hyped cases of split or multiple personality, consciousnesses appear not simultaneously but in sequence. And this of course is a fact which sits easily with the idea that the entity we are dealing with is a *state* of the brain—which just can't be in two of them at once! Indeed, an appreciation of this 'one-state-at-a-time' characteristic of consciousness is important to an understanding of much more than just multiple personality disorders. Consider, for example, the so-called 'binding problem': that is, the problem of how, even though they are processed in different parts of the brain, the various components of an experience (red, small, sweet-smelling) are 'bound' into a single experience (rose). It seems to me that this is likely to turn out to be a problem considerably more apparent than real. For just as Galileo pointed out that there is only a problem of motion if one assumes non-motion in the first place, so there is only a problem of binding if one assumes non-binding to begin with. But why do that? Surely the essential singularity and unity of a nervous system is something already built into it as a product of its evolution. A brain, after all, evolved as that part of a *single and unified living body* which would monitor the importance for that *single and unified living body* of the stimuli impinging upon it, so that that *single and unified living body* might respond appropriately to them. Indeed, much more likely than that there should be a problem of assembling various disparate brain processes together is that there might be a problem whereby accident or disease might *dis*assemble what was a single and unified process to start with.

But the essentially singular nature of consciousness (sentience? the muddle?) is not just a matter of our not being able to be in two conscious states at the same time. When we think not of the afferent processes whereby input results in one brain state rather than another brain state, but of the efferent processes whereby output results in one action rather than another action, the singularity and the unity become even more obvious. As the eminent biologist C. S. Sherrington was wont to insist, the only action available to us is to move things; the only way we can move things is via muscle contraction; and the most important thing to remember about muscles is that they cannot be in their two different states simultaneously. Think about that *muscular* 'one-state-at-a-time' reality in relation to consciousness (sentience?), and what it seems to me to do is to lend some support to the theory of consciousness (sentience?) put forward by Nicholas Humphrey.

As I understand his argument, he is suggesting that because natural selection can only act on those features of an organism that make a difference to its chances of survival, and because such features have to be biologically public in ways that sensations seem not to be, it follows that, in order to be selected, the ability to experience must once have been something done: that is, must once have been an active (that is muscular) response to stimulation happening on the surface of our bodies which only with time became 'privatised'. Using the example of seeing red, he argues that the *activity* 'sensing red' would once have involved responding to stimulation by red light in a particular way: with a particular sort of wriggle, he suggests, which we might think of as 'the red wriggle'. To sense red would then have been to issue the efferent commands which would bring about 'the red wriggle'. On becoming privatised, on becoming what he calls 'the wriggle within', the response would cease to be subject to selection. But not before its historical form had become established in such a way that, having had the same evolutionary history, my sensation of red and your sensation of red would be the same.

It is possible, I think, to pursue this line of thought a little further. In the brain of the developing foetus, the first 40 hertz range oscillations, thought by Francis Crick and others to be characteristic of consciousness (sentience), emerge at about the twenty-first week of pregnancy: that is, at the same time as the first muscle contractions characteristic of movement (quickening). In addition, retardation rates for immobile infants are, in the absence of active therapeutic intervention, considerably greater than for, say, visually impaired infants. These are the sorts of findings which, among others, suggest to me that it is not at all obvious that J. B. Watson was so far up a gum tree when he characterised thought as anyway maybe beginning as 'a sub-vocal tremor of the larynx'. The work of Vygotsky illustrating the way in which, between the ages of two and seven, small children spend much time talking aloud to themselves—rehearsing, he argued, what eventually they would internalise as thought or 'inner speech'—lends support to the idea that thought, at least initially, involves some sort of simulation of muscle movement: as also perhaps does the fact that at both the individual and the historical beginnings of literacy, reading aloud precedes reading silently to oneself (with St. Ambrose, Julius Caesar and Alexander the Great being among various contenders for the honour of being the first person to have been recorded doing the latter). And muscle movement takes place, of course, not in the dead entity which is a corpse (there are no conscious *spirits*); not in the never-alive entity which is a computer (there are no conscious *word processors*); but uniquely in *a single and metabolically unified living body.*

Third, we know that there is no 'hard problem' of consciousness: or rather, that what has come to be called 'the hard problem' can only be given the status of a serious problem if those who propound it can offer us just some idea as to what kind of thing, what kind of explanation, they would accept as its solution. For serious problems are like that: while 'the hard problem' seems to comes formulated in such a way as to *preclude* solution, they come formulated in such a way as to *entail* solution—anyway in principle.

At least in so many words, it was the philosopher David Chalmers who introduced 'the really hard problem of consciousness' as the problem of experience. But I would suggest that he makes things 'really hard' for himself in a totally unnecessary fashion. By dismissing not only all attempts that have *actually* been made, but also all attempts that could *possibly* be made, to solve the problem using some form of identity theory—that is, some form of the theory which works by correlating physical states of the brain with experiential states of the animal—he virtually rules out in advance any answers to his questions. And is left with only what Dennett calls—quite rightly I think—the 'drastic move' of adding experience to mass, charge and space-time as one of the fundamental givens. At this point, 'the problem of consciousness' really does seem to have become one which warrants instead Dennett's own solution: that as a pseudo-problem, it is one which should be put on the back burner.

In developing his case for the 'drastic move', Chalmers asks us to consider, as analogy, the way in which Maxwell had to add electromagnetism to the list of basic elements of physical reality because attempts to explain it in terms of underlying mechanical processes failed. But—as David Papineau points out—he does *not* ask us to consider, as *dis*-analogy, the way in which nineteenth century scientists did *not* have to add heat to the list of basic elements of physical reality because attempts to explain it in terms of molecular motion succeeded. Indeed, Papineau's position in general—as anyway I heard it further elaborated at a March 1999 conference on 'Consciousness' held at King's College, London—seemed to me to make much better sense. If I understood him aright, what he there wanted to insist upon was that theories of consciousness (sentience?) are trying to answer a question that doesn't exist; that *feeling* pain, *tasting* chocolate and *seeing* elephants exist, but that *consciousness* (yet another reification) does not; that conscious properties are going to turn out to be identical with physical properties; that while anti-physicalists think the existence of phenomenal concepts means that there must be a distinct phenomenal reality in addition to physical reality, physicalists think that the existence of phenomenal concepts just means that there are two ways of referring to a single physical reality; that most theorising about consciousness is motivated by residual dualism (hear, hear!); and that 'if anything, we need a theory of selves, not a theory of consciousness' (hear doubly hear!).

It seems to me that it should not be easier to think of consciousness, à la Chalmers, as always and everywhere already existent (for which there is *no* evidence) rather than as sometimes and somewhere gradually emergent (for which there is *much* evidence), when *that* the physical and the experiential somehow combine is still a problem—if one *thinks* it is a problem—whichever position one holds. As has several times been pointed out, Newton famously refused to put forward hypotheses to tackle what could have been called 'the hard problem of motion' (*why* does gravity behave in accordance with the laws he proposed). And it will probably turn out to be sensible to do the same with 'the hard problem of consciousness': especially since, if Chalmers were to be construed as right, we would either have to overthrow the whole edifice of physical cause and effect, or consciousness would have to be thought of as totally without consequence: both of them seriously 'drastic moves' which can't begin to accommodate any sort of reading of Occam's razor.

It seems to me also that the hard problem is only really a problem to those of a religious turn of mind. As much that it has since contained has I think only served to substantiate, the editorial in the first (1994) issue of the previously mentioned *Journal of Consciousness Studies* actually identified its authors as such. So that when they then went on to say that 'it really is not good enough just to say that it is a "brute fact of nature" that certain physical processes, in particular certain brain processes, are accompanied by a stream of consciousness', I found myself wanting to ask them what then they thought would be 'good enough'. For I got the distinct impression—from reading their articles, from discussing with some of them—that 'we don't understand it, so it must be God', or 'we don't understand it, so it must be a forever un-understandable mystery', or 'we don't understand it, so it must be a 'consciousness force' (or some such) might be all they would have to offer. And for me, such offerings—various versions of mysterianism I would say—are very far from 'good enough'. 'God', or 'a mystery', or 'a consciousness force' are not good enough— because they don't explain.

Fourth, we know that there is no homunculus. The more we come to understand the way in which the brain actually works, the less it seems in any way to resemble a machine moved by a ghost, or a vehicle driven by a self. Consciousness, whatever it may be, does not seem to be situated at some point in the brain. As neither does an I. The old idea of a something called me viewing the images conjured by 'my mind' and issuing executive commands for action to 'my body' receives less than no support from current neurological research. Indeed the very idea of a *centre* of consciousness, of a *centre* of any sort—as opposed to massive parallel processing —would seem to be thoroughly in question.

Unsurprisingly given the many and awful things which flesh has historically been heir to, almost every religious persuasion has an image of the body as some sort of container for the soul: a container which, when no longer needed, can be sloughed off, to leave the soul disembodied, and ready for something better. It is so widespread an idea as to make it reasonable to suppose that it must correspond with some equally widespread feeling about the human condition. And indeed it does. It corresponds to this illusory homunculus feeling: the feeling that between our ears there squats a little something called me, looking out on the not-me of the world and interacting with it, or not, as I choose [see 'acts *and* omissions']. It is a feeling which compels even when the not-me with which I interact is of that part of the world which I call 'my stomach' in which there is 'a pain'. Even more bizarre, it is a feeling which compels when the not-me with which I interact is of that part of the world which I call 'my mind' which I am about to 'change'. This is all Cartesian dualism. And no matter how compelling it feels, we now *know*, I contend, that it just isn't so. With the work of Libet and others, we now *know* that the homunculus picture is wrong: that there is no ghost in the machine.

Fifth, we know that consciousness is not a property of the inorganic: that is, that computer consciousness is either a metaphor or a category mistake. To suggest, as did a letter to the *New Scientist* (7.12.96), that robots don't yet exhibit consciousness because they are made of too few processors, but that if we could make one with something like one hundred thousand million processors, we would not have to argue about whether it was conscious or not we would just have to ask it, is to make such a mistake somewhat crudely. But more sophisticated versions are probably going to turn out to be just as much in error. In *Kinds of Minds* (1996), Daniel Dennett, in his inimitable style, suggests that 'to put it vividly, your great-great- ... grandmother *was* a robot!'. With such an opener, of course I want to read on. But when I do, there is only disappointment. I find an argument which runs something like this: that because 'your body' is composed of 'machinery that dumbly does a wonderful, elegantly designed job'—he refers to your haemoglobin molecules, your antibodies, your neurons, your vestibular-ocular reflex machinery, and so on—he claims that that shows that 'something made of robots *can* exhibit genuine consciousness, because you do if anything does'. But this, surely, is just hopelessly to confuse levels The 'you' that is *made of* robots is an organic entity: is an organism. The 'you' that not so much '*exhibits* consciousness' as '*consists in* consciousness', is a personic entity: is a person.

Consciousness, I am suggesting, is that phenomenon which emerges as the result of the linguistic structuring of the functioning nervous system of a *Homo sapiens*. And the crucial difference between the inorganic machine and the *Homo sapiens*, indeed between the inorganic machine and any living organism, is that, via metabolism, the latter has interests—at base, the interest to remain living—which the former does not. Given that interests only emerge at the level of the organic, anything which might be *described* as interest in an inorganic machine is being described so only analogously: which means that anything which might be *described* as consciousness in an inorganic machine is also being described so only analogously. And whether or not an analogy is appropriate always depends on context. In this case, when some of the most common errors of thinking involve the (mis)application of the concepts of one level of reality to the problems of another and different level of reality, it seems to me to be an analogy not only inappropriate but fundamentally misplaced. If, with Jerry Fodor, we describe the analogy between the workings of the mind and the workings of computer software as 'the big idea', we deal in such inappropriateness. For it was the minding of persons which produced the non-minding of the soft-ware. And because I think that that makes all the difference to the way we need to conceptualise the relationship between them, it is worth spelling the point out just a little.

In academic places 'artificial life', 'A-life' as its practitioners call it, is studied by a world-wide research community established after a founding conference in Los Alamos in 1987. It includes people from a considerable array of specialisms: from biology, chaos theory, computer science, ecology, genetics, neurology, philosophy, psychology. And for the most part, the 'life' they study 'lives' in their computers and is basically simulation. But in philosophical conversation, while this is of course recognised as one sort of artificial life, the sort which arouses considerably more interest, and often acrimonious debate, is different. It is rather more like that encountered in the film *The Stepford Wives* in which robot-women, more or less indistinguishable from the carbon-based variety, serviced their apparently delighted troglodytic men without a feminist 'thought' in their heads. This is artificial life as science-fiction: as walking, talking, zombie computer. And it is whether consciousness can be thought of as, potentially, a property of *this* artificial life which exercises the acrimonious ones.

Now all organic forms are carbon-based. Indeed, in discussions of the 'what-is-life?' kind, this is often taken, if only in part, as definitional (see Chapter 5). Of all the elements, carbon is unique in the huge range of bonds it is able to make. And given the chemical complexity of organic systems, it is perhaps such a range which is essential. Thinking then of whether or not non-carbon based entities, perhaps silicon-based entities, could come first to live, then to exhibit consciousness, since the bond-making capacities of silicon are, by comparison, minimal, it would be possible to argue that such an idea is just a non-starter, and that nothing more need be said. Possible: yes. But not necessarily wise. It could perhaps be wiser to recognise that everything here depends on what 'life' or 'living' is to be taken to mean. And then try to say something useful about some of the definitional niceties.

To be organic, and thus alive, let me again suggest that what we should perhaps home in on for our definition is 'interest': that is, on the fact that because they have metabolisms and are thus capable of ceasing to be alive, things that *are* alive have the interest of not being dead. Now some wit has suggested that if you want to know which is intelligent, your computer or your dog, you should try throwing both into the river and see which one scrambles out first. But it seems to me that that quip would have contained an even more telling point if the question had concerned life rather than intelligence. On the standard definition of life—that it is a carbon-based complex dynamic system maintaining a stable configuration far from thermodynamic equilibrium—the dog is alive and the computer isn't—as neither would be a Stepford robot. And this is a standard definition which has considerable consequence for our thinking about consciousness.

Humans have not made life, it has evolved. Computers have not evolved, they have been made by humans. And while I confidently expect that at some time in the future we almost certainly will make life—the number of amino acids in Stanley Miller's more recent test tubes considerably exceeds the number in his earlier ones—*when* we do so, it will surely have to be just a definitional requirement that it be based in carbon, that it have a metabolism—and so on. That which we have made already, based in silicon, is *an original*: is *artificial*. That which then we will have made, based in carbon, will be *a copy*: will be *synthetic*. Indeed, if the former were not artificial originals, if they were synthetic copies, they would be much less use to us than they are. It is precisely the fact that they are not like us which means that they can do things which we can't. Let me spell that out. Just to *be* synthetic, each of 'synthetic life', 'synthetic intelligence' and 'synthetic consciousness' (sentience) would have to be not-more-than, and not-less-than, *copies* of their non-synthetic originals. While just to *be* artificial, each of 'artificial life', 'artificial intelligence' and 'artificial consciousness' (sentience) would have to be (as two of them are already) no such thing. And what that means is that the initial question as to whether consciousness (sentience) can be thought of as potentially a property of artificial life can not be answered until it has been sophisticated. A full answer, it seems to me, would need to be this one: that from evolved life, consciousness (sentience) *has* emerged; that from synthetic life, synthetic consciousness (sentience) *could* emerge; and that from artificial life, artificial consciousness (artificial sentience) *might* emerge—*if such could be made a coherent concept.*

Sixth, we know that there is nothing it is like to be a bat. For as long as I have been aware of Nagel's now famous formulation of the 'what-is-it-like-to-be-a-bat?' question (in his 1974 paper of the same name), I have also been aware of an answer given once by Princess Anne to a probing journalist. He asked her what it was like to be a princess. And she replied that she didn't know because she had never been anything else. In idle moments, I have sometimes wished that the philosopher and the princess might meet and discuss these things.

Speaking first for ourselves, there is something it is like for a human to *be* a Hungarian, to *be* a diabetic, to *be* a cricketer or to *be* a princess. And that is because there is something it is like for a human *not* to be a Hungarian, *not* to be a diabetic, *not* to be a cricketer or *not* to be a princess. But there is nothing that it is like for a human just to *be*—and for the same reason: that is, because there is nothing that it is like for a human *not* to be. Speaking next for the bats, there is something that it is like for a bat to *be* flying, to *be* injured, to *be* captured because there is something that it is like for a bat *not* to be flying, *not* to be injured, *not* to be captured. But there is nothing that it is like for a bat to *be*—and again for the same reason: that is, because there is nothing that it is like for a bat *not* to be.

But where being is *like* nothing to either of them, being *is* something to one of them. Being non-linguistically structured experience, being a non-person, the bat has no means of comparing its current state, and those of its fellows, with its past or future states, or with theirs. But being linguistically structured experience, being a person, the human has some means of comparing its current state, and those of its fellows, with its past or future states, or with theirs. That 'being' which is no more *like* anything to the human than it is to the bat nonetheless *is* something to the human only. It is that which will end in nothingness. And the fact that we refer to ourselves as 'human *beings*' but not to bats as 'bat-*beings*', indeed not to any other creatures as 'some-creature *beings*', suggests, I think, a recognition. It suggests that we recognise the sense in which bats—and the rest—don't really 'be' at all. To repeat: being is *like* nothing to either bats or humans, but *is* something to humans alone. It is that which will end in nothingness.

The 'what-is-it-like-to-be' confusion embedded in Nagel's formulation flows, I think, from adopting what I am calling the 'having' model of experience. Taking as I do the position that the truly basic point about experience is that it is not something which we *have* but something which we *are*, my response to Nagel and his bat is this one. The main thrust of his piece was to argue that qualia, the raw feels of experience, must for ever be beyond scientific account: that is, that no scientific account, no third person account, can say what it is like to have an experience. And the main thrust of my reply is that the reason why there cannot be a third person scientific account of what it is like for a person to have an experience, or even a first person non-scientific account of what it is like to have an experience, is because a person does not have an experience in the first place. There cannot be an account, non-scientific or scientific, of that which does not happen.

But there *can* (by contrast) be an account, and it *can* (by contrast) be scientific, of a person *as* experience: of a person *as* linguistically structured experience. It's just that while the combined efforts of the already existent proto-sciences variously mis-named 'human' or 'social' or 'psychological' have certainly started to deliver such an account, such effort will only be able to result in substantial progress when a change of paradigm something on the lines of the I AM to I IS shift is made. In the course of such a shift, what are still the traditional almost folk divisions between human-science disciplines will need to go. Such disciplines will need to crack open and their contents be re-aligned—or junked—to allow the emergence of a category equivalent to its inorganic and organic precursors: a category which here I am calling the personic.

Finally, and seventh, we know—and this has emerged across our consideration of the previous six negatives—that 'consciousness' as a compendium concept embracing both the experience (non-linguistically structured) of bats and the experience (linguistically structured) of humans is not coherent: it will not allow us to theorise experience coherently. What has now fully emerged, or so it seems to me, is the necessity of making the sort of distinction I earlier suggested between sentience and consciousness. So let me now attempt to spell out the essence of just such a distinction.

In *The Miraculous Birth of Language* (1937), R. A. Wilson put the point rather graphically this way:

> This new world of conscious mind is, first of all, not a mere addition or expansion of the sensuous world of nature. In its elaboration there is nothing added to the sum-total of the natural material world. Nor is there any modification of the natural world. It involves no change in that world. It is, on the contrary, a new and distinct world now come for the first time

> into actual existence, an invisible supra-sensuous world rising above and as a counterpart of the visible sensuous world.

My terminology would be different. For me, although it would indeed be 'a new and distinct world now come for the first time into actual existence', I would be for calling it *the* world, where previously there was only the universe and the biosphere. But the really important point which arises from this extract is the one contained in the phrase 'an invisible supra-sensuous world rising above and as a counterpart of the visible sensuous world.' Wilson's 'sensuous world' is of nature: in my terminology, it is the *sentience* which is characteristic of (some of) the organic entities of the biosphere. But his 'supra-sensuous world', the one which rises above it and as a counterpart to it, is not: in my terminology, it is the *consciousness* which is characteristic of (all of) the personic entities of the world. In other words, *experiencing* is the process whereby the complex dynamic system which is a biological nervous system changes its state as the result of (external or internal) environmental input via a set of state-to-state transformation rules; *experience* is the nounal reification of the process; *sentience* is experience non-linguistically structured; and *consciousness* is experience linguistically structured. Almost all the technical detail of the means by which language use makes the difference between such sentience and such consciousness remains to be discovered. But its main consequence is already clear. It is already clear that, being non-linguistically structured experience, being a non-person, the bat has *no* means of comparing its current state, and those of its fellows, with its past or future states, or with theirs'. But that, being linguistically structured experience, being a person, the human has *some* means of comparing its current state, and those of its fellows, with its past or future states, or with theirs. It is the 'con', the 'with', which makes the difference. It is in discourse 'with' the other, both the other which was my own former self and will be my own future self, and the other which was your past self, is your present self and will be your future self, [see 'I *and* thou'] that I am constituted as person. Consciousness, the 'with' phenomenon which emerges with language use, is not an *inorganic* phenomenon, not an *organic* phenomenon, but a *personic* phenomenon: uniquely.

On my hypothesis, these I suggest are the seven things we already know about what consciousness is not. So what then might it be? Well, in turning my attention now to consider more specifically the possible role of language in providing an answer to that question, I am of course turning my attention to where many before me have also turned theirs.

The general approach which argues that while the non-human animal foundation of mind (that which I am calling sentience) is a biological phenomenon, the specifically human development of mind (that which I am calling consciousness) is not, is one which sees the development of speech, more specifically the development of the ability to internalise speech so that it becomes a potentially controlling inner voice, as the key. And it is among psychologists that this approach, one sometimes called 'social constructionism', has been pioneered. It is an approach which sees consciousness as a product of that inner speech which is achieved via childhood socialisation into specifically human culture. As already mentioned, the work of the Russian psychologist Vygotsky has been seminal here. He argued that what he called 'self-addressed speech' was the basis of our ability to remember; that it is our immature sense of self which is responsible for infantile amnesia; and that the very process of learning the vocabulary of a natural language wires in some kind of structure to that memory. While accepting the general drift of this, I would want to talk in terms not of self-*addressed* speech but of self-*creating* speech: not of an immature *sense of* self before language, but of a non-existent *actual* self before language. But either way, what is being stressed here is that humans, by contrast with animals, use their inner speech (or thought) to re-present actual previous consciousnesses, and to pre-present possible future consciousnesses. And that it is these processes which create what someone has called 'the eerie illusion that consciousness is also conscious of itself'.

The main consequence of this approach—in my terminology, of the recognition of sentience and consciousness as different phenomena—is that the task both of the organic sciences in explaining the first, and of the personic sciences in explaining the second, becomes that much easier: not easy, but that much easier. And that is because the questions can now be posed in a way that does not from the outset preclude the very possibility of answering them. It will very probably be technical advances, such (I think) as those represented by the newly sophisticated scanners which can make images of functioning brains, which primarily will lead to more satisfactory theories of sentience. But it will almost certainly be conceptual advances, such (I hope) as those represented by the I AM to I IS shift of the candidate theory I am proposing, which primarily will do the same for theories of consciousness.

It is a candidate theory taking its origins from two different research areas: generally, in the research tradition which has produced what is sometimes called the social intellect theory of consciousness; and specifically, in the work of Benjamin Libet. I am suggesting that it is that syntactic manipulation of symbols called language, specifically as internalised to inner speech, which creates consciousness; that the consciousness it creates is essentially an inhibitory rather than an excitatory phenomenon; and that the resulting emergence of the dual construction of 'I' and 'other' [see 'I *and* thou] facilitates that *co*-operation which is the hallmark of the human. The emergence involved here is of that which folk psychology calls 'imagination', and the philosophers call 'theory of mind' (the ability of one I to attribute mental states to another I) or 'second order intentionality' (the ability to think that you think that I think). It is studied in children as their developing linguistic ability correlates with, for some examples, their developing ability to achieve object permanence (12 months), to recognise themselves in mirrors (18 months), to indulge in make-believe play (24 months), to abandon animism (48 months), to take the point of view of another (6 years)—and so on. And it is posited by me as being a two-stage emergence whereby the earlier I AM thinking (that of primary and secondary first persons relating normally) is replaced by the later I IS thinking (that of equal, just persons relating morally).

Such a candidate theory must of course be predicated upon some concept of 'I' in the first place: that is, some concept of that which in I AM consciousness is wrongly posed, in I IS consciousness is rightly posed. It is my claim that that 'I' is the moment to moment, linguistically structured experience which patterns and re-patterns the state of the brain of a *Homo sapiens*. Such an 'I' is not a continuous entity, but rather a set of discrete moment to moment entities. The semblance of continuity comes from the fact that linguistic structuring enables patterns constitutive of previous 'Ies' to get something of that previously mentioned 'aufhebung' treatment of Hegel's: that is, to be annulled while yet being preserved, and thereby to be lifted to a new level. Since it is only when the sensory activity of 'now' is hooked up to the memory of 'then' that such a notion could arise, this is the process, described so well by William James, which Wittgenstein saw as the source of meaning. As it is also the process which produces Dennett's formulation of the 'I' as the main fictional character at the heart of our every life-story: though this last formulation, because it implies that 'fictional' is to be equated with 'non-matter-energetic', and thereby also implies that things like democracy, love, peace, goodness and almost everything of greatest consequence to persons are also all 'fictional', is not, it seems to me, one to be recommended! Indeed, consistent with the fact that in my scheme of things reality is all that there is and ipseity is, one of my aims in turning my attention directly now to the role of language will be to demonstrate the thoroughly *non*-fictional way in which the emergence of *consciousness*, as the emergence of *language* is quintessentially related in the production of a thoroughly real 'I'.

When considering the nature of the human, almost every philosopher, both the professionals and those taking part in philosophical conversation, gives considerable attention to the role of language. So we find the American philosopher Donald Davidson following Descartes in denying thought to 'dumb' animals, and the Canadian philosopher Charles Taylor arguing that humans are 'beings who exist in, or are partly constituted by, language.' And neither is this a contemporary phenomenon only. In his

Essay on Man (1944), Ernst Cassirer was already illustrating the fact that what has now come to be called 'the linguistic turn' – the phrase is probably that of Gustav Bergmann – was well underway some sixty years ago:

> Man has, as it were, discovered a new way of adapting himself to his environment. Between the receptor system and the effector system, which are to be found in all animal species, we find in man a third link which we may describe as the *symbolic system*. This new acquisition transforms the whole of human life. As compared with the other animals man lives not merely in a broader reality: he lives, so to speak, in a new *dimension* of reality.

In questioning this turn to the linguistic, Bertrand Russell was wont to respond by insisting that philosophy was about the world rather than about language. But Wittgenstein's approach whereby 'the limits of my language' are to be thought of as 'the limits of my world' already seems to be suggesting that such a language/world difference is perhaps rather more apparent than real. And once the distinction which I am defending is in place – the distinction between universe, biosphere and world – the difference between that which is 'about the world' and that which is 'about language' is actually no difference at all. Indeed, in what I now want to say about language, the main question I must address is that of how such a 'difference which makes no difference' should be formulated.

About forty thousand years ago, in the period of the so-called 'upper Palaeolithic', archaeologists have established what appears to have been a considerable and relatively sudden improvement in both the tool-making skills and artistic abilities of humans. And given the general drift of the idea that it is language which makes us different, it is tempting to suggest that, despite the fact that most historical linguists would agree that the reconstruction of past languages can only take us back to about ten thousand years ago, it might have been that forty thousand years date which saw the emergence of language use. Before then, perhaps it was such adaptations as the changes (consequent upon bipedalism) in the anatomy and physiology of throat, tongue and larynx, and the changes (consequent upon what?) in the physiology of hearing and noise-making which were taking place. And if they were, fossil remains suggest that they were absent in *Homo habilis*, perhaps emerging in *Homo erectus*, quite advanced in *Homo neanderthalensis*, and full-blown in *Homo sapiens*. Anyway until recently, it is this I think which has been the orthodoxy.

But what it ignores is that language and speech are not the same thing. All the emphasis on the anatomical and physiological changes which would permit enhanced vocalisation has perhaps obscured the fact that vocal language is in no way the only possible language, and was maybe not the earliest language. Such an idea is not one which gives any credence to the notion that the vocalisations of animals – birdsong, howls, alarm calls and the like – constitute languages. On anything remotely like the criteria suggested by Hockett (1960) or Aitchison (1983), clearly they do not.

For these now widely accepted yardsticks include such requirements as that a language needs to be handed down from generation to generation (the 'traditional transmission' criterion), and that it needs to enable its users to produce and understand an infinite number of novel utterances (the 'productivity' criterion). And birdsong, howls and alarm calls, rather more likely to be the precursors of our sighs and laughter and sobs than of our vocal language, cannot pass any such muster. Instead, it is an idea which might give some credence to the notion that, prior to its existence vocally, language might well have existed first gesturally.

This is not a new idea. It was put forward—by Condillac—as early as 1746. And in more recent times it has been developed by several people: during the 1970s, for example, by the anthropologist Gordon Hewes of the University of Colorado at Boulder. That gesture and speech are somehow linked will come as news to almost nobody: think only of all the hand-waving we do when we are talking on the phone, or of the gestures with which infants communicate, culminating perhaps in the all-important pointing. But that they are linked as parent and offspring, so to speak, is a rather more specific claim.

One of the more prominent exponents of such a claim is Michael Corballis of the University of Auckland in New Zealand. He is particularly interested in the evidence to be gleaned from the congenitally blind: particularly in the fact that the gestures that accompany their speech accompany it not only when it is directed to the sighted but also when it is directed to other blind persons. With the first gesture probably being that archetypal pointing, he hypothesises that it might have been as the result of the formation of the Rift Valley in Africa that a full-blown gestural language developed. First, the ancestors of the great apes remained on the western side of the rift, while the ancestors of the hominids were separated from them to the east. Then, the protective forests of the eastern side gave way to exposing savannah, meaning that silent communication about the presence of predators, followed by silent cooperation in dealing with the threat they posed, was at a premium. He thinks—though I wonder about the time-scales involved—that this offers some explanation as to why it is vision not hearing which is our most highly developed sense. And he raises again the bipedalism issue—though not in the way which meant that earlier researchers got their fingers burned by associating it with tool-use rather than protection against the heat of the sun on the savannah. In effect he suggests that, whatever its cause, its effect might well have been that our ancestors not only had hands free to carry spears and light fires but also to make a wider range of gestures than would have been available only to the face. As to the eventual shift from gesture to vocalisation, he suggests that, once the first was in place, it would have been the pressure to communicate even when it was dark, or even when the speaker was hidden, which could have been that which pushed language to its vocal form. Indeed, by transferring language to the voice, early humans would have been able to communicate and to perform manual tasks at the same time. He tries further to establish his claim by pointing out that there are strong neurological links between speaking and movement. The language areas of the brain are active not only when we speak but also when we gesture. And the pre-motor and motor areas are active when we use language, particularly when that language is that which is associated with the hand, as for example is that concerning tool use.

What Corballis's suggestion amounts to is that the emergence of language, as identified by criteria such as those of Hockett or Aitchison, is a two-stage process: first the gestural linguistic, then the vocal linguistic. That the gestural can be just as much a natural language as the vocal, that it can have more or less the same linguistic structure, is confirmed, of course, by the current existence of the various languages used by the deaf. In *I See a Voice* (1999), for example, the philosopher Jonathon Ree concluded that the differences between gestural language A and vocal language B need be no greater than the differences between vocal language B and vocal language C. And in *Hand and Mind: What Gestures Reveal about Thought* (1993), David McNeill argued that the popular idea that something called 'body language' is a mode of communication separate from and additional to verbal language is strongly challenged by the fact that there is a detailed and complex correspondence between the two, both temporally and iconically.

Building on such ideas as these, what my suggestion amounts to is that the emergence of consciousness might well be a three-stage process: first, the emergence of the gestural linguistic (as suggested by Corbalis); from that, the emergence of the vocal linguistic; and from that, the emergence first of the sub-vocal linguistic (as identified by Watson) and eventually of the non-vocal linguistic (as identified by Pavlov, Vygotsky and others). It is in this sense that I am suggesting, as also are the very roots of the word, that fully-fledged con-sciousness, 'knowing with' past Ies of the self, past and present Ies of others, just *is* fully fledged language use. To be fully conscious (to be awake and alert) *is* to be fully connected to the world-wide-web which is language. To be partially conscious (to be dreaming, by day or by night) *is* to be partially connected to the world-wide-web which is language. And to be not at all conscious (to be under a general anaesthetic or in a coma) *is* to be not at all connected to the world-wide-web which is language. The I AM paradigm, misreading as it does all these 'being' phenomena as 'having' phenomena, reading every 'is' as a 'has', posits not that language use *is* consciousness but that language users *have* consciousness. And thereby creates the terrible homuncular muddle which results, amongst much else, in the phenomenon of there being too many poorly differentiated words for ourselves: we become conscious 'beings', with self-conscious 'minds', making us 'persons' with 'selves' and even 'souls'.

What I am claiming, by contrast, is that the I IS alternative, the alternative that consciousness just *is* language use, allows us to escape such muddle. It has been a commonplace to think of such language use as in some way miraculous: think only of the very title of the already-quoted Wilson book *The Miraculous Birth of Language* (1937). And there are perhaps two main reasons for such a thought.

Firstly, there is the apparently totally transformative nature of its use. The sequencing of the chimpanzee genome, recently accomplished by scientists working in the Max Planck Institute at Leipzig, has shown it to be one minimally different from our own. And that is a finding which can only lay further emphasis on that transformative power. For there does, after all, have to be some reason why species so similar genetically should be so different. And the use of the thereby 'miraculous' language cannot help but offer itself as a likely candidate.

Secondly, there is the incredibly prolific nature of the phenomenon itself. At one level of such richness, a consensus figure among linguists is that there have probably existed, at one time or another, some ten to fifteen thousand different languages. At a different level of such richness, it is estimated that, among English speakers, the average six year old recognises perhaps thirteen thousand words; the average adult without a higher education, about sixty thousand; the average adult with a higher education, about eighty thousand; and the highly literate person, about one hundred and thirty thousand. And anybody being around a small child as he or she is acquiring her or his language, apparently so effortlessly and at such a rapid rate, might well be inclined to talk in terms of 'miracles'.

But miracles don't happen. And that being so, it's just as well that the scientists who are linguists and psychologists—and others—are on the case. Among the mass of literature in this fascinating and rapidly growing research area, what I want to do here is to focus on just one issue. I am offering the hypothesis that it is going to be some lingual informational device which, in all language-using humans, structures brain states into minds—or better, into minding persons. The variables that *do* matter to the identification of such a device are currently little understood. But some of those that *don't*—as for example the use, in English, of 'ain't got no' rather than 'doesn't have any', which to a linguist, as Steven Pinker insists, makes no difference—are perhaps becoming clearer. In the interests, therefore, of trying to lay the 'miracle' ghost, let me pursue a little further the question of which are likely to be the variables relevant to such identification.

The linguist Noam Chomsky has many times expressed the opinion that a linguist visiting from Mars would very quickly become convinced that, notwithstanding their mutually unintelligible vocabularies, all worldlings speak the same language. He of course had a particular story to tell, which later I will consider, about just what it was that accounted for that sameness. But first, the very sameness itself is interesting. If it is going to turn out to be 'some lingual informational device which, in all language-using humans, structures brain states into minds' then it cannot be one which differentiates Hindi from Chinese, Polish from Portuguese or American Sign Language (ASL) from Mandarin. Which then might be the features of that 'world-wide language' which, according to Chomsky's visitor from Mars, all worldlings already use, which could be candidates for providing at least some clue as to the way in which such language might structure the experience of all humans into personness?

It would seem to me appropriate that we anyway begin our search for such features conservatively: that is, that we look at the things which are already known to be common to all natural languages. And such things are of course many. We know that there are invariant stages in the acquisition of language by infants: that cooing, babbling, morpheme production, holophrases, two-word sentences—for some examples—emerge in the same sequence in the acquisition of language in all speech communities. We know that, in all speech communities, the child's ability to walk unaccompanied, usually somewhere around fifteen months of age, is followed by both a massive vocabulary spurt and the beginnings of more sophisticated grammar. We know that, among most researchers in the area, the criteria as to what constitutes a language offered by Hockett and Aitchison—to which I have already referred—are ones which would seem to command widespread acceptance. We know that many have been the theorists, perhaps beginning with Frege, who have argued that it is the sentence which is the basic unit of meaning in all languages. We know from such studies as the seminal 'wugs' work of Jean Berko Gleason, that it is not the culturally-linguistic specific but the universally-linguistic general which structures novel output. We know that, as Steven Pinker puts it, a grammar is 'a discrete combinatorial system' (a finite number of words which can be strung up into an infinite number of sentences), which means that that other 'discrete combinatorial system' which is the DNA genetic code (four nucleotides combining into sixty four kinds of codon, which can be strung up into an infinite number of genes) almost inevitably offers itself as some sort of suggestive comparison. We know that a word, in whatever language, is almost always all symbol: that is, that the relation between the way it sounds and what it means is almost always fully contingent. And so on. This is a list which could grow quite extensively.

So if such facts as these suggest that the basic design features of all human languages are the same the world over, why might that be? Famous now, not to say notorious, Chomsky's nativist explanation of such regularities used to be that members of the species *Homo sapiens* were just born with what he called a 'language acquisition device', or LAD. That was taken to mean that the capacity to use natural language was hard-wired into only human brains; that 'linguistic universals', as he called them, were to be thought of as properties of language that were pre-formed in the nature of an already existent and organic entity; in other words, that language was to be understood as a biological phenomenon. His critics have argued something different. Taking off only from such simple observations as the fact that English speakers correctly guess that in Chinese *ch'ing* means light and *ch'ung* means heavy – the observation is Pinker's, the interpretation of its significance is mine – it is entirely possible to argue instead that it is because the basic design features of the human environment are the same the world over that linguistic universals emerge. Lifting something heavy produces an effortful 'u', something light an easy 'i', universally. For those taking part in philosophical conversation, it is still the work of Chomsky, together with that of other Chomsky-influenced linguists, which is best known and most often drawn upon. But I think that when in such conversation we borrow his ideas, when we use for example the concept of a 'universal grammar' to think with, we need to be very careful about the level at which we want the 'grammar' word to operate. Let me try to make clear what I mean by that.

Since it seems clear that vocabularies (in the form of single morphemes, for example) *precede* rule-based sentences, it would seem to be just a given that, at one level anyway, semantics *precedes* syntax. Whatever a universal grammar is going to be able to be made to mean, it has to be able to accommodate that reality. Similarly, that a natural language should make a syntactically structured unity called a sentence out of something standing for a thing (noun, noun phrase) and something else standing for an action (verb, verb phrase, predicate) hardly needs the more complicated explanation suggested by there being an in-built universal grammar when the simpler explanation suggested by there being an out-built universal universe, biosphere and world will surely do. It seems to me that in such debate [see 'nature *and* nurture'], those defending the 'nature' corner frequently conflate culture-independent and culture-invariant: that is, they argue that because a phenomenon is culture-invariant, that means it must be culture-independent, or natural. But this of course is not so. There are going to be huge swathes of behaviour which are culture-invariant not because they are culture-independent or natural, but because they are the same cultural responses to those features of environments which are also the same across all of them.

It is this reality which leads me to offer an explanation for the existence of linguistic universals rather different from that of Chomsky. Where his position seems to require us to think of them as properties of language that were pre-formed in the nature of an already existent and organic entity, I am suggesting that we think of them instead as properties of language that are emergent from the nurture of a thereby created and personic entity. And this is an explanation in which every word and phrase used to express its alternative nature—not 'pre-formed' but 'emergent from'; not 'nature' but 'nurture'; not 'already existent' but 'thereby created'; not 'organic entity' but 'personic entity'—is to be regarded as crucial. It is an explanation both obviously of a kind with, but clearly somewhat different from, those of many of Chomsky's critics. And that being the case, I must deal with both of those realities: that is, first with its 'of a kind' features; then with its 'somewhat different' ones.

That of my formulation which is common to many Chomsky critics is that which is emerging most clearly from the work of those frequently labelled 'connectionists'. Inheritors of the centuries-long debate between rationalists and empiricists, while the Chomsky side puts its emphasis on in-built genetic programmes which innately give us our capacity to manipulate symbols, connectionist modellers (those building neural-networks) stress the capacity to generalise from past experience 'on the hoof' as it were. Of course, in 'the linguistic wars', as some journalist somewhere has no doubt called them, each side often recognises, if sometimes only formally, that the other perhaps has some of the truth. Thus in his *Words and Rules* (1999), Steven Pinker works through a possible compromise on the specific issue of the past tenses of verbs in English. He suggests that each individual human infant comes with an in-built, innate genetic program to deal with the regular ones: the rule is to add 'ed' to the present tense. But for the irregulars, for the need to transform 'give' into 'gave' or 'catch' into caught', there are no rules. And small children just have to memorise and thus learn. If you think it seems odd, offensive perhaps to Occam's razor, that we should posit two such different mechanisms when one will probably do, I am with you. And if you were to read the book, you would also notice that while its author does a pretty good job of undermining exclusively rule-based accounts, his real interest, the one that produces most of the passion, is in bashing the connectionist accounts. So what might a connectionist say about the acquisition of the correct forms of the past tense of verbs?

The mainstream connectionist position within neuroscience—there are variations on the theme—is the view that neural networks in the brain change at the synapses with learning: that the more stimulation a particular network receives, the more are those pathways (learnings) strengthened. And at the very start of that research tradition, Rumelhart and McClelland, who created their network in the mid 1980s at the University of California, set out to study just that. Their relatively simple network had just 920 'neurons': a layer of 460 input units connected in parallel to a layer of 460 output units. Each unit could be in one of two states: 'on' or 'off'. And with their patterns of activity acting as representations of the phonological structure of verbs, and the whole system having the in-built capacity to adjust the strength of the connections among the units, the whole network was able to produce correct past tenses without the use of rules, and without the need to treat differently verbs those plurals were regular and verbs whose plurals were irregular. The learning algorithm they used is that known as 'back-propagation of error', and has two stages: first, an example verb is programmed into the input units and signals spread from these to the output units; second, the resulting past tense produced at the output is compared with the correct answer from a 'teacher'. The weights are then changed in such a way as to ensure that the next time that input arrives at the output, a more correct form is produced. After training, the network resulted in an almost errorless performance. In one run, for example, it gave the correct past tenses to 90% of a batch of eighty-six unfamiliar regular verbs. These then are clearly powerful models which require some further attention.

For a succinct summary of their workings, we could do much worse than turn to two of their most ardent supporters. In *The Churchlands and their Critics* (1996), Patricia and Paul Churchland summarise what they take to be the nature and significance of such models. They describe representations as 'patterns of activation (or sequences of such patterns) across millions of neurons'. The computations which are then the basis of their model are 'synapse-driven transformations of such patterns (or sequences thereof) into further such patterns across further neuronal populations'. That which 'steers the network between alternative interpretations of the same data' is its 'prior dynamical state', which they also refer to as its 'frame of mind'. The basics of this which they refer to as 'contemporary connectionist theory' are the transformations effected at the synapses by 'learned patterns of synaptic connection strengths'. In such models, 'each element of the network contributes such a tiny amount to the overall process' that the failure of just one unit or just one synapse, while it will mean that 'the quality of the network's responses will be degraded slightly', is no more likely to change what they call a 'behavioural profile' than is the failure of just one pixel on a television screen to make much difference to the quality of the picture. The models they are defending here are those based in parallel distributed processing: that is, processing whereby 'input-output functions are executed by configurations called "nets" that consist of many *units* that admit of varying levels of *activation*, and many activation-transforming *connections* between those units with stable but modifiable *weights*.' This is the 'framework', as they call it, which they want to contrast with the 'propositional-attitude' framework of folk psychology. And most important of all, anyway to my argument, they want to suggest that 'the bottom line claim of the eliminative materialist is and always has been that the content and character of our social practices in the domain of mutual perception, explanation, anticipation, and behavioural interaction are going to change, and change substantially, with the dawning of a truly adequate neuropsychology'. Why? 'Current practices will change'—they continue—'because, with a transformed understanding of what we are and how we work, we are almost certain to develop different and much *better* practices.'

This is the type of model which I think makes best sense of the learning of a language. And more generally, it is precisely their 'transformed understanding of what we are' to which I am appealing with my hypothesis of a transition from I AM to I IS persons: with no idea as to whether or not they would agree with me, I have no hesitation in saying that the 'better practices' to which they refer will be those based in equality. But beyond the basics of how we *learn* a language, we are still left with the question as to what its transforming role might be once learned. After all, the models as here the Churchlands describe them are those which, in general outline, are deemed to be applicable to learning of *any* sort, and in *any* creature with a neuron-based brain. The idea of persons as entities emergent only from brains where a natural language is what has been learned needs further explication.

It seems to me that on top of some sort of 'connectionist model' base, posited as explaining the micro-level of mind formation, there has to be erected some sort of 'constructionist model' superstructure, posited as explaining the macro-level of mind formation. And by 'constructionist model' I mean that which I can most easily summarise in the words of John McCrone. Writing in the *Journal of Consciousness Studies* (Vol. 6, No. 8-9, 1999), he begins by observing that it is 'a commonplace that the human mind is shaped by culture'. But he then points out that the usual assumption is that the consequence of this is just 'horticultural': that is, given that the really important things (in the words of a now dead psychology, 'the faculties') are actually a function of genetics, and therefore 'present in seed form at birth', what culture provides is 'only a modicum of watering and perhaps some later judicious pruning'. As alternative to such a picture, he then outlines 'a more radical line' that has been adopted by 'a series of authors': he mentions Muller, Vygotsky, Luria and Mead as its possible founders. And suggests that it has its more recent expression in what he calls 'the socio-cultural school of cognitive development' (he mentions the work of Zivin, Wertsch, Diaz and Berk); 'the social constructionist movement' (he mentions the work of Coulter, Berger, Luckmann, Gergen, Davis and Burr); and 'most recently in the philosophy of language and thought' (he mentions the work of Dennett). And what do all these have in common? He suggests that their common argument is 'that we are born with naked animal brains', but that then, 'through the learning of language', we are able to 'internalise an apparatus of socially-formed thought habits', and that these 'language-enabled skills' are what give us 'all our distinctively human abilities', including such things as 'self-awareness, recollective memory, socialised emotions and structured thought.' And then—the seemingly inevitable computer analogy!—we 'become loaded with a mental software that can drive our brains to places they would not normally go.'

This is a loading which it is perhaps most easy to see in non-human creatures for which the 'not normally' which is involved is quite spectacularly 'not normal'. Chimps are 'not normally'—and in the full-blown Hockett or Aitchison sense I would say 'not ever'—language users But as is well known, with massively attentive training from humans who are, some have been able to get to the edges of some sort of symbolic functioning. One such animal study, one to which I have briefly referred already (see page 235), will make my point very nicely.

When studying deception, Sarah Boyson required a chimpanzee to select one of two portions of food. The portions were of unequal size; the animal was hungry; it always chose the larger portion. It did this repeatedly. And even though the experimenter then gave the smaller and rejected portion to the experimental animal, the larger and chosen portion to another chimp, it did not seem to be able to learn to do otherwise. However, when she replaced the food items with a food symbol, a symbol in whose use the experimental chimpanzee had been trained, she found it could cope. The symbol used was numerical and the food was presented in discrete and thus countable units. And in that circumstance, the animal was able to chose the symbol which indicated the smaller portion in order to be rewarded with the symbol which indicated the larger portion, which symbol could then be exchanged for the actual larger portion. What appeared to be happening was that the use of the symbol enabled the animal to distance itself somewhat from the evidence of its own eyes—'that's more, I'll have it'—and behave somewhat more efficiently. One commentator on the work suggested that the symbol seemed to enable the animal to 'keep in mind' that what was chosen was *not* what was received. I am suggesting a more parsimonious explanation. I am suggesting not that the symbol enabled the chimpanzee to 'keep it in mind', but that it actually *created* the (incipient) mind (or minding), or better *constituted* the (incipient) mind (or minding), in which it might be kept. I am suggesting that the mind (or minding) *is* the working of a brain structured by a symbol system.

It could be wiser to leave things here. For while, with an hypothesis like mine, I clearly cannot have *nothing* to say about the mechanisms whereby language might structure experience, and thereby create consciousness, that in no way means that the something which I *must* say need be substantial, still less decisive. My baseline answer to the question as to just how language structures those 'patterns of activation' is that I don't know: as neither indeed does anyone else. But perhaps precisely because of my distinctly amateur status—I have no reputation to lose—I can maybe more easily than some afford to fly a few kites. So I will.

Kite One: the chunking kite. I remember as a student being for some long time puzzled as to how the notion of '*a* gene' (discrete) could be made coherent with '*a* continuous stream of As, Cs, Ts and Gs' (non-discrete). It could at the level of a single base. But I knew already that that was not what was meant by 'a gene'. A continuous stream of bases can only carry information as to how to make a body if it is not continuous: if it breaks in some way. And in time I came to understand that the relevant way in which it breaks, the way which ensures the emergence of those 'chunks' called genes which are capable of being selected, is into triplet codons (and ways of signalling 'stop'). My first thought on what is likely to be the nature of the linguistic structuring of a nervous system is analogous. Here also, a continuous stream of symbols can only constitute information as to how to make a mind if it is *not* continuous: if it also breaks in some way. Now we are already well aware of some of the ways in which language breaks: into phonemes, morphemes, sentences, for example. But whether it is one of these ways, or one as yet unrecognised, which is the way relevant to its function of providing those 'chunks' capable of linguistically structuring the nervous system of a *Homo sapiens* to make a mind (a person that *minds*) we don't know. Anyway for now, the hypothesis that it is *some* chunking capacity of language which carries the information necessary to the making of a mind (a person that *minds*) has to remain neutral with respect to which of various candidate capacities that might be. Or neutral at least with respect to all but one of them.

For it does seem to me that the one that can probably be ruled out of court reasonably unceremoniously—and it's the one, wouldn't you know it, which has received most popular attention—is that which proposes 'the meme', or even 'the idea' as a candidate chunking mechanism. This would seem to me to be a notion at a totally wrong *level* of analysis: to be far too macro to be of any use at all. Putting it crudely, the structuring of a sodium-potassium pump mechanism by 'an idea' seems just a little bit off the wall. Even 'a sentence', in other respects an interesting possibility, would seem dubiously wrong-levelled for such a role: though at least in its case, the possibility of a tighter, rule-based description of what it is means that I at least can just about imagine a way in which such a constrained entity just about might have a discrete informational read-out.

Kite Two: the crayfish kite. In humans, transmission at the synapse is chemical transmission. Throughout its characteristic many-inputs (dendritic) one-output (axonal) structure, what happens at the synapse either increases or decreases the likelihood of the next cell firing: that is, each individual synapse is always either excitatory or inhibitory. The most common excitatory transmitter is glutamate (the charged form of the amino acid glutamic acid), and the most common inhibitory transmitters are gamma-aminobutyric acid (GABA) and glycine (the simplest amino-acid of all). But in the *Journal of Physiology: London* (1959), E. J. Furshpan and D. D. Potter reported what, anyway back then, was a surprising discovery. Working on crayfish, they found a huge synapse where transmission was electrical rather than chemical. And in the almost fifty years since, it has been well-established that electrical synapses are very common in invertebrates; less common in vertebrates: and much less common in mammals. Now this was a distribution which of course suggested that electrical synaptic transmission evolved first, chemical synaptic transmission later. But in that case, given that electrical transmission is considerably faster than chemical transmission, is indeed almost instantaneous, the question arises as to why the *slower* form of transmission should apparently be more appropriate to the *higher* animals? A candidate answer to that question can be found in the fact that while electrical synapses are necessarily excitatory, chemical synapses can be either excitatory or inhibitory. By extension therefore, might it not be that it is inhibition which is necessary to greater cognitive capacity? And thus that when we get to the level of the very highest cognitive capacity of the lot, to the level of the human, it is a yet greater inhibitory development which enables its emergence. Made available by linguistic structuring, my earlier interpretation of the work of Benjamin Libet as demonstrating a qualitatively new level [see 'quantity *and* quality'] of inhibition of the natural, a level we call consciousness, a level which Ramachandran characterised as that which gave us not free will but free wont [see 'free-will *and* determinism'], is that which I have in mind. But maybe that is not all. I could perhaps play out this particular kite to fly even higher by suggesting that maybe some interesting experimental possibilities are offered here. If it is hypothesised that it is the linguistic structuring of the brain which is responsible for what Libet's work suggests is the inhibitory nature of consciousness, then given that *both* the 'transmitter source' (an excitatory or inhibitory synapse) *and* the 'transmitter substance' (excitatory glutamate or inhibitory GABA or glycine), are clearly distinguishable and thus unambiguously definable, might that not mean that experiments necessary to the exploration of the connection between language and inhibition *at the micro level* could soon come within range of 'the art of the possible'? As illustrated earlier by reference to Sarah Boyson's chimpanzee experiment, work at a macro level already exists. But micro-level additions would indeed be splendid. There would of course be the most complex ethical issues to consider. In the case of the full-blown use of the symbolic in humans, such studies would of course have to consist in the 'natural experiments' offered by the study of human disease processes and incapacity rather than the ablation experiments possible with other creatures. And even in the case of only the incipient use of the symbolic in other animals, given that the other animals concerned would have to be the higher apes, the constraints would maybe still be such as to rule out much that might suggest itself. But such problems notwithstanding—and despite the fact that I don't have the knowledge or competence to fly it any less clumsily—this is my crayfish kite.

Kite Three: the not-only-at-the-synapses kite. As outlined above, the connectionist modellers have (probably entirely appropriately) concentrated their efforts at what constitute the 'synapses' of their systems. But brains are more than synapses. And it might not be only there that language accomplishes, however it accomplishes it, its restructuring job in the brain. To illustrate this point, consider the work of Jacobs and Scheibel (1992). They studied dendritic length in twenty human brains, making their measurements in Wernicke's area which is known to play an important part in language comprehension. They found variations of up to forty per cent. Fibres were longer in female brains, shorter in older brains, longer in the brains of those who had been most intellectually active. And checking out that last finding against animal enrichment studies, they concluded that it was probably the intellectual life style which produced the longer fibres rather than the longer fibres which enabled the intellectual life style.

So my kite on this one is as follows. Might it not be fruitful to try to discover rather more about what, in addition to dendritic *length*, might correlate first with degrees of linguistic development, second with degrees of linguistic enrichment? Might it not be that dendritic *abundance*, for example, would so correlate? After all, the work of such as John Dobbing has already shown that, as long as it is given at the appropriate growth-spurt times in the development of brains, such abundance does correlate with nutritional enrichment. Or what about axonal length? Or the extent of glial support? Or limen levels? In short, while the states of synapses are probably going to be crucial, other variables can surely not just be ignored.

Kite Four: the computers-operated-by-the-spoken-word kite. Such machines are of course now a reality. But the techniques which go into their construction are not those into which most of the early research funding was poured. Such early research drew its inspiration from the already mentioned 'LAD' ideas of Chomsky. The programmers reasoned that since, according to him, all languages had the same deep grammatical structure in common, all they had to do was model this structure, and thereby build rule-based systems that would mimic what was taken to be the human. Given the myriad idiosyncrasies of the speech of myriad individuals, the alternative idea—the 'pattern-matching' approach—was thought to be impossible. But this turned out to be wrong. It is in fact sophisticated pattern matching which has won the day. The programs of such machines analyse the frequency and volume of the speech sounds, and then compute phonemic and morphemic probabilities using a statistical technique (called Markov modelling) to identify the words spoken. Would-be users have first to read some text to the machines so that they can become familiar with individual pronunciations. That done, they can then reach speeds of seventy words a minute, and have a ninety five to ninety-nine per cent success rate with vocabularies of at least around twenty thousand words. The Markov modelling systems have required that many more than the forty or so phonemes traditionally identified by linguists as being the building blocs of the English language be identified: a typical system might use something more like forty thousand 'phonemes-in-context' (PICs), as they are called, to identify all the detail of variously spoken words. It is perhaps sobering to note that it was probably because the linguists who began the research just did not have the necessary mathematical expertise to realise the full potentialities of Markov modelling that pattern matching was for a time (and wrongly) ruled out in favour of Chomsky-led ideas. As the mathematician Ian Stewart is fond of pointing out, there is a mind set almost ubiquitous amongst the mathematically unsophisticated majority which is far too ready to see un-caused and/or un-understandable mystery—in this case in the guise of something like innate ideas—where in reality there is only that where the mathematics of its chains of causation are so complex as not yet to be understood.

My fifth and last kite—if I toss up any more, then if they are not so already they are likely to become quite hopelessly entangled—is my favourite: the one that seems to me to be the most suggestive of all if it is the consciousness/language-use interface we are interested in. It is the 'articulation capacity' kite. And it flies because the conscious processing of information would seem to involve a very interesting limitation. Expressed informally in the everyday complaint 'don't talk to me while I'm on the phone', and more formally in the experimental work of Anne Treisman in the 1970s, we don't seem able consciously to process two linguistic inputs at the same time. The best we can do is to flip back and forth between them. And only if there is sufficient redundancy in one or both of the messages conveyed can we 'get' both of them. In the *Journal of Consciousness Studies* (Vol. 6, No. 2-3, 1999), Bruce Mangan from the Berkeley Institute of Cognitive Studies sees this 'operative limitation' as one of 'articulation capacity'. And thinks that what it suggests is that '*at the deepest level, consciousness IS the limited but infinitely plastic capacity to articulate experience*' (his emphasis). He elaborates: 'this overall capacity is conserved during a huge number of phenomenological transformations' such that 'when something becomes clear something else becomes vague' but 'total articulation remains at least roughly constant'. If consciousness is the capacity, at any one time 'operatively limited' but across time 'infinitely plastic', to articulate experience, then what the I IS hypothesis is adding is that it is not that persons uniquely *have* that capacity but that persons uniquely *are* that capacity: that attained only with struggle, and over a long developmental history, that is the capacity which constitutes the new personic reality called 'the world'.

So at the end of my chapter, and with five kites flying, how might I try to draw together what I take to be the essence of the theory I have now tried to spell out? To do so, I want to return to the image I used in Chapter 6 [see 'freedom *and* determinism'] of a cursor crossing a computer screen. The cursor had width. But as with all the terminology of space used in that image, that 'width' (a space word) was to be thought of as a 'duration' (a time word). The person imaged by the cursor was imaged as a duration three hundred and fifty milliseconds 'wide'. And its life-line along 'the arrow of time' was imaged as a movement across the screen. A *non*-person which was a dog would be 'narrower': wide enough to learn to fear that its owner will beat it, but not wide enough to fear that its owner will beat it next week. A *non*-person which was a mollusc would be 'narrower' still: maybe wide enough to learn to fear a moving predator, but not wide enough to learn to fear that same predator when it was stationary. What I am suggesting is that for all non-human sentient entities, the relative 'width' of their present is a variable which correlates with cognitive capacity. But that for the one and only human sentient entity, the linguistic structuring of its brain results in a width sufficiently enormous for a *totally new entity*, a *conscious* person, to emerge as a consequence. And it is as inappropriate to study that newly emergent entity which is a conscious person as if it were, or could be reduced to, the entity from which it emerged which is a human animal, as it would be to try to study the emergent entity which is a smile as if it were, or could be reduced to, the entity from which it emerged which is a face. Persons no more have metabolic rates than smiles have lips.

It is entirely possible to have the best of motives for attempting such inappropriate study. In *The Illusions of Postmodernism* (1996), this is what, in his own inimitable style, Terry Eagleton was pointing out. He was writing about the then current ubiquity of 'back-to-the-body' disquisitions, spattered throughout with the then buzz-word 'embodied'. And he observed that 'the post-modern subject, unlike its Cartesian ancestor, is one whose body is integral to its identity': that 'from Bakhtin to the Body Shop, Lyotard to leotards, the body has become one of the most recurrent preoccupations of post-modern thought'. He thought that it was because 'bodies are ways of talking about human subjects without going all sloppily humanist', that what this represents is 'the final expulsion of the ghost from the machine': a stance which allows us to avoid what he called 'that messy interiority which drove Michel Foucault up the wall.' Well of course the motive of trying to articulate a scientific approach to the study of persons is entirely appropriate. But it is my claim that the way to avoid Foucault's 'messy interiority' is not to jettison the 'interiority' but to jettison the 'messy': that is, to construct a *non*-messy interiority instead. A *non*-messy interiority is certainly a scientific one. It's just that the science concerned needs to be one which has as its object of study not the non-mysterious material *thing* which is 'that part of the body called the brain', but the equally non-mysterious *happening* which is 'a sequence of linguistically structured states of that part, called the person'. So with what the I IS hypothesis takes to be the relationship between consciousness, language and mind now outlined, I can turn to the third and last part of my book. In Part Three, I will try to make good my claim that formulations such as those I am proposing make possible the creation of the moral sciences of the personic.

PART THREE

Chapter Ten

SCIENCE AGAIN

Of that which is good

Moral Philosophy exists. In many of the institutions of higher education of anyway the Anglo-American world, it is pursued as an academic discipline. And many a fat tome, with considerable but not total overlap with many an other, will tell you just what its author takes the subject matter of the discipline to be. Moral Science, by contrast, does not exist. Nineteenth century writers—John Stuart Mill, for example—habitually used the phrase 'moral science'. And in Britain, one university—Cambridge—for years hosted a Moral Sciences Tripos, and, famously, a Moral Sciences Club. But while the second survives, still with its singularly inappropriate 'science' title, the first, established in 1851, has gone. Coming into line with most other institutions, Cambridge now too offers a degree just in 'philosophy': a move made increasingly inevitable as the subject matter of what formerly had been mis-labelled 'moral science', by being now able to be studied scientifically, leached steadily away into the burgeoning social sciences. And today, the content of courses in the one case and debates in the other is nothing more than philosophical: a situation which anyway much more accurately reflects the very widely accepted premise of much of mainstream Anglo-American philosophy that since natural science and moral philosophy are so conceptually distinct, it therefore follows that a stance which takes the 'science' from the one and the 'moral' from the other and comes up with 'moral science' is, in some fundamental sense, just a contradiction in terms.

But all is not well in Moral Philosophy. It exists. But probably without exaggeration, it could be said to be at least thoroughly unsatisfactory, at most in some sort of crisis. Writing in the *Times Literary Supplement* (3.7.92), the American philosopher Martha Nussbaum outlined what she called 'one story that is often told about the current state of things in Anglo-American moral philosophy'. It was a story which suggested that it was 'turning from an ethics based on enlightenment ideals of universality to an ethics based on tradition and particularity; from an ethics based on principle to an ethics based on virtue; from an ethics dedicated to the elaboration of systematic theoretical justifications to an ethics suspicious of theory and respectful of local wisdom; from an ethics based on the isolated individual to an ethics based on affiliation and care; from an a-historical detached ethics to an ethics rooted in the concreteness of history.' And she added that 'this new anti-theoretical ethics, being rooted in the traditional understandings of virtue of particular communities, is likely to become linked with some sort of political conservatism, and with the abandonment of the Enlightenment's radical demand for human equality across differences of ethnicity, nationality, class, gender and race.'

It is my impression that the reason why this is a story so often told, is that this is a story substantially true. For while it was of the essence of an Enlightenment approach to morality to seek foundations on which universal moral principles could be built, it seems equally clear that one of the main thrusts of much recent moral philosophising has been strictly anti-foundational. Thus in considering the debate between Habermas and Derrida as to whether there is any human universal which could provide a basis for universal moral principles, the self-styled pragmatist Richard Rorty, for example, sides with Derrida in finding Habermas's universalism just 'hopelessly metaphysical'. And in philosophical conversation anyway, the conviction that what is good has to be just a matter of opinion rather than actually a matter of fact is ubiquitous.

Though there is a single exception. Almost as widespread as the 'says who?' relativism which tends to challenge any attempt at the construction of moral universals, there is a different position: one which defends a universalism alright, but one based in biology. The only anyway widely canvassed and current proposition about what just might serve as foundation, as *scientific* foundation, of universalist principles for morality is evolutionary-theory. Thus in *Darwinian Natural Right: The Biological Ethics of Human Nature* (1998), Larry Arnhart suggests that where 'a naturalist explanation of morality would assume a moral universalism founded on the unity of human nature, a culturalist explanation of morality would assume a moral relativism founded on the diversity of cultural traditions'.

Now as I've already made clear—and in several places—this biological and thus evolutionary approach to the moral seems to me to be just a total non-starter. And by leaving out of account that of which, in my opinion, we are most in need, Arnhart's way of formulating things here makes that abundantly clear. Where he counterposes 'a moral universalism founded on the unity of human nature' with 'a moral relativism founded on the diversity of cultural traditions', I want to ask: whatever happened to 'a moral universalism founded on the unity of human culture'? Isn't it surely just an error to argue, this time in the words of the ethologist and zoologist Frans B. M. de Waal and Jessica Flack in a co-authored paper in the *Journal of Consciousness Studies* (Vol. 7, No. 1-2, 2000), that 'the degree to which the tendency to develop and enforce moral systems is universal across cultures—suggests that moral systems—have biological origins'? Surely that which is 'universal across cultures' can just as readily be universal for cultural as for biological reasons. For just as the universe and the biosphere present similar species with problems which they tend to solve in similar ways (anti-gravity reflexes, for example, are much of a muchness across many animal species), so the world presents similar selves with problems which they tend to solve in similar ways (where younger children make moral judgements based on consequence, for example, older ones make them based on intention). The fact that such moral judgements are pretty similar across many young selves no more means that they did not develop culturally than does the fact that such anti-gravity reflexes are pretty similar across many animal species means that they did not evolve biologically. This is my position. But I need to say a little more about *why* it is my position.

On March 12 1953, Crick and Watson announced to the world that they had cracked the double-helical structure of DNA. On March 12 2003, we celebrated the fiftieth anniversary of that wonderful breakthrough. Its application to the already well-established sciences of the organic has been spectacularly enriching. It's just that its *mis*application to the only embryonic sciences of the personic has been just as spectacularly dire. For where the application *gives us* (as human animals evolved through speciation) the actuality of our origins, the misapplication *takes from us* (as human persons developing through selving) the possibility of our progress beyond them. By returning right back to Darwin himself, let me try to show how this confusion comes about.

In *The Descent of Man* (1872), he claimed that 'any animal whatever, endowed with well-marked social instincts—would inevitably acquire a moral sense or conscience as soon as its intellectual powers become—nearly as well developed as in man'. And there it is: the confusion. It lies in his use of the word 'man' as if it delineated a non-problematic concept. I am wanting precisely to problematise that use: to say that 'man', or its now rather less sexist counterpart 'human', is one of those weasel words which, if we are not careful, leads us wildly astray when we think with it. The problem is that 'human' delineates both a biological and a non-biological entity. And it is of the essence of my hypothesis that, when we use the word, we need clearly to distinguish between them. I do so distinguish. But since it is as rare to do so as it is crucially appropriate to do so, let me detail just two of the things I am not doing, but could perhaps be thought to be doing, when I so distinguish.

First, I am not using the now familiar genotype/phenotype distinction. In a 1909 paper, the Danish botanist Wilhelm Johannsen suggested that, from a Greek word meaning 'to give birth to', the factors of inheritance discovered by Mendel be called 'genes'. And in a 1910 follow-up, he proposed the terminology 'genotype' and 'phenotype': the first to indicate the genetic composition of an individual; the second its apparent, visible and measurable characteristics. This is not my distinction here. Both genotype and phenotype refer to biological characteristics: the first to those genetically derived, such as being an hydrangea not a lupin; the second to those environmentally derived, such as being a pink hydrangea (because growing in acidic soil) rather than a blue hydrangea (because growing in alkaline soil). And to the task of weakening the knee-jerk geneticism which speaks, carelessly, of a genetic disposition to behave in way X rather than, more carefully, of a genetic disposition to behave in way X in environment Y, this genotype/phenotype distinction is not unimportant. But my distinction is more fundamental. It is that between *both* the genotypic *and* the phenotypic characteristics of a human animal on the one hand, and *neither* the genotypic *nor* the phenotypic characteristics of a human person on the other.

Second, I am not being wild and wacky. The tradition within which I am wanting to place myself, although it would not necessarily be recognised as being a tradition by those thinking outside the bounds of my candidate theory, is long and I think honourable. Perhaps just a scattering of not-quite-but-almost randomly chosen milestones across its history will be sufficient to make my point. *John Locke* suggests that it is experience as recorded in memory which provides the basis for personal identity: that I am what I remember myself being. *David Hume* argues that 'the soul so far as we can conceive it is nothing but a system or train of different perceptions'. *Auguste Comte* argues that since 'the thinking individual cannot split himself in two, one part of which would think while the other would watch the former thinking', it follows that the psychological methods of introspection are therefore 'radically faulty in principle'. *William James* adds that 'the thoughts themselves are the thinkers'. *Henri-Louis Bergson*, thinking in terms of something like a flow of presents, sees each of the present states of an individual as being the reflection of the entire history of preceding states. *Schopenhauer* sees the world as 'representation' for all living and knowing beings, but as what he calls 'reflective representation' (in their capacity for the abstract) for only humans. *Radcliffe-Brown* thinks of every human being as a duality: as individual, 'a biological organism'; as person, 'a complex of social relations'. *John Macmurray* argues that 'the proper relations between human beings' are those in which 'organic differences have no essential standing'. *Susan Greenfield* talks about the mind as 'the personalisation of the brain'. *Charles Taylor* maintains that we become persons, and remain such, only as interlocutors. *George Pattison* insists that 'the person doesn't exist outside the social process of which dialogue is the articulation'. *Galen Strawson* points out that when philosophers talk about a 'subject of experience', they do so in such a way that said subject is to be thought of as existing even in the absence of any of said experience—as just sitting around waiting to be 'subjected' to it, I thought, when I came across his comment—when it is much more parsimonious to think of there being 'no subject of experience if there is no experience.' And so on, and so on, and so on: this could become a very long list indeed. But in relation to such a list, short or long, what I here want to claim is this: that while, if my hypothesis were to be accepted as useful, of course it would entail a further *development* of any such tradition—after all, it could hardly justify several hundred pages of explication if it were merely a *restatement* of it—nonetheless it would in no way be an out-of-the-blue heresy with no foundation at all in the history of such things.

That which I think *is* new in what I am proposing is that which I also think represents some sort of resolution of what I take to have been one of the main problems with any such tradition: that is, its ever present tendency, much more obviously marked amongst those who have contributed to it from an idealist rather than a materialist position but present to some extent almost everywhere, to slide off into mysterianism. It is most commonly the hard-nosed, biological reductionist who gets irritated by talk of entities called agents, persons, selves, subjects, even souls, without any too clear an identification of what they are, or even what they might be. But just because hard-nosed biological reductionism is not the solution to the problem of what they are (which it isn't), that doesn't mean that it's not a problem to say clearly what they are (which it is). When in *Consciousness Reconsidered* (1992), Owen Flanagan says that 'persons experience their experience as their own', then along with the biological reductionist, I am wanting to mutter: 'in which case, what are persons, if not Cartesian homunculi'? But when in the *Times Higher Educational Supplement* (29.5.98), Mary Warnock says that 'there is everything to be said for reminding us that genes alone do not constitute a person', then this time across the bows of the hard-nosed biological reductionist, I am wanting to shout: 'but genes don't constitute a person at all'! In taking such a position, what I am wanting to do is to flesh out, but in ways different from those in which he fleshes them out, a couple of points made by Dan Dennett in his book *Darwin's Dangerous Idea* (1995).

First, a point on genes. Dennett suggests that our relationship to our genes is 'importantly different from the relationship of any other species to its genes'. Why? Because 'what *We* are is not just what we as a species are'. Indeed. But let's be clear. My position is that when 'I' is read as 'a member of the species *Homo sapiens*', it has exactly the same relationship to its *Homo sapiens* genes as 'a member of the species *Wug*'—where *Wug* stands in for any other species you care to name—has to its *Wug* genes. It is when 'I' is read as a person that things are different. When I is read as a person, as the linguistically structured experience of a *Homo sapiens*, it's not that it has a *different* relationship to its genes, but that it has *no relationship at all* to its genes—*because it doesn't have any*. With apologies for the clumsiness of the formulation, it is my position that persons just don't have genes to have a relationship to! What we are in need of here, or so I anyway maintain, is a thoroughly non-genetic hypothesis as to the nature of persons: an hypothesis at one and the same time both thoroughly non-biological *and* thoroughly scientific; an hypothesis the formulation of which could provide us with a foundation on which could be developed the moral sciences.

Second, a point on language. I would argue that the enablements and constraints of millions of years of pre-linguistic evolution have been for far too long pre-wired for the internal environment so made by the genes to be significantly modified by the genes: a position which means that any such significant modification would have to come from elsewhere. Now assuming that the human way of being can appropriately be characterised as a very significant modification of all other animal ways—and who among even the most dedicated of evolutionary psychologists would want to suggest otherwise?—then what might have been the non-genetic source which, from elsewhere, might have brought such modification about? In one way or another, wrapped in one or another of many specific hypotheses about just how it might have happened, by far and away the favourite candidate as the source of such modification is language. It is Dennett's candidate too. He writes of the use of language being able to submerge 'the old hard-wiring under a new layer of pre-emptive control'. He suggests that in 'verbally embodied surrogates that could be reviewed in any order at any pace', it creates 'a whole new class of objects-to-contemplate'. Instead of all this, what I am suggesting is that what the use of language created was *the contemplator its-self*: a formulation which parsimoniously avoids the 'contemplated by what?' question which otherwise so readily arises.

It is worthwhile pausing here to spell out that last idea in a little more detail. Amongst linguists and philosophers of language there is a theory of meaning commonly called the 'Fido-Fido' theory. It suggests that the meaning of a word is the thing it represents. And it is a theory now almost universally rejected. Instead, so the orthodoxy now goes, words as such don't refer to anything. It is the *uses* of words which refer. And the things to which such uses refer are identified by knowing about usage rules: rules, as sometimes it is put, which are intrinsic in the language, not extrinsic in that to which it refers. It seems to me that this emphasis on intrinsic rules is entirely appropriate. It is an emphasis, for example, which leaves us able to be very interested in the work of Luc Steels at the Free University of Brussels, who has modelled the invention of language using rules which are gradually invented, negotiated, and built upon, by pairs of talking robots which learn their grammar from one another. But I also think that mine is a formulation which, by delineating the world *in terms of* the linguistic, avoids any such intrinsic/extrinsic dichotomy in the first place. And in so doing, aids the search for what we need: which is an hypothesis at one and the same time both thoroughly non-biological and thoroughly scientific as to the nature of persons. For it is the formulation of such an hypothesis which could provide us with a foundation on which could be developed the moral sciences.

In its most general form, the search for such an hypothesis has been with us for at least a couple of centuries. Perhaps essentially initiated by Hegel, the general idea that the personal is something structured inter-subjectively—as a linguistic hypothesis requires—is well established. But it is my contention that the single most important error which has prevented it from becoming more than just such a general idea, which has kept many of its many protagonists slopping about in various forms of mysterianism, has been what might be called the error of the subject. We have conceptualised a person as that which is the subject of physical and experiential states: as that which, as subject, has a body and has experience. But as the I IS hypothesis I think makes clear, so conceived there is a totally intractable problem. If it has a body, it must be something other than that body: if it has experience, it must be something other than that experience. But what else is there, other than a body with experience? And how could there be anything which could evade the problem of infinite regress? There is only the body and the experience of the body. If the person is to exist, it must *be* one or other or both of them: without becoming some third entity, it cannot *have*—as agent, self, subject or even soul—one or other or both of them. The I IS hypothesis informally expressed is that it *is* the experience of the body. The I IS hypothesis more formally expressed is that it *is* the linguistically structured experience of that part of the body called the nervous system. And that there is no *has* involved at all.

With these few introductory pages, I hope I have laid the groundwork both for the rest of this chapter and much of the final two. In the rest of this one I want to turn my attention more directly to the question of what the moral sciences, founded on such premises, might look like: a task which will involve a brief look at what, in my candidate theory, is meant by 'experience' (and thus by 'person'); a more lengthy analysis of my idea of 'the self', and of the process I am calling 'selving'; a brief illustration of what a moral science constructed within such parameters might have to say about just a few of the most discussed practical philosophical problems of today; and, to conclude, a first very brief spelling out of the role of equality in the project of their construction. In the next chapter, in a return to the idea of God, I will deal more extensively with the theoretical underpinning of what I am calling 'the having ideology', especially as it impinges upon our ability, as persons, to love; concluding (once again) with a consideration of what I take to be (once again) the centrality of the idea of equality, this time to the construction of a God-as-love entity. And in the final chapter I will at last let that 'sovereign virtue' have the centre stage thus justified for it. And in so doing, try to summarise as simply as possible the whole thrust of my text.

As conceptualised by the I IS hypothesis, the moral sciences would have as their object of study a particular category of experience. In his *Truth and Method* (1975), Hans-Georg Gadamer argues that the concept 'experience', as used by philosophers, is one as frequently employed as it is infrequently explained. So: let me try to be as clear as I can about the way in which I am employing the concept. Mine is an hypothesis which makes six claims about the nature of experience:

(1) that the ability to experience is a characteristic only of the nervous systems of living things;

(2) that it consists in states of those nervous systems brought about by input into them from their environments;

(3) that their environments are both internal and external, with the rest of the body of which the nervous system is a part constituting the internal environment, and the parts of the universe, biosphere and world outside the body but available as input to its senses constituting the external environment;

(4) that because it consists in receiving input from the internal and external environment, 'experiencing' is always 'experiencing something', and that therefore the verb 'to experience' cannot be intransitive;

(5) that the type of experience unique to *Homo sapiens* is linguistically structured experience;

(6) that it is this category of experience which constitutes personness.

And thus the I IS hypothesis is one which suggests that it is 'person-ness' so characterised which would form the subject matter of the moral sciences. Persons would be posited not as material things but as happenings: entities which exist not as stuff in space but as durations in time. The point of view *at* which a human animal stands is not really a point but an area: organisms *take up* space. The moment of comprehension *as* which a human person understands is not really a moment but a duration: persons *consist in* time. The 'point' is actually a square half-metre or so in area: the 'moment' is actually some three hundred and fifty or so milliseconds in duration. But while the human animal is always *at* such a point (area), the human person is always *as* such a moment (duration). That means that, because it is *at* such a point (area) that the human animal exists, it *is* coherent to postulate it as an entity which can move from that point (area) to another point (area). As it also means that, because it is *as* such a moment (duration) that the human person exists, it is *not* coherent to postulate it as an entity which can move from that moment (duration) to another moment (duration).

This is a formulation of the nature of persons which solves, or rather just *dis*solves, the so-called problem of personal identity. For if we now ask what are the necessary and sufficient conditions for person X at time 't' to remain person X at time 't + 1', our answer is that it is the same person when it is the linguistically structured experience of the same *Homo sapiens*. It is sometimes suggested that this one-person-per-body principle is violated by multiple personality syndromes. Are there not two persons at home in the one body which hosts both Dr. Jekyll and Mr. Hyde? But the I IS hypothesis avoids any such problem. Since it is just definitional that a system cannot be in two different states at the same time, when 'person' is operationalised as the (linguistically structured) experience of a *Homo sapiens*, and 'experience' is operationalised as the (dynamically shifting) sequence of states of its nervous system, the unity of the person becomes a given of the theory: a conclusion thoroughly consistent, for example, with the experiments of Anne Treisman in the 1960s and 1970s which suggested that Dr. Jekyll-ness and Mr. Hyde-ness simply oscillate in the single nervous system of a single *Homo sapiens*.

Conceiving then of experience in the way outlined above, the I IS hypothesis is suggesting that while the non-linguistically structured experience embodied in the nervous systems of all but one animal is *not* such as to constitute the emergence of persons, the linguistically structured experience embodied in the nervous system of that one animal *is* such as to constitute the emergence of persons. And as already discussed in the previous chapter, while we already have some understanding of the way in which the information as to how to make a body is carried genetically, we as yet do not have an equivalent level of understanding of the way in which the information as to how to make a mind (or person) is carried linguistically. But such discrepancy notwithstanding, I am nonetheless proposing that it is the search for such understanding which is the appropriate search if we are to construct the moral sciences.

And some few things we know already. We know, for one example, that all the thousands of phonemes of all the hundreds of languages of all the peoples of the world can be uniquely described in terms of the shapes and movements of what are sometimes called the six 'speech organs' (the larynx, the soft palate, the root of the tongue, the body of the tongue, the tip of the tongue and the lips). We know, for another, that the written symbols of all writing systems each represent one of only three types of linguistic structure: a phoneme, a syllable, or a morpheme. We know, for a third, that in severe cases of the loss of language as the result of injury to the brain (aphasia), family and friends often feel that it is the whole person which has gone: which is maybe because it is. And we probably know, for a fourth, that at the linguistic outset it was not that humans were becoming smart enough to make language, but that they were becoming linguistic enough to make smartness.

Mine is an hypothesis putting heavy emphasis on human animals becoming human persons not by being generally socially situated with respect to one another, but by being specifically linguistically situated with respect to one another. It is not *just* saying—though of course it *is* saying—that it is the wiring of the interpersonal in the world which produces the wiring of the personal in the brain. It is also saying that each of us becomes a person not so much by freely joining the human conversation, but rather more by necessarily being one of its con-scripts. And such an emphasis raises an important conceptual problem which needs to be resolved. It is the question of the point at which experience ceases to be non-linguistically structured and becomes linguistically structured. And thus the point at which a non-person becomes a person. Even within the constraints of the I IS theoretical position, there are still two possible answers to the question rather than one. And what is at issue in the choice between them is whether it is afferent structuring by *other*-generated language or efferent structuring by *self*-generated language which matters.

There is a good case for saying that *self*-generated language is in place only from about eighteen months of age with the emergence of Roger Brown's eight types of two-word sentence. On the other hand, *other*-generated language is part of the experience of most human animals at least *from* birth, and maybe from *before* birth. At which of these points it becomes appropriate to say that experience is from then on linguistically structured is therefore debatable: one could stress the significance of efference, of output, and plump for the first possibility; one could stress the significance of afference, of input, and plump for the second. One possibility which I find suggestive is again based on the idea that, when offered a choice, it is often best to take both. Could it be that the point at which ones own efferent output (voice) becomes available as afferent input to ones ears is the point at which language use results in consciousness? The reflexiveness of such a proposition sounds about right. In the last analysis, of course, the only possible stand to adopt on this question is that it does not yet have an answer. It will be only when we have a greater understanding of the *mechanism* by which language structures experience that we will be able to answer it. In the meantime, I will sit firmly on the fence: that is, when I want to refer to humans between birth and their *production* of language, that is to infants, I will use the concept 'proto-person'. And will do so because, provisional though this one is, some concept rather than no concept is needed: and that for at least two reasons. First it is needed – by all of us – in order to be able to talk about some current moral debates, such as the one posed by Peter Singer's proposition that parents should have the right to kill severely disabled babies up to the age of one month. And second it is needed – by me – in order to be able to present my concept of a self more fully. It is to this second that now I must turn.

My basic proposal is that where *evolving organisms* are particular 'ensembles' (with thanks to Marx) or 'structurations' (with thanks to Giddens) of the genetic relations which constitute the biosphere *actualised in space as species*, so *developing persons* are particular ensembles (ditto) or structurations (ditto) of the linguistic relations which constitute the world *actualised in time as selves*. And via what will be a further explication of the selves concept as it is used in this formulation, I want now to defend an extension of that basic proposal to encompass the idea that where the organisms constituting the biosphere *live* as one or another type of species, so the persons constituting the world *love* as one or another type of self. It is that 'one or another type of self' which will get its main treatment here: persons as entities capable of loving will be a major theme of the next chapter.

In a personal conversation, an Oxford philosopher of my acquaintance once said, beautifully succinctly I thought, that 'while science has given us a wonderful picture of how the world is, we just can't find a way of putting ourselves into it'. The 'science' he was talking about was what I have been calling natural science: the 'world' he was talking about was what I have been calling the universe, the biosphere and the world combined. What I want now to illustrate is the way in which we can create that science which can give us his 'wonderful picture' of the world – not of the universe or the biosphere, sciences of which already exist, but of the world – into which we will not need to find a way of 'putting' our selves *because those selves are what constitute it*. Such a science will be moral science.

I am arguing that, as the linguistically structured experience of a *Homo sapiens*, a person – and to the extent to which the yet-to-be-acquired evidence requires it, a proto-person also – always exists as a type of self. And that what I am calling the selving process is one whereby, across a lifetime and within a person, one self follows another self sequentially. As I have already spelled out, the process is one which, at its most micro level, involves one state of a linguistically structured nervous system succeeding the immediately previous state of a linguistically structured nervous system nanosecond by nanosecond. But there are two important qualifications to make about such a conception.

The first is that the passing of each self, and its replacement by its successor self, is a passing which also entails a non-passing. William James puts it this way: he says that that which 'owns the last self owns the self before last', for 'what possesses the possessor possesses the possessed.' And while his 'owning' and 'possessing' language would not of course be mine, it is just such a realisation I am after. I am wanting to say that given every subsequent unit of experience is the product of every prior unit of experience plus new input, one self following another self sequentially is not a process involving discontinuities.

And the second is that while the ultimate self unit might be the nanosecond-long state of such a system, the units with which it is, anyway as yet, operationally possible for a science of selving to function—and which anyway may turn out to be the ones which eventually we need—are units considerably more gross-grained than that. The various already existing stage theories of the development that takes place across a lifetime within persons, which development results in the sequential series of selves which I am delineating as the products of what I am calling the selving process, delineate stages not nanoseconds but years long. These are the theories of such as Vygotsky, Piaget, Erickson or Kohlberg which, as I have already argued, might turn out to be the precursors of that more fine-grained theory of the selving process which in time, the I IS hypothesis would predict, will emerge.

Such emergence, depending as it does on on-going empirical research—and even on *not-yet* on-going empirical research—is almost entirely a development of the future. But at whatever level the selving process comes to be understood as happening, mine would be a formulation of it, as process, which would be couched in terms of 'progress': if not a simple and linear 'progress' coming with some sort of teleological guarantee of its satisfactory outcome, then at the very least, a negative impetus, whereby structures stunting of the selving process are 'progressively' eliminated across both historical and personal time. And that being the case, it behoves me to have at least something to say about what I mean by such 'progress'. In the course of saying it, what's more, I will be able not only to deal with the substantive, specific and controversial issue of 'progress' itself, but also to illustrate the more general way in which an insistence on the difference between the organic and the personic can be such a powerful conceptual tool in helping us to solve philosophical problems.

In *Complexity* (1993), Roger Lewin quotes Roger Thomas, a palaeontologist from Franklin and Marshall College, Pennsylvania: 'The most significant pattern in the history of life is the progressive net increase in complexity of structure and dynamics that has occurred in organisms and the ecosystems in which they participate'. At first flush, that would seem to be a reasonably unexceptionable idea: there is pretty obviously something rather more complex about a *Homo sapiens* than there is about an amoeba. But as Lewin then goes on to discuss, the question as to whether evolution necessarily generates ever more complexity, especially as expressed in Thomas's 'progressive' word, would seem to be one of the most contentious issues in biology. Indeed, the current orthodoxy among biologists is that the notion of progress 'up' the phylogenetic scale is hopelessly anthropomorphic. So *at the level of the organic*, can increased complexity (a) be shown to be the case, and (b) be used as a measure of 'progress'?

Well although, as Lewin himself points out, 'complexity' in this context is a concept very difficult to define, there have been a number of candidate attempts to do so. Some biologists have tried to base a measure on the number of anatomical parts an organism has. A proposal very well received by many of his fellow scientists was that made in 1984 by John Tylor Bonner, a biologist at Princeton, to the effect that a measure be based on the number of different types of cell an organism has. John Maynard Smith's observation on the problem was that when you start simple, 'there's no way to go but up': that while the direction of change is constrained in one direction (towards the simple) it is open in the other direction (towards the complex). And that anyway, as prey develop ever more sophisticated ways of staying alive, competition between predator and prey is going to tend to produce increased rather than decreased complexity. Putting in his oar, Norman Packard, a pioneer in the development of chaos and complexity theories, argues that it is increased computational ability which drives the evolution of living organisms: that is, that it is the ability to process information that 'progresses'. And this is an approach which sounds somewhat similar to that taken by Dobzhansky et al (1977) in their classic textbook on evolution, where they argued that the ability to obtain and process information about the environment increased through evolutionary history, and was thus a mark of 'progress'.

But others—among them, and notoriously, Stephen Jay Gould—insist that since evolution adapts species to local environments, and local environments are no more likely to demand *increased* complexity than they are to demand *decreased* complexity, this is an approach fundamentally in error. Gould speaks of 'progress' as 'a noxious, culturally embedded, untestable, non-operational idea'. And suggests that the reason why there is nonetheless what he calls 'a profound unwillingness to abandon a view of life as predictable progress' is because 'to do so would be to admit that human existence is nothing but an historical accident'.

It is my own suspicion that, Gould notwithstanding, a perfectly respectable notion of 'progress', *even at the level of the organic*, could readily be arrived at using some combination of ideas such as those to which I have briefly referred. But even supposing that none of these 'progress' arguments could be made to stick; that Gould is right to excoriate the whole notion of progress at the level of the organic: what then? It would be my claim that, if we unpack Gould's word 'human' in the way that I've been proposing, a way which insists on the difference between a human animal at the level of the organic and a human person at the level of the personic, something not only new but almost diametrically different emerges.

The organism 'human-animal' has evolved via a speciation process: a process whereby those species which, as living things, constitute the biosphere, become, via evolving genetic competition and across time, differentially successful. *By absolute and word-by-word contrast with such a formulation*, I am wanting to suggest that the person 'Ann Long' has developed via a selving process: a process whereby those selves which, as loving things, constitute the world, become, via developing linguistic cooperation and across time, similarly successful. The essence of my case here is that it will be only in getting fully hold of the total and complete difference between these two processes, between the organic and the personic, that we will give ourselves a conceptual tool sufficiently powerful to enable us to start to construct the moral sciences. And since what therefore needs to be stressed is *difference*, what needs to be made clear is that, even if students of the organic should find it more appropriate not to talk in terms of progress in their domain, that is *no sort of reason at all* for students of the personic to follow suit in theirs. Progress at the level of the personic is an absolutely central feature of that level. For moral *development*, the steady creation of that which is more fully good, is quintessentially a one way process. To use (without accepting) Kohlberg's (alias Kant's) terminology, it's not that having achieved autonomous morality a self would no longer *choose* the heteronomous kind. It's that the heteronomous kind is no longer available *to what has become a different self*. Development, as opposed to just growth or maturation, is like that. It is progressive. So how then is this 'progress' concept, one essential at the level of the personic *whether or not* it is useful at the level of the organic, to be conceived?

Across both the collective history of our societies and the individual history of our persons, perhaps the most important thing which progresses is our ability to decentre: that is, our capacity increasingly to develop selves unafraid of notions of process and progress, both of which necessarily carry with them the realisation that our own small mortality can play only an historically contingent and limited part in the whole. Let me illustrate this point with what I take to be a very poignant example of its misunderstanding from the poetry of W. H. Auden.

A much quoted and much discussed stanza from his poem *September 1st, 1939* reads:

All I have is a voice
To undo the folded lie,
The romantic lie in the brain

Of the sensual man-in-the-street
And the lie of Authority
Whose buildings grope the sky.
There is no such thing as the State
And no one exists alone;
Hunger allows no choice
To the citizen or the police;
We must love one another or die.

The poem was one which very quickly became one of his most popular. But perhaps when he quite rapidly realised just how significant and demanding a message it maybe half inadvertently contained, he disowned it: in particular, its last line. He defended his distrust of what he had written by arguing that 'or die' should have been 'and die'. But he was wrong. It should not. The younger, braver, less successful Auden had it right: the older, more frightened, too feted Auden got it wrong. Auden must die, and did. I must die, and will. But whether or not *we* die, whether or not the whole human *enterprise* dies, is a totally different issue. And there is a compellingly strong case to be made for saying that the successful resolution of that totally different issue does indeed depend on our learning to love one another: part of which learning means our coming to understand, contra the post-modernists, that it is *progress* towards such loving which is of the essence not of human animals but of human persons.

In his essay *Tradition and the Individual Talent* (1922), T. S. Eliot wrote of the progress of the artist as 'a continual self-sacrifice': as 'a continual surrender of himself as he is at the moment to something which is more valuable'. This is selving. It is the continuous making and remaking, across personhood, of a sequential series of selves. Each such self acts in what it perceives to be its own interests: that is definitional. But since each succeeding self is developmentally *progress upon* the previous one, its perception of its own interests is also developmentally *progressive upon* its previous perception. Its acts are not merely different from those of its ancestor selves: they are progress upon those of its ancestor selves. Of course such selving—such 'menting', if alternatively it is thought of as the continuous making and remaking of the mind—is a process subject to dislocation. One such dislocation occurs when menting becomes *re*menting: when previously made selves are simply recycled, and development is stunted. Another such dislocation occurs when menting becomes *de*menting: when even previously made selves begin to disintegrate and development goes into reverse. Dementing is usually seen only in the old, and in only a minority of them. But rementing, by contrast, is widespread. And its avoidance—that is, the nurturance and thus preservation of the ability to 'ment' or selve—is possibly the biggest single problem of personness. To start to show what I mean by that, let me now turn to a consideration of the selving process in a little more detail.

At the level of the organic, speciation is the evolutionary process whereby, via natural selection amongst genetically structured bodies, new species emerge. By analogy, I am suggesting that, at the level of the personic, selving is the developmental process whereby, via moral election amongst linguistically structured minds, new selves emerge. And by deliberately 'comparing and contrasting' the two processes in this way, I am wanting to draw attention to two things: the one very simple; the other very complicated. The simple thing (which needs no further comment here) is the fact that they actually are two processes, not one. The complicated thing (to which now I must turn) is that when either is thought about from within what I am calling the I AM paradigm—and to date that has meant for ninety nine per cent of the time that either has been thought about at all—each process, the organic and the personic, is thereby open to serious misrepresentation. To illustrate the way in which such misrepresentation comes about, I want to unpack the consequences of the use of the two phrases 'natural selection' and 'survival of the fittest', as they occur in evolutionary theory.

First, 'natural selection'. Sourced in its positing of its *own* self as agent over against the world, as chooser vis-à-vis that which is chosen (or not chosen), the I AM person has a marked tendency also to posit a particularly interesting *imaginary* self in much the same way. The imaginary self is the one called Mother Nature. And the tendency is to posit her, analogously, as agent over against the biosphere, as selector vis-à-vis that which is selected (or not selected). What is going on here is that in much the same way as the I AM person sees chooser and chosen as *externally* related the one to the other in the first case, so he or she also sees selector and selected as *externally* related the one to the other in the second. This is the similarity.

But there is also a difference. The difference is that while the animism involved in the selector case, the Mother Nature case, is now pretty readily understood to *be* animism, and thus inappropriate, amongst a large and ever growing number of people, the animism involved in the chooser case is hardly understood at all. More and more of us now appreciate the fact that, in the organic case, the selection process is one *internal to* the system of competitive relations between genetically structured organisms living as many types of natural species, and that Mother Nature is an animistic fiction. But in the chooser case, this is not so. The homuncular animism involved in the posit of an I AM self is still only minimally appreciated. In philosophical conversation anyway, it is still extremely rare to encounter the realisation that just as the natural selection process is one internal to the system of competitive relations between genetically structured organisms living as many types of natural species, so what I am calling the moral election process, the process of selving, is one internal to the system—*the no less real system*—of cooperative relations between linguistically structured persons loving as many types of moral selves.

Second, 'survival of the fittest'. Anyway *some* of the confusions which this phrase fosters have, of course, been very widely recognised. Many have been those who have lamented the fact that Darwin, apparently at the suggestion of Herbert Spencer, introduced it into later editions of *The Origin of Species*. For one thing, it is a phrase which can encourage the (faulty) criticism of evolutionary theory mounted by those who argue that because 'fit' can only be defined as 'that which survives', the whole idea is just circular. But for another -and this is the case that concerns me—it is a phrase which, spectacularly as used in an I AM context, lends itself to a thoroughly interested reading of just what 'fit' means.

It was because I had long recognised this double potential for confusion that I was interested to somewhere come across the advice of Fred Leavitt, professor of psychology at California State University, that we should mount the case for evolution using neither phrase. His proposal, as I remember it, was that we should adopt a four-stage delineation of the process something like this: single organisms vary—an empirical statement; they live (survive and reproduce) with varying degrees of success—an empirical statement; some of the variations which affect success in living (in surviving and reproducing) are transmitted genetically to offspring—an empirical statement; therefore populations of organisms will come to consist in more of them resembling those living more successfully, fewer of them resembling those living less successfully—an empirical statement and a logical deduction from the preceding three. I liked it. And I liked it even more when I effected its translation to that which, by contrast with the evolution of the organic, I am calling the development of the personic.

Such a translation would run something like this: single persons vary—an empirical statement; they love (actively pursue the well-being of the loved) with varying degrees of success—an empirical statement; some of the variations which affect success in loving (in actively pursuing the well-being of the loved) are transmitted linguistically to others—an empirical statement; therefore societies and communities of persons will come to consist in more of them resembling those loving more successfully, fewer of them resembling those loving less successfully—an empirical statement and a logical deduction from the preceding three. What this means is that the development of love across the series of selves which constitute the personic, just as surely as the evolution of life across a series of species which constitute the organic, is a process: and a process which, like its natural counterpart, needs not an 'animation' nor an 'ought' in sight in its conceptualisation.

Delineating the most appropriate way to characterise the developing sequence of selves which, across time, constitute a person is almost entirely a task for the future. Developmental psychology has made a start. The theories of such as Vygotsky, Piaget, Bruner, Donaldson, Freud, Erikson, Kohlberg, Bandura, Wright, and many others, have begun to give us a rough outline of possible sequences. Whether any of them will turn out to be the most appropriate characterisations remains, I think, to be seen. As earlier I spelled out, Erikson, for example, delineates eight ages across which, in discretely identified stages, an ever widening circle of persons become significant Others, and thus of moral concern. I like it. But whether or not it will turn out to capture the essence of the sequential selving process I am talking about will depend, among other things, on the extent to which its stages turn out to be correlative of any increasingly well understood linguistic and neural correlates. Perhaps we can already guess that such broad categories as 'infancy' (or proto-personhood), 'the child until it is seven', and 'adolescence' are going to turn out to be sufficiently well established as to be unlikely to shift dramatically. But even these are in no way absolutely clear. And much of the rest, particularly in adulthood, is positively murky. When we can say, as perhaps we now can, that with the emergence of two word sentences, wilfulness and mirror recognition at somewhere between twelve and twenty-four months of age, the first self, reaching its apogee with the notorious 'terrible twos', is in place, it is not that we can say nothing. Far from it. But it is that we can say nothing as yet fine-grained enough to satisfy entirely appropriate scientific standards. We will. But to do so, one of the things we have to do is to tackle head-on the naturalistic reductionism which tries to tell us that we need not bother.

For we are at the moment living through a period of what might appropriately be called rampant pseudo-Darwinism. An initial sociobiology has become a more sophisticated and therefore more dangerous evolutionary psychology. Particularly perhaps since the demise of Marxism, the reduction of the personic to the organic is everywhere: and everywhere a menace. In such a situation, the task for an hypothesis such as mine is to try to underpin an approach to the specialness of the person, its language, its morality, which refuses any such reduction. But which does so in a way which nonetheless whole-heartedly welcomes as progress, rather than erroneously laments as doom, the fact that such underpinning *must* now be done scientifically: *can't* now be done mysteriously, or emotionally, or religiously, or dualistically. If sometimes unnecessarily unpleasantly, especially I think in relation to persisting religious beliefs, the rampant ones are entirely right to rule any such mysterious, dualistic approaches totally out of court. And wishing that anyway some of them could muster more compassion as they do so does not make it any less right that they do so. They are entirely right to do so. Where they are wrong, and entirely wrong, is in their choice of the science with which they think they can achieve their 'ruling out'. Let me illustrate this point by reference—once again—to just one of their number.

In *Philosophy Now* (Autumn 1996), Les Reid reviewed Dan Dennett's *Darwin's Dangerous Idea*. He suggested that Dennett presented all our history as 'one long story of evolution, starting with chemical reactions, evolving to simple organisms, evolving to conscious organisms, evolving to language users, and so evolving to all the peaks of culture as we know it today'. And when I came to read the book for myself, I saw indeed what he meant. There is just too much 'evolving' going on in its pages. Darwin himself claimed—and it might be a good idea to listen to what he claimed—that 'one general law leading to the advancement of all organic beings is to multiply, vary, let the strongest live and the weakest die'. But 'chemical reactions', 'language users' and 'peaks of culture' are *not* 'organic beings'. According anyway to its principle author, the theory of natural selection is *not* a general theory about the way in which anything and everything *changes* across time, but a specific theory about the way in which organic-beings *evolve* across time. He after all gave *The Origin* the alternative if unfortunate title *The Preservation of Favoured Races in the Struggle for Life*. And things that don't so live can't so struggle. Whether we work, as initially, with the organism as the unit of selection (and Darwin would have had no trouble at all with the current doubt about the inviolability of the 'species' concept for he always saw species not as real entities but as aggregations of individuals), or, as currently, with the gene as the unit of selection, is here (but not elsewhere) irrelevant. For 'chemical reactions', 'language users', and 'peaks of culture' correspond to none of these: they are *not* species; they are *not* organisms; and they are *not* genes. And just as evolutionary theories about how species, organisms, genes change expression across time were made out of observation of, hypothesis about, and experiment on, species and/or organisms and/or genes, so *other* theories about how chemical reactions, language users, and peaks of culture change across time must be made out of observation of, hypothesis about, and experiment on, chemical reactions, language users, and peaks of culture. There is no scientifically legitimate route whereby theories of things can be derived from theories of other things, rather than from observation of, hypothesis about, and experiment on the things themselves. Theories about chemical reactions, language users and peaks of culture must be made from observation of, hypothesis about, and experiment on chemical reactions *as* chemical reactions, language users *as* language users, and peaks of culture *as* peaks of culture, not as pseudo-species, pseudo-organisms or pseudo-genes. And concerning the latter two—language users and peaks of culture—that means they must be made in the domain of what are now called (I think erroneously) the social sciences: what will come to be called (I think more accurately) the moral sciences.

This of course in no way means that the as yet unformulated foundational theory of the sciences of the personic will *necessarily not* have features in common with the already existent foundational theory of the sciences of the organic. It just means that it will *not necessarily* have features in common with the already existent foundational theory of the sciences of the organic. Short of its foundation, we cannot know whether it will or not. There are reasons to expect that it might. There are reasons to expect that it might not. And to illustrate what I mean by that, let me speculate, no more, on some of each. Of reasons to expect commonality, I can immediately register four.

First, they might well share what has come to be called the 'falsifiability' criterion, which is anyway a fairly well respected test of the scientificity of at least some levels of theory. While there is considerable debate as to which such levels are—I find the approach appropriate at the level of very specific theories, inappropriate at the level of more general theories—it can be nothing but bonus for a theory that its proponents should be able to specify the evidence they would take as sufficient to make them at least question its validity, at most reject it outright. The easiest way to mount such evidence is for the theory to make predictions which can then be tested empirically. Now whether or not the theory of evolution, explicitly or implicitly, makes such predictions is a not an uncontentious issue. In *The Language Instinct* (1994), Steven Pinker not only argues that it does himself, but claims that Darwin thought so too. To falsify the theory of evolution, he suggests, 'all it would take is the discovery of a trait that showed signs of design but that appeared somewhere other than at the end of a lineage of replicators that could have used it to help in their replication.' The examples he uses to flesh out his argument are 'the existence of a trait designed only for the beauty of nature, such as a beautiful but cumbersome peacock tail evolving in moles, whose potential mates are too blind to be attracted to it'; 'a complex organ that can exist in no useful intermediate form, such as a part wing that could not have been useful for anything until it was one hundred percent of its current size and shape'; 'an organism that was not produced by an entity that can replicate, such as some insect that spontaneously grew out of rocks, like a crystal'; and 'a trait designed to benefit an organism other than the one that caused the trait to appear, such as horses evolving saddles.' OK. So how might an as-yet-unnamed underlying theory of the moral sciences measure up to criteria such as those here applied to its earlier and simpler counterpart? How about judgement by intention rather than judgement by consequence arising in two year olds? Or identification with all persons emerging when prior identification with none or some could not have been experienced?

Or an advanced I AM self arising spontaneously in a giraffe? Or authoritarians found to be setting up consciousness-raising groups?

Second, just as there are no essences in the sciences of the organic, so there will probably be no essences in the sciences of the personic. The Platonic notion that there is an essential hippo, an essence of hippo-ness, is just wrong. There is this hippo, that hippo and the other hippo, each of them a variation on the theme hippo-ness if you like, but no essence of hippo. And this sort of essentialism is shaping up to be a notion just as wrong in the sciences of the personic as it is in the sciences of the organic. We are steadily recognising that there is no essential man-woman relationship; no essential democracy; no essential way of growing crops, making babies, deciding what is to count as literature, being fulfilled or playing cricket, but again just variations on these various themes. Of course, just as in the first case, that does not mean that a hippo comprising one set of variations is not better equipped to flourish in its hipponess in the biosphere than would be a hippo comprising a different set, so neither in the second case does it mean that a person comprising one set of variations is not better equipped to flourish in her or his personness in the world than would be a person comprising a different set.

Third, just as the theory underpinning the organic sciences came into existence slowly over time, so the theory underpinning the personic sciences will do so too: indeed, is doing so too. While the Great Man, Whig history of the nineteenth century has meant that it is Darwin's name which is anyway for now attached to the former, even the most elementary scholarship reveals that this is not entirely appropriate. Dogs were domesticated, and thus notions of inheritance probably born, around about 8000 BCE. And long before Darwin wrote, many educated people thought that the living things of the present had evolved as the result of natural causes. As is evident not only in the later *Critique of Judgement* (1790) but already in the earlier *General History of Nature and Theory of the Heavens* (1755), Kant, for example, was of their number. And just as the concepts of adaptation, environment, selection and inheritance are all to be found—though not yet purged of the teleological notion of purpose—both in his work and in that of many an other, so the concepts that will come to constitute the terminology of the yet-to-be-formulated theory of the personic sciences will already be 'around and about' now. Indeed, I like to think that some of them might even be 'around and about' in this book.

Fourth, just as an organism can be conceptualised as an hypothesis about its environment, so too can a person. The most significant environment of the organism is other organisms in the biosphere. The most significant environment of the person is other persons in the world. And this is a similarity which is embedded in the very 'oikos' (house) names given to the current sciences of the two. That which is current 'oiko-logy' (ecology) deals with relations between organisms which, in a particular habitat, constitute an eco-system, within which they gain differential access to its resources. And that which is current 'oiko-nomics' (economics) deals with relations between persons who, in a particular society, constitute an economic system, within which they do the same.

These then are just four of the ways in which the foundational theory of the sciences of the organic (already in place) and the foundational theory of the sciences of the personic (as yet to be formulated) might well be similar. But we need to be careful even with these. Each will have to be *shown* to be a similarity after the new theory is in place, rather than just *assumed* to be a similarity at the outset. That which Dennett labelled as Darwin's dangerous idea was not, contrary to his reading of it, a general idea of any type of change over time in any medium, but a very specific idea of one type of change over time in one medium. It is only if, with Dennett, we posit that which Darwin did not—that is, that evolution be read as any type of change over time in any medium—that 'dangerous' becomes an appropriate epithet: though in ways neither intended by, nor friendly to, Dennett's usage. There is nothing remotely dangerous about Darwin's specific idea: that the origin of species is by means of natural selection. What is however dangerous is the thoroughly unscientific extension of the specific idea such that it is read as the more general. This is an extension, scientifically totally illegitimate, which tries to get theory on the cheap: to get a theory of the personic not by observation of, hypotheses about, experiments on its own object of study, but by lifting a theory from elsewhere and simply plonking it down in the alien domain. But theories are not like that. They are only ever won—and very hard won—as the result of very extensive work on their own object of study. A mode of change demonstrated to be valid as explanation of phenomena in one domain has to be *demonstrated* to be valid, not *assumed* to be valid, as explanation of phenomena in another. To test that second validity—to see if what one has is anything more than an analogy—one has to generate one's theory the hard way: out of observation of, hypotheses about, experiments on its own object of study.

This I think is the serious issue of substance in this area: the importance of realising that you can't get theories from theories. But there is also an issue, to some extent distinct, of terminology: of what we are therefore to do with just the word 'evolution'. I think there are three choices, all freighted with problems of prior and inappropriate reductionisms. We could go with its most specialist meaning and restrict the word to the organic, where it would be a shorthand term for 'evolution by natural selection'. We could go with its less specialist but not yet fully generalist meaning, and use the word for 'evolution by natural selection' *and* 'other types of change in other areas sharing some (specified) similarities with evolution by natural selection'. Or we could go with its most general meaning, and use the word for 'any type of slow change over time'. The first option would mean having it as a tightly defined technical term in the natural sciences; the second having it as a more loosely defined technical term applicable across both the natural and the moral sciences; and the third, losing it as any sort of technical term at all. I would oppose the third on the grounds that technical terms are useful. I would oppose the second on the grounds that while it might eventually prove appropriate to have one term to apply to processes sufficiently similar to warrant it across both the natural and the moral sciences, we cannot know, in advance of the very existence of the latter, that it will. Given the as yet critically under-developed state of those latter, we just cannot yet know that it will. So if only by default, and without being very sanguine as to the possibility of its ever losing its reductionist baggage rather than just dropping out of our technical vocabulary altogether, I would support the first. It is anyway the first which has been, and will continue to be, my own preferred usage.

When turning next to thinking not about possible features which might make it the case that, when the foundational theory of the moral sciences is in place, it *will* be found to share characteristics with the theory of evolution, but about possible features which might make it the case that it will *not*, it is again four such which primarily occur.

First, there are vast differences of time scale. The period over which the organic change which is captured by the theory of evolution has taken place is huge: spectacularly lengthier than the period over which the personic change not yet captured by theory at all has done so. For example, we have to get our heads around such differences as that while dinosaurs lasted for more than one hundred and fifty million years, modern humans have been around, at the most, for perhaps one hundred thousand or so; that while in palaeontology the emergence of a new species within a one hundred thousand year time span is referred to as 'instantaneous', in psychology the process the result of which is the emergence of the sixth and last self of the six-stage selving process envisaged by Kohlberg's theory of the development of moral reasoning, for example, is referred to as 'slow'. It is at least plausible to argue that the vast stretches of time over which natural selection has held sway have produced differences not half as spectacular as have emerged in the (by comparison) tiny moments of time over which the as-yet-untheorised personic has existed. Evolution has to explain a biological process taking place over maybe more than three thousand million years. A theory foundational of the moral science of which I speak, by contrast, will have to explain a personic process taking place over not more than a few thousand: as also of course it will have to explain the very coming to known-ness—known-ness being personic not organic—of the biological process itself. It is therefore not an obviously crazy hypothesis which suggests that these spectacular time-scale differences are quite likely to mean that we will need substantially different theories to explain the organic and the personic.

Second, there are considerable differences of what might be called 'style'. As expressed originally I think by Monod, but since by many an other, evolution is not a designer but a tinkerer. And to get a flavour of the significance of that distinction, nowhere can its huge significance be more vividly illustrated than in the field of human reproduction.

Think first of human animals. By showing it to be a veritable battle ground for control of the female body, modern evolutionary theory has revolutionised our understanding of the relationship between a pregnant female and the conceptus-embryo-foetus that she carries. No more can romantic nonsense about mystical bonds, maternal instincts, symbiotic accord stand up in the face of the much less cosy evidence that, given the huge metabolic cost of pregnancy, any mutation which would pass the buck to the other sex would be as biologically beneficial to females as it would be biologically costly to males: a cost-benefit analysis which would figure hugely in the resulting tinkering battle for its success or failure. This is the merely organic: this is the only natural.

But—and by contrast—think, second, of human persons: that is, about entities dealing in realities way beyond the mindless and thoughtless ken of the merely natural. Here things are considerably different. Even within the constraints put upon straight thinking by the ideological I AM paradigm, a variety of stories about love, about 'the four loves' of the Greeks for example, permit thinking about relations between mother and child, father and child, offspring and parent, sibling and sibling, man and woman, friend and friend, which already are far better than such totally conflictful stuff. And when the shift to a scientific I IS paradigm comes to offer not a variety of stories about love but a single explanation of love, of love as 'the active pursuit of the well-being of the loved'—as in the next chapter, I will defend its being—then that which is not the tinkering of the natural, not the partial (because ideological) designing of the normal, but the complete (because scientific) designing of the moral becomes available. In the reproductive example I am currently considering, such would not be found in a tinkering passing of the buck of pregnancy from women to men. For if you distend *them*, can *they* not also ache? If you assault *them*, can *they* not also die? Instead it could be found in a completely designed and moral passing of the buck from women to the fully non-natural, non-organic 'wombs' of a minding, thinking, loving and making personic.

As tinkerer rather than designer, evolution is concerned only with that which is slightly more adaptive now, not with that which might be considerably better then. Natural selection does not work to ensure a high degree of fit, just a high enough degree of fit: to borrow a horrible word from economics, fitness is not maximised but 'satisficed'. For the personic, by contrast, we need a theory not of tinkerer but of designer and maker. And given that it is the designing and making of the good which is quintessentially of the personic, it is particularly the designing and making of the good which we need such a theory to explicate. It is such a candidate theory which the I IS hypothesis offers.

Concerning such designing and making, what it is suggesting is that we have been unable coherently to say what the good *is* within the I AM paradigm because the good *isn't* within the I AM paradigm: that from within the I AM paradigm, that which is the good just cannot be coherently formulated. Within that paradigm, we can (and have) come up with coherent theories of the biological phenomenon 'life', amongst which the theory of evolution by natural selection is one of the most important. But within that paradigm, we have not (and cannot) come up with coherent theories of the non-biological phenomenon 'love'. To do that, a total shift of paradigm is what we need. And my claim is that, via its designation of a coherent theory of the good as that which emerges with the emergence of the I IS self, the I IS hypothesis is one which offers just such a shift. As is the case with all selves, an I IS self is (just definitionally) selfish. But this totally new type of self is one which can only be selfish, can only look after its own totally new interests of the disinterested, by also and necessarily looking after the interests of all other selves at the same time: that is, by actively pursuing the well-being of all other human persons; that is, by loving. When Walt Whitman says that he will have nothing fine which cannot be had, and in equal measure, by each and every Other, that is the loving good of the I IS self in the making. And such new and emergent realities have absolutely no parallel in the organic.

Third, there are differences of scope. Success for the tinkerer means that each individual tinkering has to be viable: to be naturally selected, it is each individual tinkering, rather than the whole and eventual package of tinkerings, which has to confer reproductive advantage. In 1997, for example, palaeontologists for the first time found, as part of a dinosaur skeleton, a wishbone. This was a find lending some support to the idea that birds might have evolved from small dinosaurs. But to take such an idea further, it just won't do to explain the evolution of feathers in previously feather-less creatures by recourse to the idea 'they evolved for flight'. To have been naturally selected, each tiny increase in featheriness, each increase way short of the level eventually permitting flight, would have had to have conferred on its bearer some reproductive advantage. The hypothesis 'feathers might have evolved for incubating eggs' has a chance of being validated. The hypothesis 'feathers might have evolved for flight' does not. But for the designer and maker, where each individual 'design' and 'make' is morally created rather than naturally selected, it is not self-evidently obvious that the creation would have to proceed so. It is at the very least a reasonable conjecture that via not individual agency but collective origination, persons, as opposed to organisms, can come collectively to determine determinations, and thus collectively to make the good. Again, such realities have absolutely no parallel in the organic.

Fourth, there are differences of instantiation. Organisms are instantiated in space: persons are instantiated in time. It is not, at first flush, immediately obvious why entities consisting in stuff in space would be likely to need theorising in the same way as entities consisting in states of stuff in time. Indeed, why should one even begin to suppose any such unlikely thing?!

In summary, then, of these likely differences, it is my contention that when in *The Origin of Species* Darwin argued that 'the one general law leading to the advancement of all organic beings is to multiply, vary, let the strongest live and the weakest die', he meant exactly what he said: not approximately what he said, but exactly what he said. He said 'organic beings' and he meant 'organic beings'. His theory is a theory about organic beings. And the quite cosmic error of sociobiologists, evolutionary psychologists and all their ilk, is that they take that word 'organic'—and ignore it. For whatever are the difficulties of delineating the concept 'person'—and they are many!—one of them is *not* trying to decide what sort of an organic being it delineates. It does not delineate *any* sort of organic being. And that is a reality such that, were one to rewrite Darwin's sentence above to *try* to make it appropriate not to the organic (to which it pertained) but to the personic (to which it did not), one would have totally to reverse its meaning. For persons, it would have to read something like that 'the one general law leading to the advancement of all persons is so to develop, so to unify, that none can be strong (because none are weak), none can be weak (because none are strong).' Hardly a difference that makes no difference.

As I indicated towards its beginning, I want to end this chapter with a first and brief spelling out of the role of equality in the project of the construction of moral science: a one-paragraph spelling out, to be more fully fleshed out in my final chapter. But before I do that, and again as already indicated, I want just to outline a few of the more interesting practical problems which the emergence of such a science might help to resolve. For an insistence upon a clear distinction between the concept 'human animal', with which such a science would not be concerned, and the concept 'human person', with which it would, leads to interesting approaches to several of today's 'hot topics': approaches which would enable also *applied* moral philosophy to become *applied* moral science. A brief look at each of four such examples should suffice to suggest what I mean.

First, cloning. The arrival in 1997 of Dolly the sheep occasioned much press coverage. But in that coverage, and even in serious places, the distinction between an animal and a person was frequently lost. Thus an editorial in the *Times Higher Education Supplement* (28.2.97) referred to the possibility of 'copying people'. But cloning is a biological technique. Only if people were biological entities could they be cloned. And they are not. A member of the species *Homo sapiens* could be cloned: and probably will be. But the linguistically structured experience which is one person is not and cannot be the different linguistically structured experience which is another. A non-scientifically based opposition to the cloning of *Homo sapiens*—and there's plenty of that about—could turn out to be thoroughly unhelpful: there is a very respectable case to be made, and in *Remaking Eden* (1998) Lee Silver makes it, for the hypothesis that a new *Homo sapiens* resulting from cloning, far from having many more genetic problems than one resulting from natural conception, is quite likely to have many fewer. And anyway, some problems being better to have than others, the real problems of such an eventuality, or so it seems to me, might well turn out to be those which it would be rather good to have. It would be good not bad if patriarchal (and matriarchal) rights to assume parentage were to become even further problematised; if images of parents as in some sense the owners of children were to become even more questionable; if traditionally based ideas of what a family has been were to become even more extensively challenged by new ideas as to what it might better be. But this all concerns the organic. To see the personic as that which, being something different, just cannot be cloned, we need to listen to voices rather less careless than such 'copying people' journalists. We need to listen, for example, to the presidential professor of philosophy at the University of California in San Diego, Philip Kitcher.

Writing in *New Scientist* (13.3.97), he argues that while clones would be as closely related genetically as monozygotic twins, they would be less similar than such twins, and for three reasons: first, they would have developed from eggs with different cytoplasmic constituents; second, they would not have shared the same uterine environment; and third, they would experience after birth environments which would diverge considerably, especially where the generation gap resulted in changes of such things as diet and education. He went on to point out that, in contrast to the few cases where a single gene appears to be responsible for dire genetic diseases, such as is the case with Huntington's chorea and Tay-Sachs, the characteristics we most often value in people result from very complex interactions between their genetic endowment and the environment in which they function which cannot be replicated. His conclusion was first that rather than relying on cloning, 'demented dictators wanting armies of *Ubermenschen* would do better to fund military academies'. And second that 'those who think that cloning portends more great artists and scientists would be advised to invest in scientific education and support the arts.' Of course the detail of this, particularly what I would see as a failure carefully enough to distinguish between organisms and persons, would not be mine. But its general drift, its realisation that cloning human animals is not and cannot be cloning human persons, seems to me right.

Second, genetic engineering as it could extend our reproductive technologies. Robert Sinsheimer, the molecular biologist who initiated the human genome project, was quoted in the *New Scientist* (31.10.98) as saying that while the 'old dreams of the cultural perfection of man were always sharply constrained by his inherited imperfections and limitations', the 'horizons of the new eugenics are, in principle, boundless.' It was his opinion that 'for the first time in all time, a living creature understands its origin and can undertake to design its future'. To some this will sound like scary stuff: like hubris on a very grand scale. Especially where such designing might involve altering DNA in sperm or egg, in the germ line, there are well recognised and serious questions to raise: and answer. Among the most commonly heard are those which suggest that we might dangerously reduce of the size of the human gene pool; encourage a black market in desirable qualities; create the possibilities of abuse by Hitlerist types; further marginalise those who nonetheless are born less than 'perfect'; or create the danger of moving the goal posts as to what 'perfect' means. But it seems to me that, with the possible exception of the first, these are all questions not about the use but about the misuse of a technology: not about the existence of the natural sciences which can offer us the opportunities, but about the non-existence of the moral sciences which can educate us in how to use them. In the *Times Higher Educational Supplement* (14.3.97), James Watson argued—concerning the misuse of genetic data in combating schizophrenia—that all information can be misused. And continued: 'We are trying to get this information so we can combat schizophrenia. I think the real ethical issue is not its misuse but the possibility of *not* using it.' This I think is the sort of approach that we need. But consistent with my hypothesis concerning the nature of moral science, I would want to try to spell it out a little differently from the way in which he does.

The I IS hypothesis would require us to recognise four different scenarios. First, there is the case where the processes of natural procreation result in the birth of a child. In such a case, the evolutionary process is one flowing along lines of least resistance, along pre-wired 'always already' existent synaptic connections which send the instinctive procreative current flowing one way. And in such a case, whether that child is well-formed or de-formed is not a moral issue at all: it is amoral, or maybe pre-moral. As about all that is merely natural, there can be nothing that is either good or bad about it. It just is. Second, there is the case where the processes of making, the making of a well-funded health service for example, result in the birth of a well-formed child. That is the moral. It is good. Third, there is the case where the processes of making, the making of medicines intended for the control of morning sickness for example, result in the birth of a de-formed child. That is the absence of the moral. It is the absence of the good, or the presence of the not yet good. And fourth, there is the case where the processes of making, the making of genetic change which could countermand the processes of evolution and protect against the birth of a deformed child for example, are available but are not used. That too would be the absence of the moral. That too would be the absence of the good, or the presence of the not yet good.

Third, our attitudes to neonates. The philosopher Peter Singer, at the time of writing the professor of bioethics at Princeton, argues that the parents of severely handicapped neonates should have the right to euthanasia for them. Because he argues that before then infants have no human consciousness and so do not have the same rights as adults, his cut-off point would seem to be one month of age. It would be interesting to debate this choice of 'one month': there would appear to be other possibilities. But that aside, what seems to me excellent is that he should be requiring us to think in terms of choosing at all. Quite indefensibly, it seems to me, pro-life activists argue that aborting a foetus is killing a child. They maintain that it's rather odd to say that it's OK so to kill before birth, but that after birth it suddenly become wrong; becomes murder. In arguing so, they are trying to make the line drawn at the moment of birth less salient. And what interests me about Singer's 'one month' position—and my own—is that so too are we. Those adopting a pro-life stance try to push the line backward in time. They want to say that it is the point at which the new organic is conceived which is critical. But apart from anything else, that's just incoherent. When egg meets sperm, what is conceived is both that which will develop into a *Homo sapiens* and that which will develop into its placenta (and other supportive material) which at birth is merrily burned. By contrast, those adopting what I would call a pro-love stance want to push the line forward in time. They—we—want to say that it is the point at which the new personic emerges which is critical. This is what Singer is after. And this is what my division between a human animal and a human person, while it does not automatically provide a definitive answer to the point at which the killing of neonates becomes the killing of persons, does suggest some rather more compassionate and loving possibilities than those offered by pro-lifers.

Fourth, animal experimentation. My hypothesis would suggest that in doing experimental work on animals, two things need to be given equal (and considerable) attention. One is the fundamental continuity between other animals and *Homo sapiens*, based on the existence of the phylogenetic scale. The other is the equally fundamental discontinuity between other animals and *Homo sapiens*, based on the emergence of persons from among the latter but not the former. The continuity should make us aware of different levels of sentience (sentience being that which *Homo sapiens* shares with most of the others), and therefore different kinds of approach appropriate to work with different species: fruit flies are not bonobos. The discontinuity should make us aware of consciousness (consciousness being unique to human beings), and therefore of the fact that harm caused to a person is of a different moral order from harm caused to a non-person: bonobos are not persons. One of the few things on which I am in agreement with Roger Scruton is that in philosophy, the language of rights—if it is to be useful at all, as perhaps he thinks it is and perhaps I think it isn't, but no matter—needs to be accompanied by (and usually *is* accompanied by) the language of duties or responsibilities. It seems to me that it is no more morally defensible to try to elevate animals to the level at which they could be said to have rights than it would be to try to elevate them to the level at which they could be said to have duties or responsibilities. Cats cannot appropriately be put on trial for torturing mice. As the only entities which can deal in such things, we can give ourselves (or *not* give ourselves) rights, duties and responsibilities in relation to animals. And if we are to use such concepts at all, I certainly think that we should. But it is just surely patently obvious that they cannot either give themselves (or *not* give themselves) rights, duties and responsibilities in relation to us. They cannot give them*selves* such things because they are not *selves* in the first place. As stages in the development of persons, selves are entities which become capable of the moral: capable, for example, of settling their disputes by negotiation; capable, if we want to use such concepts, of the exercise of rights and duties. And non-human animals, since they are not persons are also not selves.

If space permitted, these four examples could be joined by many more. Two others that interest me, for example, concern zombies and sex-change individuals. On the first, the I IS hypothesis makes it quite clear that all the philosophical thought experiments based on the idea of a zombie are just ridiculous. They involve our trying to think of a person physically exactly the same as ourselves yet with no experience. But since a person is not a physical entity, it can't be physically identical to anything. And since a person *is* experience, it cannot be *without* experience. And on the second, the I IS hypothesis serves to remind us that sex-change people are not person-change people. But rather than go on accumulating examples, I want to end this chapter, as promised, with as clear and as simple a one-paragraph statement as I can muster about the connection between the establishment of the moral sciences and the making of equality. It is this.

The making of science requires the making of disinterest. We have to be able to stand before that which we wish to comprehend, disinterested. The making of the personic sciences, the moral sciences, the sciences of the good, requires such a stance before not the universe, not the biosphere but the world. To make those sciences, we have to stand disinterested before the personic relations which constitute the world. The necessarily unequal relations of I AMs offer no such vantage point. From out such relations, all points of view are skewed by being those of the *ad*vantaged or the *disad*vantaged. No sciences of the good can be made from such places. To make the sciences of the good, we need the disinterested stance available only from out the necessarily equal relations of I IS selves. And to begin to make such relations, we need first a thorough unmasking of that *having* to which I have been referring when I have used the phrase 'the having ideology', and second a more complete understanding of that *loving* to which I have been referring when I have used the phrase 'the active pursuit of the well-being of the loved'. It is to these two needs that I will turn my attention in the next chapter.

Chapter Eleven

GOD AGAIN

From I am that I am to I is

For many persons, of course not yet for all persons, the God concept is one which no longer compels. Where Kant saw what he called 'the emergence of man from his self-imposed minority' as an emergence from 'his incapacity to make use of his own understanding without the guidance of another', many have been those who have since maintained that by far the most important 'other' whose guidance must be shed is that ultimate other called God. In *Taking Leave of God* (1980), for example, Don Cupitt argues that because 'government by personal rule, however enlightened, is always in the long run morally intolerable and spiritually oppressive', and because 'good government is government by general laws and not by personal rule', it therefore follows that 'if God is good, God must fade out'. The title of Cupitt's book comes from Meister Eckhart, a Dominican friar, whose phrase, in one of his sermons, was to the effect that 'Man's last and greatest parting is when, for God's sake, he takes leave of God'. And given that Eckhart lived between 1260 and 1327, we could perhaps say that that 'fading out', forever increasing the size of 'the God-shaped hole', has been taking place across anyway no fewer than seven centuries.

But hold. There is a story – apocryphal it may be – that in a college essay a Cambridge philosophy student once suggested that God was so omnipotent that, even if he didn't exist, he'd be able to make himself do so. That's nice! And in a way, what I will be doing in the rest of this chapter is defending the sense in which that student's comment is spot on. It's a defence which will put me at some odds with a seemingly widespread agreement, at least among many academic philosophers in mainstream parts of the Anglo-American tradition, that it was that pesky Christianity (and its monotheistic ilk) which, by getting in the way of *proper* thinking about the moral which was *Greek* thinking about the moral, held up the development of moral philosophy for centuries. By contrast with such a stance, I am arguing two things. First, I am suggesting that it is no coincidence that the main problem of all thinking about the moral, the problem of the non-relativistic grounding of the concept of the good, is one which has seen its most hegemonic solution, to date perhaps its only hegemonic solution, formulated from within the God concept. And second, I am arguing that it is a development of that concept which can supersede the various relativistic stories currently constituting moral philosophy, and replace them by a single universal explanation which can form the basis of moral science.

There is, of course, a long tradition of God-talk being actually about something other than an outside-the-universe supernatural entity who somehow made it all out of nothing. Thus when Wittgenstein says that 'to pray is to think about the meaning of life', and that the meaning of life is 'the meaning of the word we call God', it is these days a commonplace to understand such talk as having little to do with any such entity. Thinking about the meaning of life is something we all do. And that such thinking is frequently done by exploring 'the meaning of the word we call God' has been more or less a given of philosophical conversation for centuries. I think the militant atheists of this world, the Richard Dawkins' of this world, would do well to understand this better. For by trying to sever all their thought processes from those of their ancestors who thought with God, and so often thought such spectacularly liberating, radical and progressive things with God, they are often cutting themselves off, I think, from vital understandings. After all, shouldn't a scientist, of all people, know that an hypothesis does not have to be shown to be true to be nonetheless able to be fruitful? To illustrate what I mean by this, let me pick up on what just a handful of poets and a single scientist have come up with when thinking with the God idea.

One. When, in *The Divine Comedy*, Dante 'thinks with God', the Dante whose exceptionally brave hubris took him not only to heaven, not only into the very sight of God, but back again to write about it, he comes up, in the last canto of his poem, with the spectacular. First he addresses the Virgin, she who is the last screen before his face-to-face encounter with God, as 'Virgin Mother, / daughter of your son'. Having then passed beyond her to come finally into the presence of that ultimate visage, he writes of it as one which 'seemed to me painted with our human semblance / And for that reason my sight was set wholly upon it'. And anyway for this reader, the startling conception of an entity which is the creator of its own Creator, which Creator is 'painted with our human semblance', prompted much of the thinking which resulted in this book.

Two. When somebody said, of Milton, that had he been in the Garden of Eden he would have eaten the apple and then written a pamphlet to prove that it was his duty, then albeit humorously he or she is speaking of a man whose 'thinking with God' took him inexorably towards a God-portrait which can only lead us to question that 'justification of his ways' which the poet maybe only half thought he was undertaking. Milton's was really a search for truth: for what he called, in *Areopagitica*, that 'Truth which is strong next to the Almighty'. And it is because such as he went boldly to all the thoroughly contradictory places to which such a search necessarily took them that we are now in a position to add: 'and there *is* no Almighty!'

Three. When, in *Jerusalem* (section 96, lines 23–28), Blake understands that 'Man is Love / As God is Love: every kindness to another is a little Death / In the Divine Image, nor can Man exist but by brotherhood', his 'thinking with God', a thinking with the very God-concept which had contributed so much to his 'brotherhood' idea, is enabling him to reach after an understanding of God's death long before Nietzsche was to do so.

Four. When, in *Collected Poems* (1954), Wallace Stevens—gesturing to Heidegger—writes that 'we are too late for the gods and too early for Being / Being's poem, just begun, is man', his 'thinking with God' has taken him first to Protagoras's understanding that 'Man is the measure of all things', then to Feuerbach's formulation that 'theology is really anthropology'.

And five. When, in *What is life?* (1943), Schrödinger 'thinks with God' he produces the following notorious, faulty, but surely productive two-step (and a jump!) argument:

(1) My body functions as a pure mechanism according to the laws of nature. Yet I know by incontrovertible direct experience, that I am directing its motions, of which I foresee the effects, that may be fateful and all-important, in which case I feel and take full responsibility for them.

(2) The only possible inference from these two facts is, I think, that I—I in the widest meaning of the word, that is to say, every conscious mind that has ever said or felt 'I'—am the person, if any, who controls 'the motion of the atoms' according to the laws of nature.

(3) Hence I am God Almighty.

And while it is perhaps not surprising that his book was rejected by its original (Catholic) publisher, neither is it surprising that it soon found another. For as he himself was quick to point out, its author was in thoroughly venerable company. Having been found over two thousand years previously in the *Upanishads*, this 'grandest of all thoughts' was hardly a novel idea.

But enough of examples. Suffice it to say that, considered historically, much of our 'thinking with God' has been thoroughly productive. Although it is usually only the first that is quoted, when Marx said that religion had been 'the opium of the people' and immediately added that it had also been 'the heart of a heartless world', he was as much right on the one count as he was on the other. In particular, the 'thinking with God' which has pictured him as *person* has contributed directly and hugely to our understanding of ipseity in general. And because it can protect us against that reduction of the personic to the organic which, in the form of the 'just so' stories of much of so-called evolutionary psychology, is currently ubiquitous, this is an understanding which could not be more important.

When defending my formulation of a shift from an I AM God to an I IS God, it is probably important to stress that mine too is a 'thinking with God' which posits that entity as something considerably other than an outside-the-universe supernatural being who made it all out of nothing. In short, it is probably important to stress that the tradition of God-talk on which I draw is one that insists upon a very particular premise: in essence, the premise that that which is apparently God-talk is *and has always been* actually us-talk. The entities which create the world out of no-thing, which create it out of that 'logos' of the first verse of St. John's gospel, are us. And one of the reasons why the stress on this premise matters is because it is important that I clearly distance myself from a stance currently fashionable, but considerably different. The physicist Paul Davies, believing it would seem that the very universe is in some way rational, and thus that there is some sort of meaning to brute physical existence, sees himself as a member of what he calls 'an alternative academy'. With fellow members being, he thinks, such as James Lovelock and Roger Penrose, he sees himself as trying to construct a view which, while not religious in the conventional sense, is nonetheless in some (mysterious?) way different from that of modern science. Such a position is certainly not mine. What I want to argue—instead—is not only that it is we who make the world out of no-thing, but also that, as part of such making, it has been our making of God which has been our single most important creation to date.

Fully to appreciate the importance of that making, some of the things it is crucial to understand are its timings, its moments. Anthropologists have of course long pointed out that the emergence of God comes at just that historical moment when human groups become too big to be controlled by one powerful individual to whom all those he seeks to rule could be individually known. And in *After God* (1997), Don Cupitt points out that the emergence of God is one which 'coincides with, and indeed is the very same thing as, the emergence of a more individuated human selfhood standing in a new relation to the world.' These are timings indeed important. But it is another and different timing, one also inherent to God-talk, which is critical to my hypothesis. Indeed, it is maybe no exaggeration to say that, in order fully to understand the us-talk which has been God-talk, it is this particular timing, a timing which contrasts the first with the eventual, which is by far and away the most important to appreciate.

For in just the same way as it has put I first (as freely originating first-person), so the mainstream tradition has put God first (as freely originating first-cause). Both stances, each actually a part of the same stance, lead to dualism: to the idea that mind, spirit, thought, is something separate from and prior to that which, as action, it causes. Just as surely as I cause my arm to rise (to hit my attacker), so God causes his arm to rise (to smite the Israelites). In fact, it is because he does the one that I can do the other. For is it not in his image that he has created me? Thus is the I AM God an originator God: one who is an originating first cause in just the same way as his I AM makers posit themselves as originating first persons.

But even within the constraints of the I AM paradigm, a different tradition has also been accumulating. As well as such mainstream stuff, another variety of the God-talk which is us-talk, a variety which talks of a God who comes not first (as first cause), but eventually (as eventual outcome) has long been around. It is commonly suggested that it was perhaps Hegel who gave this entity its most completely formulated first outing. But there has been many an other. Simone Weil, for example, suggests that we have to believe in a God who is like the traditional God in every way except for the fact that he does not exist *because we have not as yet reached the point of his existence*. Thus when the I IS hypothesis posits persons as the *outcomes* of the linguistic structuring of experience, and God as the *outcome* of the search by such persons for an idea of the good which can be immune to a relativising critique, it is this non-mainstream, second tradition with which it can claim continuity. Indeed, one way of formulating the hypothesis of a shift from I AM to I IS thinking, about persons as about God, is as such a shift in timing. I am suggesting that where both the I AM person and the I AM God have in the past been understood as before-the-event causes, both the I IS person and the I IS God will in the future be understood as after-the-event outcomes.

It is my claim that these are timing differences which could not make more difference. It is my claim that the I AM God, made by I AM persons, is *ideological* (because interested), and thus unable eventually to resist the uprooting of its concept of the good by *relativistic* moral philosophy. And by contrast, that the I IS God, made by I IS persons, will be able to be *scientific* (because disinterested), and thus able increasingly to ground a new concept of the good in *universal* moral science.

My aim in the rest of this chapter will be to substantiate such claims. To do so, I will look first at those differences between I AM and I IS persons, as makers, which are relevant to their differential making of their respective Gods. And second, at those Gods themselves, at the I AM God and I IS God, to demonstrate the way in which the *ideological* nature of the one makes it unable to resist mounting *relativism* about the nature of the good, and the *scientific* nature of the other makes it able to ground a new *universalism* about the nature of the good.

As makers, both I AM persons and I IS persons make their God idea in their own image. But their images of themselves are substantially different. In particular, seeing themselves as first persons, I AM persons want to *have* differentially and *be loved* exclusively. So they give their I AM God those characteristics too. By contrast, seeing themselves as just persons, I IS persons want instead to *have* equally and to *love* inclusively. So they give their I IS God these characteristics as well. It is these two differences, differences in the way in which *having* and *loving* are understood by the two sorts of maker, which I think are those most relevant to their making of their Gods. So it is these two differences which next I must consider.

First, *having*. Their different understandings of having can be encapsulated by a further examination of what I have been calling 'the having ideology'. This is an ideology, I want to suggest, which has its *natural* origin in an organic source whereby all entities with metabolisms are necessarily 'havvers', and its *normal* development, first flourishing then faltering, across the whole history of I AM persons. It is that eventual faltering, into widespread relativism, which gives rise first to its exposure *as* ideology (less-than-moral ideology), then to its replacement *by* science (moral science): a process which consists in the emergence of I IS persons and the creation of their I IS God.

The life of things organic is instantiated in partly closed systems: partly closed only, because an organism must always be open to that metabolic relationship with an environment which is, quite literally, its life-line. What this means is that, for an organism, 'to have or not to have' is the critical issue. In relentless competition with other organic entities, having (or not having) its sustenance is a matter of life (or death). Only those which succeed in getting for themselves the means of flourishing—the nutrients, the shelter, the mates—will survive. For as long as it has that from the environment which enables it to continue scrambling up the gradient against the second law of thermodynamics and negative entropy, it lives. The failure of that metabolic relationship is the end of its living: it dies. And this, of course, is as true of the human organism as it is of any other. In *Science in History* (1954), Bernal, in paraphrase, puts it this way: that all disease is ultimately either *not having* the chemical constituents that a healthy body needs (being starved), or *having* the chemical constituents that a healthy body does not need (being poisoned).

But what then of things personic? When first they emerge from out the human organic, what then of persons? What I am suggesting is that when first we emerge as persons, as animal-human normal persons, we do not do so 'trailing clouds of glory' and 'from God who is our home'. This is how Wordsworth has it. But he is wrong. In language maybe less felicitous but considerably more accurate, it is my case that we come instead 'trailing all that is gory', and 'from nature which has not yet ceased to be our home'.

And that we come that way not only individually and developmentally as infants, but also collectively and historically as kind.

In nature, there is a life or death competition played out among its constituent organisms to have. It is 'red in tooth and claw'—just necessarily. And the first persons steadily to emerge from such gore, I AM persons, are able to challenge such origins, at first hardly at all; increasingly as they develop; but never completely. As first we emerge from our natural non-person origins, as first we become normal I AM persons, we trail considerably more than a little of that from which we came. In our first instantiation as persons, as animal-human normal persons, we are indeed 'havvers'. We cannot be otherwise.

We are havvers of things: with agriculture, we come to *have*, some more some less, property. We are havvers of other persons: with systems of governance, we come to *have*, some more some less, slaves, servants, children, spouses, lovers. But most fundamentally, we are havvers of ourselves: I AM persons are those who posit their very experience as something which they *have*, and thus their very selves as possessions or properties. In academic philosophy, for example, the idea of 'self-ownership' is a perfectly respectable notion: in *Self-ownership, Freedom, and Equality* (1995), the philosopher Gerry Cohen has written a substantial text with that concept in its title. Indeed, once alerted to its existence, it is hardly possible to stop noticing examples of 'the having ideology' riddling all philosophical thinking. Directly analogous with the self-ownership case, for example, it was once a commonplace, and is still I think widespread, for academic philosophers to think of entities as things which *have* their properties. And although several of their number—Anthony Flew for one—have pointed out that speaking so entails the slightly barmy corollary that if each of their properties were one-by-one removed such entities, now property-less, would still be in some sense 'there', the posit of self-ownership, concerning all entities including *id*entities, is still common among them. It is my claim that it is a posit around which the whole of 'the having ideology' accumulates as buttress.

I am suggesting that in 'the having mode', I AM persons, in competition with other I AM persons, struggle to *have* as much as they think they are entitled to of whatever are the goodies going. Such goodies are land, artefacts, the services of others, capital, status, power, whatever: for to this general case, which of these 'things had' is modal at any particular historical or personal conjuncture is of no consequence. And I am also suggesting that what each such I AM *thinks* it is entitled to is always influenced by, and usually dominated by, its perception of its self as first: as first person. It is this struggle so to *have* which dominates the I AM period of history, or maybe of *pre*-history in the Marxist sense. And it is this struggle so to *have* which spawns the whole of 'the having ideology'. It spawns the bedrock idea that we *have* our experience, and thus the erroneous, massive (but declining) ideology supporting the notion of I as something that AM, as over against the accurate, tiny (but growing) science supporting the notion of I as something that IS. It spawns all the derivative ideas which flourish when we *have* whichever of the land, the artefacts, the services of others, the money capital, the intellectual capital, the status, the power, the whatever, which, in the context of the particular debate, is deemed to be relevant. And like all deeply successful ideologies, it does all this in such a way that those who articulate its doctrines do so for the most part genuinely unaware that that is what they are doing: genuinely convinced that what they say just is so. As illustration of this last, let me offer what I take to be a particularly telling example of what I mean.

In her book *The Sceptical Feminist* (1980), Janet Radcliffe Richards suggested, as many have done before and since, that 'it is entirely possible that if you insisted on absolute equality you might make everyone worse off than the worst off might have been in a carefully planned system of inequality'. And she went on to suggest, again in the good company of many a political philosopher and economist, and of course acknowledging that it was a disputed position which she was only using to make a general point, that it could be the case that 'if taxation were relaxed so that some people were allowed to become very rich, they would work with so much enthusiasm that the worst off group would actually be *better off* [my emphasis] as a result of their increased efforts than anyone would be under a system of equalising taxation'. Having read her book, and found this, I then had the opportunity to hear her speak. She was expounding John Stuart Mill's opinion—as contained in *The Subjection of Women* (1869)—that his opponents were contradicting their own liberal principles by systematically discriminating against women in both law and treatment. She insisted, rightly it seemed to me, that in not being able to see what to us is now so obvious, those opponents were being in no way malicious or calculatedly self-serving. The inferior position of women was so much a part of the natural order of things that either they literally just couldn't see it at all, or they just couldn't see it as anything that needed justification: as water is to fish, it just was. At the end of her lecture there was a coffee break, and when we all returned she reported that during the break she and the chairperson—fellow philosopher Michael Lockwood—had wondered, by comparison with Mill's critics, just which unexamined and taken-for-granted ideas *we* were all swimming in and thus totally unable to see. The gathering was one which, over a weekend, was discussing the philosophy of equality. Derek Parfit, as I remember, was one of the other speakers. And as in the quote from the Radcliffe Richards text, in the form 'better equals *having* more', what I am here describing as 'the having ideology' was *totally taken for granted throughout the proceedings.*

I IS persons, I am suggesting, are those who steadily come to question its being so taken. They are those who suggest that we learn to problematise our *having* of things, to problematise our *having* of other persons, and most importantly of all, to problematise our *having* of our experience and our *having* of our-selves. They question 'the having ideology'. And in so doing, they open the way to that *dis*possession of selves—as 'possessions' or 'properties'—which enables their *re*possession as stages in the development of persons.

I AM persons functioning in the having mode *have* (or do *not have*) those things which are deemed good. *Having* health is good; *having* happiness is good; *having* goods (no accidental terminological similarity this one!) is good. And perhaps above all, *having* love, being loved, is good. Not for nothing have so many I AM persons thought that all is well when they are loved: totally by their personal God; and as totally as possible by a personal Other whom then they can '*have* and hold'. To repeat a previous example, it is once again interesting to register, for example, that while at the level of the organic 'viable' means 'able *to* live' (it is active), at the level of the personic 'loveable' means 'able *to be* loved' (it is passive). Indeed, '*having* and holding' all possible goodies, material and spiritual, is that in which the good life consists. And as such, it is a life which can be lived by an individual I AM person singly: that is, can be lived by her or him regardless of whether others live it or not. Indeed, it is by no means infrequently the case that the very prerequisite of one person *having* such a life is that another *has* it not.

What the I IS person comes to understand—or so I contend—is that none of this is really so: that it is merely our standard of living doing (poor) service for our standard of loving. For when we acquiesce in inter-human relationships involving the necessary inequalities of 'the having mode', in inter-human relationships still structured by the animal hierarchies of Nature, we love our neighbours as ourselves alright: mine, remember, is a formulation which interprets 'the golden rule' positively rather than normatively. It's just that, because from within the having mode, we love both ourselves and them less than well. Such partialities restrict the growth into full humanity of those who *have more* and of those who *have less*. When out of a fear of freedom—and the pull of the hierarchical is always an expression of the extent to which we fear freedom—we try to cling on to our primarinesses or our secondarinesses, we love our Others and ourselves equally alright: but because from within 'the having mode', just equally less than well. For the less than fully developed loving consequent upon our acquiescing in, rather than struggling to emerge from, our primarinesses and our secondarinesses leaves us stunted in our moral development. The consciousness-raising which makes us increasingly aware of that stunting, which makes us feel less than comfortable in what we come to experience as our partialities; which makes us increasingly unable so to acquiesce, increasingly obliged so to struggle: this is selving. It is a process which, beyond the confines of 'the having mode', culminates in our recognition of the necessity that it is only in relationships of equality that we can make love; that the structures necessary to the making of that love are communal; that its making, to a direct consideration of which I must now turn, is something we do necessarily together. This is what I IS persons come to understand about *having*. And when they make their I IS God, it makes all the difference in the world to that which they make.

Second, *loving*. The different understandings which I AM and I IS persons have of *loving* are in part derivative upon the different understandings they have of *having*: as I have already suggested above, where I AM love is that which each individually we can *have* (or not have), I IS loving is that which only collectively we can *make* (or not make). But I AM love and I IS loving are also different in other ways too. In summary, I would express such difference like this: that where I AM love is passive love, of the 'falling-in' variety, and exclusive, I IS loving is active loving, of the making variety, and inclusive.

As I AM persons, we are most of us inclined, or so it seems to me, to talk about love somewhat cavalierly. We do so spectacularly in connection with our sexual relationships, where it is still a commonplace to speak of love as something which, somewhat clumsily, we 'fall into' (or don't) with individual others: a fact which like it or not—and its advocates do not—suggests nothing so much as that it is merely an adult variety of the infant's attachment behaviour, and primarily an organic rather than a personic phenomenon. It is noteworthy that, the answer 'yes' to the question 'do you love me?' seems often to be as unproblematically accepted as the answer 'yes' to the question 'are you hungry?' or even 'are you French?' And this despite the fact that it is surely easy enough to see that two people using such huge (and hugely ambiguous) language in such a casual (and casually unproblematised) way might well result, frequently does result, in their being at what are potentially catastrophic personal cross-purposes. And even when, in philosophical conversation for example, it might be expected that some attempt be made to analyse a little more closely, to distinguish perhaps this 'falling-in' stuff from what might be love's other varieties—things like loving your grandmother or cabbage or mountain climbing or algebra—there is I think a somewhat hazardous tendency to take *what we are talking about* as both given and understood. I suspect that it isn't. And because I suspect so, I must try to spell out right at the start, and as clearly as I can, just what I take I AM and I IS persons differentially to mean when they talk, in the first case about love, in the second about loving.

I AM persons are almost always talking about a particular type of emotion, feeling or sentiment: about experiencing (*having*) a particular type of emotion, feeling or sentiment. It is a stance which, in the face of attempts to tighten up on sloppy thinking, lends itself very readily to a somewhat unpleasant put-down of those so attempting. Echoing Wordsworth, it is the put-down which suggests that 'our meddling intellects' are not able to 'dissect' love; that even if they were, they would 'murder' it in the process; and that anyway *we* all know what we're talking about even if *you* don't. But put-down or no put-down, it's important to point out—so I'll risk it!—that because it's one which neither accords very well with serious evidence, nor protects very well from serious accusation, theirs is a position of dubious worth. It does not accord, for example, with the fact—as demonstrated by the experimental work of Donatella Marazziti and her colleagues reported in *Psychological Medicine* (vol. 29, p. 741)—that romantic love and obsessive-compulsive disorder are biochemically identical. And it does not protect, for example, from the suspicion—as illustrated in the old joke about the rich woman asking the butler to throw out the beggar because his stories were breaking her heart—that its reliance on only 'feelings' to legitimate it leaves its I AM champions wide open to a charge of gross sentimentality. Where loving, in I IS terms, is recognising need and acting to meet it, such love-as-a-feeling sentimentality, as recognised by Spinoza, as recognised by Rousseau, all too readily becomes either a substitute for such action, or a prompt to action totally to its contrary—as in the case of the beggar.

I IS thinking, by contrast, takes its loving to refer to action of a particular kind, to acting in a particular way. In an attempt to capture such an I IS approach, I have earlier borrowed a famous phrase from the work of Erich Fromm. In *The Art of Loving* (1956) he writes of loving as 'actively pursuing the well-being of the loved'. This is an approach which sees love not as a feeling but as a doing; not as a sentiment but as a making. And to think of love in this way is to make two claims about it: first, that as an action, it is that which results in—rather than is, or is caused by—an emotion, a feeling or a sentiment; second, that because only persons act, and loving is an action, so it is a phenomenon only of the personic, only of the world. The first is a claim to which I need give but brief attention: to treat it more fully would involve a considerable diversion into comparative assessment of various theories of emotion, which is not necessary to my case. The second needs more elaboration.

On the first: there is a theory of emotion—the James-Lange theory—which suggests that when we run away from a bear, it's not that we run because we are frightened but that we are frightened because we run. It's a theory which suggests that the subjective feeling of fear, the emotion, is just the label we give to the physiological arousal we are experiencing *as it is given meaning cognitively by context.* We would experience much the same arousal—of the sympathetic nerves of the autonomic nervous system—if we were about to have sex. But unless we were addicted to bears, context would tell us to interpret the first as fear and the second as sexual arousal. Now without going into the relative merits of this theory vis-à-vis other theories—this is a much-contested area: there are many candidate theories of emotion—it is perhaps worth nodding in its direction when thinking about what we are wont to call our loving relationships. For there, we might perhaps just countenance the possibility that it's not that we love someone and therefore we come to take care of their needs and wants, generally look out for their comfort, and share sex with them (feelings generating actions), but that we come to take care of their needs and wants, generally look out for their comfort, and share sex with them, and therefore we come to love them (actions generating feelings). Don't we all know full well just how much more we like and love those *by whom we do well* than we like and love those *to whom we do harm*? It seems not anyway immediately preposterous to suggest that perhaps it is the behaviour towards them that generates the feeling, rather than the feeling which generates the behaviour towards them.

On the second: a more extensive comment is needed. In I AM thinking, while talk of loving *cabbage*, or *music*, or *tigers* is unexceptionable, talk of loving not *one person* but *all persons*, talk of all-inclusive loving, is as often as not something to be derided. Bernard Williams, for example, so derides. To claim such loving is seen as both laying claim to the impossible, and being more than a little sanctimonious to boot. In I IS thinking things are otherwise. While talk of loving *cabbage*, or *music*, or *tigers* is regarded as just the misuse of a concept (how does one actively pursue the well-*being* of non-*beings*?), talk of loving not one person but all persons is totally unexceptionable. This is an interesting difference. As consisting in some actions rather than other actions towards *all* persons, as resulting in some feelings rather than other feelings about *all* persons, a love of just *all* persons makes perfect conceptual sense within the I IS paradigm. The difference between loving an individual, known other, as sexual partner perhaps, and loving all other persons is not a difference which means that the one is loving while the other is not: both are loving. It's not a difference which means that both are loving but of different kinds: both are the same 'active pursuit of the well-being of the loved' generating the same emotion, sentiment or feeling. And it's not a difference which means that one is any less warm or real or life-enhancing than the other: loving that is not warm or real or life-enhancing is a contradiction in terms. The difference—or so my I IS persons would suggest—lies only in what can be *done* in the name of that 'active pursuit of the well-being of the loved' in each case. I can most easily explain what I mean by way of a telling anecdote.

On visiting Bristol a few years ago, I went to look at the famous St. Mary Redcliffe church. There was a service on at the time, so I stood in the gloom at the back, waiting for it to be over. And as I stood, I slowly became aware of a difference: an awareness which, as it came to fill more and more of my consciousness, generated a rich warm *feeling* of love. The voice that was coming from the far away altar-end of things, the voice that was saying 'the magic bits', was light and soft and clear. And I did not really need the time it took for my eyes to accommodate to the gloom to realise that the minister was a woman. In a totally unexpected totally unprepared way, I had come upon the first service I had ever heard being taken by a woman. And the rich warm *feeling* of love was momentarily almost overwhelming.

What I am suggesting about such an experience is that the difference between the love I felt for her and the love I feel for a lover or my children or my friends is not in the nature of the love but in the *actions* out from which I am able to constitute it. The actions which for me make my love of unknown-to-me women ministers struggling to find their voices, *literally* to find their voices, are limited in their scope. I can challenge sexism wherever I meet it, and I do: that forwards their struggle and loves them. I can support their campaigns financially, and I do: that forwards their struggle and loves them. When I meet one of them, as on that day, I can seek her out and express my delight to her, and I did: that forwards her struggle and loves her. But that's about it. That's about all I can do actively to pursue the well-being of such a loved *directly*.

By contrast, the actions which make for us our love, for example, of an intimate, sexual partner can be so many more. They can be of the sort which cater to the lover's sexual well-being: anything from the making of a variety of mutually satisfying sexual encounters to refusing to be anything of sexual object to her or him. They can be of the sort that cater to the lover's physical well-being: anything from cooking her or him good meals to tending to her or him in terminal illness. They can be of the sort that cater to the lover's psychological well-being: anything from introducing her or him to family and friends and trying to create circumstances which will enable all to enjoy the experience to continuing to enjoy her or him as ex-lover, with now a new partner, and letting both of them see that enjoyment. They can be of the sort that cater to a lover's intellectual well-being: anything from discussing important ideas with her or him and challenging any of hers or his which seem to us wrong to celebrating each of her or his intellectual successes. And so on, and so on, and so on. This is clearly the sort of list which could run and run. In good one-to-one sexual loving, there is probably no aspect of the being of the loved other that cannot be enhanced by one who loves: and all in a delicious, intimate, taken-for-granted, protected, easy, private, warm, space.

Thus is the range and scope of the available actions in the two cases different. When one loves one pays close, curious and tender attention to the loved, and the scope for such attention is greater in the one case than the other. But the loving that constitutes each such loving action, the active pursuit of the well-being of the loved that constitutes each such loving action, is the same. Love is indivisible. In both cases, love is being made by doing. And the making of it, the making of love in the two different circumstances, is the making of that which characterises the personic. This would be the I IS position.

How then are we to understand the fact—witness the Bernard Williams position—that there are those who claim not to know what it means to love those whom they do not know? Why is it that such folk not only claim not to feel love for unknown others, but give plenty of evidence in the way they behave that indeed they don't? Why do so many of their number argue that it's just impossible so to feel: that it's just adopting a prissy pose to claim so to feel? On my I AM to I IS analysis, it is in no way necessary to argue that such folk are misunderstanding what they feel. All it is necessary to argue is that *could* they so act as actively to pursue the well-being of unknown others, they *would* so feel; that it's because they *cannot* so act that they *do not* so feel; and that were they to come to be *able* so to act, they would come also to be *able* so to feel. Notice, and absolutely crucially, that there are no 'oughts' and 'shoulds' here. This is not moralism. My statements throughout are not normative but positive. I would certainly claim that it would be a better world in which people *could* so act and therefore *could* so feel. I would certainly claim that it would be a better world in which it *was* understood that love is indivisible: for example, that you can't love children more by loving paedophiles less. But I would also claim that no way does it follow from any of this that it is appropriate therefore to blame those who *cannot* so act and therefore *cannot* so feel. Apart from anything else, such people are all of us. We are all, I am claiming, at some point or other on a developmental shift from I AM to I IS thinking. And that being the case, all that it is appropriate to do is to help to create the circumstances in which all persons *would* so feel because all persons *did* so act. Such circumstances, I am arguing, are those which involve the total shedding of 'the having ideology' via the steady making of real equality. This is what I IS persons come to understand about *loving*. And again: when they make their God, it makes all the difference in the world to that which they make.

These then are the different understandings, first of having then of loving, which inform the thinking of I AM and I IS persons. And what I want now to do is directly to compare and contrast their respective creations as God-makers: that is, their respective attempts to make that idea which, by being immune to a relativising critique, can ground an absolute rather than a relative understanding of the good.

Introduced in the companion chapter to this one (Chapter 2), and drawn upon across several other parts of the book, I have presented an image of the I AM God. It is an image, I would claim, of one stunted—*made* stunted by its makers, of course—in three ways: first, by being given four inappropriate attributes; second, by being given three inappropriate powers; third, by being given two inappropriate demands. I want now to deconstruct each of these attributes, powers and demands. And to do so in such a way as to reveal two things about them. I want, first, to show just how it is that they stunt, and finally destroy, the I AM God. And I want, second, to explain the way in which their alternatives—*different* attributes, powers and demands, given by *different* makers to a *different* God—would enable that God, the I IS God, instead to flourish.

The inappropriate attributes are (1) that he is he; (2) that he is one; (3) that he is first; and (4) that he is supernatural: all four of them attributes copied from those which, if on a lesser scale, his makers deemed to be their own. The inappropriate powers are (5) that he has the view from nowhere; (6) that he is omnipotent; and (7) that he is omniscient: all three of them powers which his makers would like to have had for themselves, indeed tried to have for themselves—to the extent to which they could. And the inappropriate demands are (8) that we love him 'with all our hearts and with all our minds'; and (9) that we understand the doing of his will as the doing of the good: both of them demands which, at their own more humble level of course, his makers tried hard to demand of their own Others too. In the interests of clarity, let me deal with each of these nine under their individual sub-headings.

(1) that he is he

One of the most important contributions of the several monotheisms of the Axial Period, especially of imagery of God as father, was that it made all men brothers. If all men have one father, they are all brothers. If they are all made in his image, they are all, in at least some sense, equal. And an offence against a fellow man, also made in God's image, becomes an offence against God. But as the consciousness-raising associated with the now several waves of feminism has made so spectacularly obvious, it is of course necessary now to register a somewhat substantial rider. Particularly perhaps associated with the most recent and veritable flood of such enlightening, it can now be seen as a rider so substantial as to make what we might call the 'first coming' of monotheism, the Axial coming, thoroughly problematic for more than half the human race. For when the 451 CE Council of Chalcedon declared Jesus to be 'true God and true man', the truth claimed for the second was at least as important as the truth claimed for the first. Adam, man, was first: and Eve was a spare rib. And as Milton's quite terrible phrase 'he for God only, she for God in him' makes abundantly clear, both the he who has been (until very recently) the unchallenged image of the I AM God, and the he who has been (until equally recently) the unchallenged image of the I AM person, have been also first in this other and terrible way: first, that is, in relation to the sex which has been second. It has meant that the 'first coming' of one-race-the-human-race, the one which arrived with the three great monotheisms, was in its partial because only male guise only a partial coming. As for the first time in the history of the world women slowly cease to be 'the second sex'—'I am not an outboard motor, I am another boat', as the poet Adrienne Rich puts it—it is now that its second coming, first and last, can becomes a reality.

> *His successor I IS God is otherwise. It is both he and she, as it is also neither he nor she. It is both he and she in being sourced in all the female and male human beings of the biosphere. It is neither he nor she in being instantiated as all the persons of the world.*

(2) that he is one

To be one *can* seem an advantage. Indeed, he was made that way precisely because it *did* seem an advantage. After all, if what is needed is the grounding of a moral absolute, things are not really helped by there being more than one of them: hence mono-theism. For as candidate sources of moral absolutes, just as surely as it is the weakness of human persons that we are many, so it is the strength of the personal God that he is one. Isn't it?

Well it can't be denied that, as any such source, the multiplicity of we worldlings is something of a problem. If we are to defeat moral relativism and find a way of grounding the good in we who are many rather than in he who was one, there does indeed have to be a way of making some 'unum' out of that 'pluribus'. The making of that 'unity in plurality' will be hard. But not nearly as hard as having to make do with such a ready-made version as is offered by the I AM God.

For if he is posited as both *one* and yet also all-*explanatory*, there is at least a case for saying that it only needs his redundancy in one of the areas for which his existence has been held to be explanatory—universe? biosphere? world?—for it also to become redundant in all the rest. For he is one. And particularly if it is creation which is the order of the day, being one can be considerable disadvantage. An entity which, like the genetic code or a natural language, exists instead as one of Steven Pinker's 'discrete *combinatorial* systems'; an entity which, being *particulate*, can accommodate Marx's 'uneven development' with ease; an entity which, because it incorporates maximum *redundancy*, can readily deal with the demise of any of its myriad (pixel) units because such demise leaves the whole still whole; an entity in which, even when the limitation of its units functioning individually tends towards faltering, still the capacity of its units functioning collectively tends towards flourishing; an entity which can function *as* an entity not only when persons can be persuaded to believe in it as such, but even when they can't and don't: such an entity, were it to exist, would win hands down in competitive stakes with what by comparison would become the inflexible not to say ossified oneness of a sole and single I AM God.

> *The I IS God is just such an entity. Instantiated as all the persons of the world, it is the ultimate in discrete combinatorial systems. In their uneven development, its persons, singularly linked but yet still singular, constitute in the linguistic web which is their world a six and a half thousand million strong 'unity in plurality'. Pixel-packed for maximum redundancy; self-assembling and self-developing; self renewing and self-organising: this is an entity consisting not in one but in many; an entity really* all-*explanatory; and an entity which functions as an entity even before the persons which constitute it believe either that they do or that it is.*

(3) that he is first

The I AM God is not just *a* first person but *the first* first person; not just *an* I AM but *the* I AM. In the words of 'the great tautology', he is I AM THAT I AM. But if one is first—in his case doubly—ones Others can only be second and third. And this is a problem. For just as I have argued that it is not only their secondariness but also their primariness which stunts the moral development of his I AM creators, so it is also not only his secondariness (when he washes our feet as servant) but also his primariness (when he rises above us as lord) which stunts the moral development of their I AM creation. What the construction methods of his I AM creators have denied to him is the possibility of ever walking beside us just as friend. The I AM God does not have friends. The I AM God cannot have friends. For as C. S. Lewis makes so movingly clear—in the 'philia' chapter of *The Four Loves* (1960)—because it is in no way biological (as are familial and lover relations) and in no way necessary (as are relations between neighbours, colleagues and the like), true friendship requires a level of mutuality in equality of the very highest order. And the poor I AM God has been made in such a way that he is not capable of reaching even the puny first-among-equals kind. As just *spectacularly* first, he is both beached in a formlessness which leaves him incapable of engagement, and marooned in a timelessness which leaves him incapable of development.

For his I AM creators there is a way out. As they selve their ways across a natural-to-normal-to-moral developmental process, their steady accumulation of experience of these crippling twin disabilities, of the Scylla of primariness and the Charybdis of secondariness, means that steadily they can come to understand—they *will* come to understand, they *do* come to understand—that it is the mutuality of equality which is necessary to loving: that to love the selves of Others is to want those selves to be stunted by neither primariness nor secondariness; that to love ones own self is to want it to be stunted by neither primariness nor secondariness; and that therefore only in the mutuality of fundamental equality can such loving be achieved. But for the formless and timeless I AM God, there is no such salvation.

> *For the I IS God things are different. In being made by I IS persons, persons for whom relations of equality are essential, it is instantiated as all of those not-first-but-just persons in all of those not primary-secondary but equal relations.*

(4) that he is supernatural

The problem here is not so much that he *is* supernatural, but that he is supernatural in entirely the wrong way. In the form of '*the* supernatural', talk of that which is 'super the natural' in an I AM context is always talk of the magical. Just as they have posited themselves magically, I AM persons have posited their I AM God magically too. They have posited themselves as magically outside a world with which, as first persons, they interact (or do not) as they freely choose. And they have posited him as magically outside a world with which, as first cause (or *first* first person), he also interacts (or does not) as he freely chooses. Thus has the magical supernatural, and in both cases, been seen as separate from, as over against, that which is natural.

But then the magical starts to fade. As the development of science comes to mean that magic is increasingly understood by more and more people to be no sort of explanation of anything, as the magical supernatural fades, there are dire consequences for the I AM God. The fading of the I AM person, anyway not nearly so far gone because in some ways even more frightening, at least leaves *something* behind. It leaves behind a natural person, a flesh-and-blood person, an animal person, a material person: the sort of person which evolutionary psychologists and their ilk can welcome and study; the sort of person which, even as my hypothesis shows it to be wrongly construed, is not nothing. The fading of a magical-supernatural I AM God, by contrast, leaves behind just *nothing*. If persons can no longer be construed as magical-supernatural, they can be natural: for them, if you like, 'being natural is something to be'. But if God can no longer be construed as magical-supernatural, what else can he be? He can't be natural. For if he were, then as an entity just as situated, just as changing, just as partial as you or I, as a source of moral absolutes he would be no more use than you or I. By being construed as magical supernatural, he becomes, in effect, just impotent. Concerning any candidate successor he might have, it would need to be one the supernatural element of which—and to avoid those 'situated' and 'changing' and 'partial' problems, it will have to *have* a supernatural element—is in no way magical.

The I IS God is just such an entity: supernatural but no way magical. It is not an entity outside a world by the origination of which it was the creator, but an entity inside a world by the development of which it was the created. It is certainly not natural. It is certainly supernatural. But it is not that which, before the natural, was its origin in the magical supernatural. It is that which, after the natural, is its progress to the moral supernatural.

(5) that he has the view from nowhere

This is the power which, more than any other, enables him to ground a universal concept of the good. The argument has been that it was his singular, impartial, whole, disinterested and universal 'view from nowhere' which enabled him to be the source of such grounding. And that it is with his 'death' that there is no way in which our multiple, partial, incomplete, interested and relative 'points of view' can ever enable us to provide such ground instead: hence relativism. [In passing, might I suggest that it would probably be a good idea for us all to train ourselves to stop talking about the 'death' of God. Only things that live in the first place, organic things, can die. And God is not of their number. God cannot die because God has never lived. God is an idea. And old ideas never die, they only fade away.] But: back to those multiple, partial, incomplete, interested and relative 'points of view'. What I want to suggest is that it is only because we have not properly understood the complete inappropriateness of the 'viewing' metaphor contained in 'view from nowhere' and 'point of view' language that the I AM God has been able to be credited (wrongly) with the power associated with having such a view, and relativism has been able to be posited (wrongly) as being the only alternative when he fades.

Entities *having* a point of view, material things existing in space, are locational: they stand *at* such points of view, and where one stands another cannot. But entities *being* instead moments of comprehension, happenings existing in time, are durational: they understand *as* such moments of comprehension, and when one so understands so too can another. Moments of comprehension are not points of view at all. They are instead what developmental psychologists have called stages of development. And as I have already tried to demonstrate, it is only the misreading of a person as an organism, only biological reductionism, which has allowed us unquestioningly to accept the 'point of view' metaphor as appropriate. It is not appropriate. And just as it is not appropriate in the case of human persons, so it is not appropriate either in the case of the I AM God.

When I AM persons make their I AM God, they make him in their own image. And it's perhaps important to register that to claim so is to claim that which has to be equally as acceptable to believers as it is, perhaps more obviously, to unbelievers. For if we ask whether, if the whole human race dropped dead simultaneously two seconds from now, God would still exist, those who answer 'yes' would still need to accept, along with those who answer 'no', that we make our God *image* anthropomorphically: that our God *image* is an invention. That which is ineffable, when nonetheless it is described, has to be described in the words of the human (all too human) world. That which is unknowable, when nonetheless it is revealed, has to be revealed through the images of the human (all too human) world. In *A History of God* (1993), this is what Karen Armstrong recognises when she comments that 'when they attributed their own human feelings and experiences to Yahweh, the prophets were in an important sense creating a god in their own image'. Indeed. But to protect him from what they saw as their own limitations, their construction methods were such as to inflate him more than somewhat: in short, to put him outside space (formless), and outside time (changeless). It seemed sensible at the time. Where we are situated, and thus partial, he was not. Where we are centred each in our own concerns, he was the circle with centre nowhere and circumference everywhere. Where our situations in a particular place and at a particular time meant that our understandings could be only 'in part', his view from nowhere meant that his could be complete. But now—it is my claim—we are coming to know rather better.

Because we now understand that existence in space is actually *necessary* to points of view, we are coming to know that what our construction methods have really meant is that, unlike even the simplest organism, our I AM God can have not all points of view but none of them. And because we now understand that existence in time is actually *necessary* to moments of comprehension (stages of development), we are coming to know that what our construction methods have really meant is that, unlike even the least-developed self, our I AM God can have the prospect of progressing not through all stages of development but through none of them. In the old 'appearance *and* reality' language, what this faulty I AM construction method has really done is to put both the *appearance* of the universe and biosphere, and the *reality* of the world, permanently beyond his ken. For the universe and the biosphere *appear* the same from the same point of view, *appear* different from a different point of view: and an entity which is outside space cannot *have* such different points of view. By contrast with which, the world actually *is* different as different moments of comprehension (stages of development), *is* the same as the same moments of comprehension (stages of development): and an entity which is outside time cannot *be* such different moments of comprehension.

The I IS God is different. It's not just that, sourced in all the humans of the biosphere, it stands at a point of view from which it has the view from nowhere by having the view from everywhere. It's that instantiated as all the persons of the world, it understands as a moment of comprehension (stage of development) as which it is the view from nowhere by being the view from everywhere.

(6) that he is omnipotent

He is not just powerful—which would be stunting enough. He is 'all powerful'. And that means that he is stunted by having the power which his others have not. In a letter to Bishop Mandell Creighton, dated 3 April 1887, Lord Acton now famously wrote: 'Power tends to corrupt and absolute power corrupts absolutely'. Almost certainly, he did not have the I AM God in mind. But he might just as well have done. After all, isn't any being which allows itself to be worshipped—worse, requires that it be worshipped—already thoroughly corrupt?

I AM agents—Gods or humans—speak always from some power-base. As I registered earlier, anthropologists point out to us that gods began to appear at just that point in the development of human societies when the human group became larger than could be 'policed' in a face to face fashion. What was then for the first time needed, and therefore for the first time created, was a source of power and authority with a voice apparently from nowhere but actually from the 'somewheres' of the powerful. And after a long creation process, after a procession of sundry lesser immortals, the I AM God of the monotheisms emerged as the hegemonic all powerful solution. It was as though the I AM secondary persons were told by the I AM primary persons: 'look, even if we can no longer keep an individual eye on what you're up to, he can—so watch out!'

But after it had been around for a good few thousand years, steadily its only apparent impartiality began to emerge: the fact that it was actually and always the mouthpiece not of a disinterested 'all' but of an interested 'some' became steadily clearer. And with that increased clarity, the idea that the power of the I AM God was legitimate power began to disintegrate, eventually to the point where what predominates now is a God-shaped hole and a lot of pesky relativistic theories of the good trying hopelessly to fill it. What these post-modern and relativistic hypotheses very appropriately and importantly highlight are what are indeed the considerable problems of defending a realism and universalism concerning the nature of the good on the basis of I AM agency. *Which* agent? Out from *which* power base? With *which* authority? Given that 'nowhere' is not available, from *which* view-point? Given that 'always' is not available, at *which* time? The great *strength* of this relativistic critique has been that, relentlessly, it has forced us up against the fact that *all* defences of a realism and universalism concerning the nature of the good which maintain an I AM agent are, in the last analysis, just authoritarian. From literary critics standing not nowhere but somewhere and defending a therefore partial and particular idea of 'the canon' to spokespersons for liberal democracies standing not nowhere but somewhere and defending a therefore partial and particular idea of 'the terrorist', in the last analysis it can only be the power and authority of the judge rather than the independent goodness or badness of what is judged good or bad *by an I AM agent* which is determinate. However, the great *weakness* of this same relativistic critique is that having forced us to acknowledge that fact, they then proceed to question *not* the validity of I AM agents but *instead* the very possibility of sciences of the human! And in so doing, they join the I AM-defended realism they are attacking in leaving the I AM God if not intact, then certainly still providing some foothold for those elements of 'the mysterious' and 'the unknowable' on which opponents of sciences of the human thrive. It is my case that, anyway in the longer term, such a position becomes just untenable. And the search for an idea which, as an impartial source of moral absolutes, would *in fact* be immune to a relativising critique rather than only *apparently* immune to a relativising critique is resumed.

> *The I IS God offers just such an idea. For where the making of the I AM God was the making of an immortal entity which was 'all* powerful, *the making of the I IS God is the making of a non-mortal entity which is* 'all *powerful. The first was a making – by would-be archetypal top-dog authoritarians – of an archetypal top-dog authoritarianism. The second is a making – by actual thorough-going egalitarians – of a genuine thorough-going egalitarianism. The I AM God once* had *a kingdom over which he ruled. The I IS God will* be *a democracy via which all rule.*

(7) that he is omniscient

And since, with Francis Bacon, we know that 'knowledge itself is power', the case here is much the same as that made already above. In Corinthians 1, chapter 8 verse 1 we read that 'Knowledge puffeth up'. Indeed. And as Lord Acton might have added, had he thought of it: 'absolute knowledge puffeth up absolutely'. To know where others know not is to be primary where others are secondary. And to want to maintain either primariness or secondariness against the pull of equality is to be stunted in the capacity to love. An omniscient I AM God is precisely so stunted.

> *The I IS God again is different. Where the making of the I AM God was the making of an entity which was 'all* knowing', *the making of the I IS God is the making of an entity which is* 'all *knowing'. The first was a making – by the few who knew where the many did not – of an archetypal know-all over-against his know-nothing others. The second is a making – by the all who know and equally together – of a thorough-going and genuine democracy of the intellect in which there will be no 'others'.*

(8) that he requires us to love him

Notice it is that *we* love *him*. His own loving is not in doubt. He loves already: on that score, nothing is wanting. What *is* wanting is that his

Others—we—should love him. He wants our love: in fact, he demands it. He wants to be loved. Thus is he the quintessential I AM lover: of course, because he is the creation of quintessentially I AM lovers. For him as for them, love is essentially a feeling not a doing. In their often terrible suffering, of course he *feels* for every last member of his creation. As did that rich woman when she heard the stories of the beggar, he weeps. But although he has the power not just to feel but also to act, he—just like the woman who turned out the beggar—does not use it. And probably for much the same reason. The woman with the power to intervene to relieve the suffering of the beggar perhaps will tell herself that it's for his own good that she does not do so: for he has to learn to stand on his own two feet, be independent, and exercise his God-given free will to choose those paths which will enable him to flourish or perish. Similarly, the God with the power to intervene to relieve the suffering of all of humanity perhaps will tell himself that it's for their own good that he does not do so: for they have to learn to stand on their own many feet, be independent, and exercise their given-by-him free will to choose those paths which will enable them to flourish or perish. The woman says so 'for real': instantiated as a particularly situated linguistically structured experience, she says so because that is the sort of real person she is. The God says so 'by proxy': created by real persons who themselves are instantiated as many particularly situated linguistically structured experiences, he says so because that is the sort of imaginary person he has been made to be.

In contrast to this, the I IS God, instantiated as the I IS persons who make it, requires nothing, demands nothing. Having come to understand the limitations of the standard of loving available to I AM persons and their I AM God, such persons no longer believe that already they and he love. They know that love must be made. They do not demand it: they make it. Making love is their mode of being. And given that, for them, such love is essentially a doing not a feeling, what they do is use their power actually to alleviate what has been the often terrible suffering of I AM humanity. Together, they do so 'for real'. As I IS persons instantiating the I IS God, they do so because those are the sorts of real persons they have become. And they do so for each and every just member of what thereby steadily becomes their creation.

(9) that he requires us to do his will

The I AM God requires that we understand the doing of his will as the doing of the good. Here, at the end of a comparative analysis of the attributes, powers and demands of the two Gods, I have reached that which, for my candidate theory, is the kernel of things. And so here, at the end of my second God chapter, I will proceed somewhat differently. My case so far has been that, as an idea providing immunity to a relativising critique, the new I IS God considerably out-performs its ailing I AM predecessor. And to under-score what I take to be the huge significance of that out-performance, particularly for the currently ubiquitous relativising of the good, I will first gather up into just one paragraph all that I have so far said about the new idea, and then end with just a question.

> *The I IS God, both he and she because sourced in all the female and male human beings of the biosphere, neither he nor she because instantiated as all the persons of the world, is the ultimate in discrete combinatorial systems. In their uneven development, its persons, singularly linked but yet still singular, constitute in the linguistic web which is their world a six and a half thousand million strong 'unity in plurality'. Self-assembling and self-developing, self-renewing and self-organising – and pixel-packed for maximum redundancy thereby – this entity, consisting not in one not in many but in all, indeed has the potential to be both all-powerful (everyone powerful) and all-knowing (everyone knowing). For in being instantiated as all I IS persons, persons for whom relations of equality are essential to their full humanity, it is instantiated as those not-first-but-just persons in those not primary-secondary but equal relations which constitute loving. Being supernatural not as that which before the natural was its magical origin but as that which after the natural will be its non-magical development, it is an entity not outside a world by the origination of which it was the creator, but an entity inside a world by the development of which it will have been the created. Sourced in all the humans of the biosphere, it stands at a point of view from which it has the view from nowhere by having the view from everywhere. And instantiated as all the persons of the world, it understands as a moment of comprehension (stage of development) as which it is the view from nowhere by being the view from everywhere. It is not an authoritarian immortal entity instantiating authoritarianism: it is a democratic non-mortal entity instantiating egalitarianism. It requires nothing. It demands nothing. Instead, it loves. Having come now to understand the limitations of the standard of loving available to its erstwhile predecessor, it is an entity which understands that love, a doing rather than a feeling, is that which must be made. And that being so, it is an entity which uses its all-powerful (everyone powerful) power and its all-knowing (everyone knowing) knowledge to make it: that is, to make that equality which is the foundation of loving.*

Would not the doing of the will of such a God be the real doing of the good?

Chapter Twelve

CONCLUSIONS

Towards a grand explanation

Lecturing on what some among the professionals call 'rigid and non-rigid designators', an Oxford philosopher of my acquaintance introduced a telling humour. 'I designates me', she said. But added: 'And funnily enough, it also designates you—you think!' At base, this is the issue which my whole book has addressed. In his *Discourse on Method* (1637), Descartes wrote that 'it is so self evident that it is I who doubt, understand and desire that there is no need to add anything here to explain it'. What I have been arguing is that what we need to add 'to explain it' is some theory as to just what 'it' is!

In my attempt to explore that issue, one of the most important questions I have raised, and must here return to, is whether an I is to be conceived of as an agent (of somebody or something other than its self), or an originator (of its own doings). The form of determinism which I have defended, a form which just might go some way towards muting Stuart Hampshire's designation of any form as 'the hideous hypothesis', argues that while in a society of I AM persons, each can be only an agent singly, in a community of their I IS successors, all will be able to be originators together. It's the crucial significance of that 'together' which now I must further explore.

The language involved in that linguistic structuring which is a person is intrinsically a public phenomenon. As Wittgenstein has convincingly argued, the notion of a private language is just oxymoronic. Language is intrinsically a 'co', 'con' or 'com' thing: a 'together with' thing. It extends greatly the possibilities of 'communication'. It brings into existence for the first time that 'co-operation', which perhaps would be more accurately labelled 'co-action'. As in 'conscience' rather than just 'science', as in 'consciousness' rather than just 'sentience', as in 'community' rather than just 'society', we have long put this 'together with' truth into our vocabularies. And I am suggesting that it is precisely the 'co' or 'con' or 'com' parts of such words, the 'together with' parts, which reach out from the un-freedom of singular free-will agency towards the freedom of collective genuine origination. There's the 'together with' unity which is 'community'. There's the 'together with' knowing which is 'consciousness', and the 'together with' consequence of such knowing which is 'conscience'. There's 'commune' and 'communism' and 'communion' and 'cooperation'. And what I'm suggesting about these 'together with' words, these 'co', 'con' and 'com' words, is that in them there sits a vital truth about persons.

Where the experience of other animals constitutes sentience (an organic phenomenon), the linguistically structured experience of human animals constitutes consciousness (a personic phenomenon). To be sentient is to be aware. To be conscious is to be aware of being aware, among others also aware of being aware. And such consciousness, I am suggesting, arrives in two quite distinct stages. First comes I AM first person consciousness, which because it mistakes *being* aware with *having* free will, wrongly takes its own self already to be a singular originator. Second comes I IS just person consciousness, which because it understands that *being* aware is just *being* aware, rightly takes all selves to be able to become originators together. Everything hangs on that 'together' phenomenon. Originators are always and can only be plural. Singly, we can each be only un-free agents of the linguistic web which is the world: nodal moments in the ensemble of linguistic relations of which that world is constituted. 'Together with', we can collectively become its free originators. But only when we are 'together with' in a very particular way.

Being 'together with' in circumstances of inequality won't do. For in such circumstances, which of course are all the circumstances in which the world so far has been instantiated, persons are 'together with' others in relationships of unequal power. And where the struggle to make the genuine origination of all out of the mere agency of each is a struggle which requires total cooperation, in such unequal and thus riven situations the cooperation is only ever partial. By contrast, being 'together with' in circumstances of equality offers other possibilities. For where persons are together with others in conditions of equal power, the total cooperation necessary to the making of origination out of mere agency becomes available. To stand in for what I think is a huge reality, let me give just a tiny example of what I mean by this. For as long as there is inequality in the relationships of females and males in the wider society, for so long a parent struggling to de-gender the upbringing of a small child can be only a less-than-free agent. Her or his making of the equal is against the grain of the more general and structural inequality. Only with the full cooperation made possible by equality can less-than-free agency become free origination. For such free origination becomes available only when the total cooperation of all gives it to each, together.

This is my position. And while aspects of it are of course just mine, in its more general articulation it is one which I am by no means alone in adopting. There are several quite well-trodden paths within philosophy which give expression to positions very similar. One such, within what is English-speaking but hardly analytical philosophy, is that exemplified in the work of the moral philosopher John Macmurray. Two others, within more recent continental philosophy, are those exemplified in the work of Giles Deleuze and Alain Badiou. To gather some support for my stance, therefore, let me, if only briefly, consider the positions of just these three.

First, John Macmurray. In his *Reason and Emotion* (1935), he suggests that 'the primary difference in the relation of persons to persons which distinguishes it *toto caelo* from other types of relationship is mutuality'. But such mutuality, he argues, cannot rationally be built in situations of inequality. This is because, in such situations, the persons are subordinated to their function. Instead of having complete integrity just as persons, they become complementary the one to the other, each having meaning and significance only in terms of the other. He works this argument through in terms of relations between men and women, but makes it clear that he sees it as pertaining to all person-to-person relations. His is a religious approach. Writing before the most recent round of feminism had alerted us to the importance of always using sex-neutral language, he maintains that 'the primary religious assertion is that all men are equal, and that fellowship is the only relation between persons which is fully rational'. By 'fully rational' he means 'fully appropriate to their nature as persons', from which it follows that 'the drive to rationality in this field is the impulse to achieve equality and fellowship in the relations of persons'. In *Conditions of Freedom* (1949), he takes this case further. Opposing them to 'primitive' or 'exclusive' or 'kin-based' alternatives, he says, of what he calls 'universal religions', the following:

> In the fullness of time, when social pressures demand it, universal religions are founded. They are founded by individual persons, of high religious insight, who seek a personal community denied or frustrated by the character of the societies of which they are members, and who seek to realise the potential universality of personal fellowship.

It is my take on this position that, with only the caveat that 'God' be understood as having been created *by* us rather than creating *of* us, I accept his case completely. Had our millennial celebrations been in remembrance of one – of course among many – whose contribution had been to seek a personal community denied or frustrated by the character of the society in which he lived, and who strove to 'realise the potential universality of personal fellowship', that for my money would have made them infinitely more telling and appropriate.

Second, Giles Deleuze. In *Radical Philosophy* (issue 81), Peter Hallward considers 'Gilles Deleuze and the redemption from interest'. He argues that Deleuze, following Spinoza, stresses the way in which we 'do not begin as God (as reasonable)'. Instead we have to develop into reasonableness.

And the 'eventually reasonable modes' are those which 'see themselves as actualisations of God's univocal power, as wholly and immediately identical to the one Real interest, the interest of divine reason itself'. He goes on to suggest that 'a conflict of interests has no more place in Deleuze's "world without others" than it does in Spinoza's "reasonable commonwealth", in Suhrawardi's luminescent "imaginal world", or in Paul's "one body of Christ".' And concludes that while, for Spinoza, 'the idea of God is the only adequate basis for the ordering of reality, it is vital to remember that it is 'a constant of Spinozism' that 'one cannot ... from the outset install oneself in God'. Although it is 'divine thought alone' that is 'authentically Real, original in both senses', it is nonetheless the case that 'we must *become* thinkers.' It is my take on *this* position that thought will become truly original rather than merely agentic, will become divine because supernatural because moral, when it shifts from I AM to I IS mode.

Third, Alain Badiou. Again in *Radical Philosophy*—this time in issue 96—the French philosopher Alain Badiou suggests that 'the vast majority of political orientations have nothing to do with truth'. Instead their role is 'to organize a repulsive mixture of power and opinions.' He goes on to suggest that what are the very rare exceptions share just one feature. They are orientations which require of people that 'they engage only their strict generic humanity: that is, they give no preference, for the principles of action, to the particularity of interests', but instead, 'induce a representation of the collective capacity which refers its agents to the strictest equality.' And what does his 'equality' mean? It means 'that the political actor is represented under the sole sign of his [sic!] specifically human capacity'. For Badiou, 'justice is the theoretical name of an axiom of equality', and 'refers necessarily to a wholly disinterested subjectivity'. In the absence of such equality, he suggests, justice becomes merely 'the harmonization of the interplay of interests': in other words, not justice at all. And it is my take on *this* position that it is precisely a political orientation which rather than organising a repulsive mixture of power and opinions engages instead with only our strict generic humanity which the I IS hypothesis requires that we espouse.

Where the objective is understood as the subjective shared, as the inter-subjective, just *how* things are shared becomes very important. As Badiou is making clear above, a sharing shaped by interest and power is crucially different from one shaped by disinterest and equality. And such interest and power is much more easily neutralised in the sciences of the natural than it is in the sciences of the moral. In the former, mutual interest is, at least to some extent, a given: a given, that is, of an 'always already' shared stance towards that which, as the inorganic or the organic, is more readily seen as other than ourselves. But in the latter, such mutuality, far from being an 'always already' given, is rather more a 'never yet' created.

Before it can be used as a base from which to study ourselves 'objectively', out of the struggle for democracy and equality mutual interest has first to be made.

Such making, however, has been widely understood as an enterprise at least hazardous, at most impossible. One of the most philosophically distinguished attempts to pull it off has been that made by Jurgen Habermas: across a lifetime, but perhaps particularly in *The Theory of Communicative Action* (1984). Along with many, he recognised that, without God, the task of a post-God modernity had to be to create an alternative grounding for the moral. Along with fewer but still with some, he maintained that without God the only possible source of such an alternative would be a rationally arrived at consensus, reached by a process of non-coercive discussion among all relevant persons: with 'relevant persons' being a category which could variously include anything from two persons to all persons. In order to establish the ground rules for such discussion, he argued that what was needed was what he called 'an ideal speech situation': that is, a speech situation completely open to all concerned, with no constraints, internal or external, on what might be said within it, and containing no imbalances of power. But such a speech situation, he stressed, did not have to exist in fact. It was only that, in the course of their deliberations, all interlocutors needed to assume its existence And in a valiant attempt to avoid the pitfalls of relativism, he then attempted to do just that: that is, to situate every actual philosophical debate he considered within a *notional* 'ideal speech situation'.

This is a stance which reminds me of nothing so much as the original Rawls: the Rawls of *A Theory of Justice* (1972), for whom decisions from behind 'the veil of ignorance', decisions concerning the distribution of economic resources and political power made in ignorance of what ones own position in the resulting hierarchies would be, might perhaps be expected to mirror decisions made by ideal speech communities containing no power imbalances. In brief: decisions made by those behind the veil and decisions made by those in the ideal speech community would both tend towards the egalitarian. Similarly, the fact that each philosopher weakens (and for my money fatally flaws) the egalitarian tendency of his philosophy serves further to suggest the comparison. Rawls does so with the now notorious 'difference principle': by claiming that inequalities are just if they are necessary to make the worst off better off than they would be with more equal distributions, it is a principle which swims up to its very neck in that 'better-equals-having-more' ideology which basically precludes a scientific approach to the problem. And Habermas does so with his stress on the point that such a speech situation does not have to exist in fact: a move which leaves him open to the claim that he is propounding nothing more than a bland, a-historical, pseudo-egalitarian liberalism.

Succinctly: these are cop-outs. What seems to me clear is that while both Habermas and Rawls go some of the way to acknowledging the need, in a just society, for disinterest, neither Habermas nor Rawls goes all of the way to acknowledging the need, in a just society, for disinterest. And what stops them is that that both, in the language of Kuhn, are doing 'normal science'. Both are trying to think the new, but from within an old and finished paradigm. And that's impossible. When even the original Rawls, the most radical Rawls, argues that the power differentials which follow upon the workings of 'the difference principle' are legitimate, he demonstrates not only that he is stuck in what I am calling 'the having mode', but also that he is, as a necessary consequence, working within an increasingly untenable I AM conception of the self. And when the Habermasian interlocutors are deemed to be capable of assuming the existence of an ideal speech situation where none in fact exists, they are being posited as those increasingly untenable free-willing, I AM, outside-the-world agents conjuring the necessary disinterest out of thin air. By contrast with both of them, it is my position that securing equality so as to get access to moral science is exactly analogous to controlling the variables in an experiment so that there should be none that confound. And confounding variables have to be controlled—for real.

So what then needs to be done? If the best way to predict the future is to make it, as a scientist at Xerox's Palo Alto Research Centre put it, what do we need to make? And if that sounds like an invitation to some social engineering, it is. For not wanting to go in for social engineering does not mean staying in some non-engineered social space: there isn't one. It just means wanting to stay in an inegalitarian social space engineered by tradition, authority and deference, rather than move to an egalitarian one engineered by reason, science and democracy. If Marx's 'end of pre-history' is to become a reality, if animal-animals (those before *Homo*), human-animals (those before language) and animal-humans (those before equality) are increasingly to be joined by human-humans (those of the future), what do we need to make?

In my first chapter, companion to this one, I suggested that at the beginning of this third millennium we need to remember and celebrate those many who initiated the first by trying to carry still further what I take to have been their project. With the making of their God, their spectacular achievement was to make an idea immune to a relativising critique: or more carefully, to the specific relativising critiques of their own time, and of most of the intervening centuries. But I have also suggested that because theirs is a making which no longer cuts the mustard, then to remember them, celebrate them and love them now, we have to make something new. What I now want to argue is that that 'something new' has to be nothing less than a whole new movement of thought: if you like, a new and more enlightened Enlightenment. To unite its supporters around a renewed struggle to make 'the simple thing so hard to achieve', it would be in essence a movement for that equality out of which moral science could then be made. Let me offer, therefore, some first thoughts concerning the possible parameters of such a movement.

It would need to be a movement in thought which gave priority to the highlighting—as almost always foundational—of economic equality. Where the various movements inspired by Marxism thought it was their strength that it was the concept of class (rather than of income or wealth) that was for them foundational, this would be a movement which saw it as their weakness, their evasion, their insufficiency: which saw that as often as class was a concept which drew *attention* to the unequal distribution of income and wealth, it was as often a concept which provided *cover* for the unequal distribution of income and wealth. But beginning, as it would, from the assumption that the promotion of all equalities would be worthy of its concern, its attention, and unless it could be shown otherwise its support, it would also be a movement in thought which would highlight each and every other inequality—personal, international, racial, sexual, social, intellectual, organisational, or any other that anybody could think of—as well. None would be ruled out only by prior prejudice, or because the world had not yet known them. All would be to be experimented with, practically and/or theoretically, to see what they could be made to offer. It would be a movement about taking the covers off: about speaking of, thinking about, challenging, all the inequalities it found unacceptable. When it found the shocking, it would be about being unashamedly shocked. It would be a movement involved in consciousness-raising. It would be about trying to raise to a conscious level a process—the process of making the more equal—which I would claim has been going on, if not always recognised, for as long as there have been persons. It would be about trying to introduce into what I have called philosophical conversation such ideas, concepts and arguments as might advance that process. To provide some flavour of what I would expect such a movement in thought to engage with, I want here to explore just four issues which would need, I think, to be on its agenda. And for the sake of clarity, I will consider them under their four distinct sub-heads.

1. Mounting a philosophically appropriate challenge to relativism

Postmodernism, as Lyotard had it, was 'incredulity before grand narratives': a 1979 delineation making it just nicely ironic that, over the period intervening between then and now, what might well be called 'the incredulity grand narrative' has itself become by far and away the grandest in town. Clearly I am no respecter of any such incredulity. Nonetheless, I think it is extremely important that the grounds on which postmodernism and its associated relativism are to be challenged are the appropriate grounds.

It seems to me that postmodernism and its associated relativism have been trying to make us incredulous before grand narratives at just that point at which the narrative of which we are most in need is one which, because it has matured into grand explanation, will be by far and away the grandest of the lot. It is my case that the various 'baby-grand' narratives now so widely found wanting—the Christian, the Kantian, the Hegelian, the Marxist, the Freudian, to name but a few—have all to greater or lesser extents served to prop up this, that or the other set of partial interests. And that that is something which it has been the great strength of postmodernism (and its associated relativism) to recognise, and the great weakness of some of its critics to fail to recognise. But beware. The critics concerned are only some of its critics. They are its reactionary critics. They are those who think that the solution to the problem of relativism is to go back to one or another of the old forms of *authoritarian* universalism. Its progressive critics, those who recognise that the solution to the problem is instead to go forward to a new form of *egalitarian* universalism, exhibit no such weakness. Their strength lies in their recognition of the fact that it is the 'grand-grand' narrative which will be the grand explanation *of* that partiality, *of* those interests, and of the possibility of making in their stead the *im*partial and the *dis*-interested, which is what we need. And it has been the great weakness of postmodernism that it has not recognised this. All of which means that the vital thing with postmodernism is to challenge it *on the right grounds*. The way to correct the errors now being recognised as contained in ideas (such as that of the I AM God) which make *invalid* claims to the universal is not to sink back into a sea of relativities but to construct ideas (such as that of the I IS God) which make *valid* claims to the universal. And a movement in thought such as the one I envisage would address itself to just such a making.

For it is a sinking which is involved in relativism. It is first a practical sinking: in the current passion for what is called 'identity politics', for example, when what we are really in need of is rather less reverence for each of many cultures and rather more love for the best moral world, then taking your political stance based in such only partial identities as 'woman' or 'black' or 'citizen' is such a sinking. But it is also a theoretical sinking. For a thorough-going relativist position—if that is not a contradiction in terms—requires the relativist to see her or his own relativism as relative. He or she must assent to the idea that no point of view is more correct than any other point of view. But he or she must also assent to the idea that that idea itself—the idea that no point of view is more correct than any other point of view—actually *is* more correct than its alternative. And that is just absurd. When Charles Taylor objects to any position which holds that human agents and/or societies can be studied objectively, that has to mean that he recognise that his own objection is not 'from nowhere' but 'from him'. And that taking his own position seriously might well require him to ask himself just what it is about Charles Taylor—his genes? his upbringing? his culture? his place in history?—which produces this particular stance in him. As in many philosophical places, in the consideration of relativism and pluralism there lurk many terminological problems and boundary disputes. In the philosophical conversations in which I anyway take part, and as I have discovered sometimes to my cost, it is possible seriously to anger a self-styled postmodernist by characterising her or his position as one of relativism or pluralism (or worse, of scepticism or nihilism). But nonetheless I'm prepared to risk it. If my project involved some study of the academic philosophy which impinges on these issues, this could be a somewhat foolhardy stance. The fine detail as to just how postmodernism (liberal-communitarian? neo-pragmatic?), relativism (epistemological? linguistic?), pluralism (political? philosophical?) scepticism (practical? theoretical?) and nihilism (ethical? semantic?) are each to be distinguished from and related to the others could fill volumes. And being only too aware that I have neither the expertise nor the inclination, neither the patience nor the interest, to try to tease them all apart, it's just as well that I don't believe it at all necessary to my current project that I do so. All I need to register is this: that anyway in philosophical conversation, postmodernism is understood to involve relativism, and relativism to involve pluralism.

Such relativisms and pluralisms are rampant. There are *political* relativisms and pluralisms: theoretically based in the democracy of diversity, they draw from the scrap-heap of what was East European totalitarianism lessons for political practice which can't stress enough the absolute impossibility of anything like a utopian or a communist or a socialist project. There are *literary* relativisms and pluralisms: theoretically postmodernist, and practically doing their very best to ensure that 'the canon' becomes an obsolete concept. There are *aesthetic* relativisms and pluralisms of a more general nature: theoretically situated in concepts of the provisional and the fragmentary, and practically giving succour to a giddy and style-oriented consumerism. There are *religious* relativisms and pluralisms: theoretically wedded to tolerance, practically resulting in both ecumenism and the splintering of received traditions into home-made sects. And there are *philosophical* relativisms and pluralisms: theoretically situated in such notions as the end of the Enlightenment project, the death of the view from nowhere, the realisation that all is words, while practically resulting in notions of the death of meaning and the poverty of any historically based optimism about the human condition. The above is of course no more than merely an impressionistic sketch of the mix-and-match of detail. But for my purposes anyway, that will suffice. It will suffice for me to be able to suggest that, in the face of such stuff, what we are in need of—and what a movement in ideas such as the one I am suggesting might begin to develop—is three things. One: some understanding as to why these particular 'isms' have emerged so ubiquitously just now. Two: some criterion by which we might sort out what of them is to be welcomed as the progress their adherents usually profess not to believe in, and what is instead to be challenged as maybe timidity, or even bad faith. And three: some idea as to whether we might move beyond them to any 'unum' which once was envisaged as being able to emerge 'e pluribus'. What follows is a brief look at each of these three points.

The reason why postmodernism-relativism-pluralism is rampant now, I think, is because now we stand at what is a very particular and heart-stopping moment in the struggle for the more equal: a moment, that is, when almost nobody can afford a notion of Truth. The idea of one Truth, the Truth, grounded as it has been in one (I AM) God, the (I AM) God, has been able to be afforded by its animal-human, normal, I AM persons only while their animal-human, normal, I AM person-ness has been secure. Only while, at least by and large, those of them who have been primary have accepted their own 'only natural' primariness (and thereby the 'only natural' secondariness of their Others), and equally those of them who have been secondary have accepted their own 'only natural' secondariness (and thereby the 'only natural' primariness of their Others), could such an idea hold. And what is peculiar to the present historical moment is that such a general acceptance *is no longer the case.* While at the same time: the idea of one Truth, the Truth, grounded as it can be in one (I IS) God, the (I IS) God, will be able to be afforded again, this time by human-human, moral, I IS persons, only when their moral-personness has become secure. Only when neither 'only natural' primariness nor 'only natural' secondariness but instead 'justly moral' equality has become that which is, by and large, accepted could such an idea take root. And what is equally peculiar to the present historical moment is that such a general acceptance *is not yet the case.* In other words, we are at a moment when the concepts of an I AM God and an I AM self have been sufficiently problematised for the issues represented by relativism to have been posed. It's just that they have not yet been sufficiently problematised for the issues represented by relativism to have been resolved.

Graphically, think of things this way. Some are defending the old: that is, trying to shore up a traditional, authoritarian universalism. But given the growing clamour of the new, they can sometimes no longer really afford to acknowledge the notion of Truth endemic to such a stance. And that is because Truth might (so alarmingly!) become what their challengers are saying that it is rather than what they are saying that it is: if you like, 'if *we* can no longer decide what is Truth, better than that *they* should is that there should be none'. Hence: relativism. Conversely, some are reaching for the new: that is, trying to come up with a scientific, democratic universalism. But given the continued power of the old, they can oftentimes not yet really afford to acknowledge the notion of Truth potential to such a stance. And that is because Truth has (so obviously!) been what those they are challenging have said that it was: if you like, 'if *we* can not yet decide what is Truth, better than that *they* should is that there should be none'. Hence: relativism again. In other words, the relativisms and pluralisms of postmodernism are flourishing because of a very particular historical conjuncture: a conjuncture where they can strongly appeal to both the passing (but not yet fully passed) old, and the emerging (but not yet emerged) new.

But if relativism flourishes, it is, to be sure, a less than robust flourishing: indeed, based as it is in a temporary loss of immunity, it is a decidedly sickly flourishing. The concept of an I AM God is now too weak to provide immunity from a relativising critique: the concept of an I IS God is as yet not strong enough to provide immunity from a relativising critique. So in the meantime—and when were meantimes anything other than thoroughly problematic?—relativism rules. But that is far from OK. It's not only that, in their more lucid moments, perhaps even its most ardent defenders can't help seeing that the now very well-rehearsed self-contradictory flaw at its heart is almost embarrassingly obvious. It's also that taking the would-be cure for relativism offered by its 'back-to' critics is a case of the remedy being worse—*much* worse—than the disease. The disease is here because the old is rotten. And in such a situation, it can be no part of any cure to try to revive the terminal. Because it is the only way in which we can 'come through' on the other side, cured, we have instead to push on into what is indubitably the crisis represented by this sickly flourishing of the incoherent. Many people have become relativists. Before we are through, many more will probably do so. In the 'two-steps-forward-one-step-back' struggle from what is an old set of universalist ideas based in elites to what will be a new set of universalist ideas based in us all, the lags and leads of uneven development will ensure that, at some stage of the process, many will be relativists of one stripe or another. For to all but the most theoretically die-hard of those yearning backward to old and 'only natural' forms of inter-human relationship, as well as to all but the most theoretically sophisticated of those reaching forward to new and 'justly moral' forms of inter-human relationship, it is a stance which can offer some albeit only temporary refuge in what is an intellectual storm. For the former, it offers such refuge while, as still they protect and defend their newly beleaguered position, they must at last concede (defensively) that the 'truths' of their Others can make some legitimate claim upon them. For the latter, it offers such refuge while, in order to consolidate and extend their newly emerging position, they must at first require (aggressively) that the 'Truth' of the old take a new place as only one truth among many. In such a situation, it could appear to be in the interests of almost nobody to defend the idea of there being a singular thing called Truth.

But not quite nobody. For egalitarian universalists, of whom of course I am one, there are three distinct challenges here: to distance oneself clearly and fully from the anti-egalitarian universalists; among the ideas thrown up by the relativist 'middle ground', to support those which are progressive and oppose those which are reactionary; and to try to develop the theories of an egalitarian universalism. The movement in ideas which I am envisioning would need to grapple with all of these things, but maybe particularly with this last. Home not for the theoretically reactionary critics of relativism but for the theoretically progressive critics of relativism, it would have as one of its major tasks the elaboration and defence of a post-relativist, universalist concept of Truth. Reactionary anti-egalitarian universalists oppose relativism because its progressive elements challenge their old and authoritarian universalism. Progressive egalitarian universalists oppose relativism because its reactionary elements challenge their new and democratic universalism. And it could not be more important clearly to distinguish the two.

For a world which, in its postmodern relativist stance, is post-Christian, post-Marxist, post-industrial, post-feminist, post-structuralist and damn near post-humous, is a dangerous world: one where we seem to know that we have left the past but not to know how to enter the future. And that is a situation which carries a serious health warning. The notion that plurality, style, difference—or even 'differance'—are good in themselves is just vacuous nonsense: a nonsense to which Terry Eagleton's telling quips concerning the legitimacy of 'a stylish rape' or 'a ceaselessly varied, robustly proliferating fascist scene' draw vivid attention. Of course the old view-from-nowhere skyhook won't do: what seemed in the past to be singular and universal Truth was all too often the 'only natural' authority of the primary. But a new view-from-nowhere skyhook can be made. The new view-from-nowhere which is available to us, anyway potentially, is the view from everywhere: a view of how things are which can command universal acceptance *not* because anybody's democratic right to dissent is removed, but because we all stand before it equally informed and disinterested; *not* because authority demands it but because understanding requires it. To wrest such a new and more securely based realism from out the morass which is current relativism, to make a *new* immunity against such relativism, we need an idea appropriate to our *new* situation. And it is my claim that such an idea is the idea of equality. We can only make rational consensus out of genuine democracy. And we can only make genuine democracy out of real equality. When we can each view the world from out the created equality of the moral, rather than the begotten hierarchical of the natural; when all interest has been removed from our inter-human intercourse: then can the circle whose centre is everywhere, because it is in each and every human person, be nowhere circumscribed. And then can Truth become the fully singular thing that only universal disinterest will allow.

Relativism has been right to draw attention to the interest of the interested. But it has been wrong to overlook the interest of the *dis*interested. It has been right to draw attention to that which makes the truth of selves selfish as animal-humans or natural persons relative to the truth of other selves also selfish as animal-humans or natural persons. But it has been wrong to overlook that which makes the Truth of selves selfish as human-humans or moral persons absolute with the Truth of other selves also selfish as human-humans or moral persons. The relativist suggests that there is not Truth, but just truth from a particular perspective. I am suggesting that truth from a particular perspective (a space metaphor) is actually truth from a particular stage of development (a time metaphor). And that while from stages of development that are within the I AM mode the relativists are right and there are indeed only truths, from the stage of development which has reached the I IS mode, it is Truth both singular and universal which can emerge. Members of all interest groups but one interest group can be only pretenders to Truth. For Truth is that which is available only to the interest group of the disinterested. With once again the caveat that in the I IS hypothesis, the disinterested and the interested are to be characterised not as the 'goody persons' and the 'baddy persons' but as the 'morally more highly developed acts' and the 'morally less highly developed acts' *of all persons*—all persons, anyway for now, being various differently proportioned combinations of the two, and in transition from the one to the other—it is only among the interest group of the disinterested that Truth can be made real. It is to the development of understandings such as these that the movement in ideas I am suggesting would address itself. And as my own contribution to the development of such understanding, the formulation of the issues which my candidate theory would involve would look something like this.

The realist says that there is 'a way that things and happenings are'; that that 'way' is one independent of any knowing that that is how they are; and that such knowing, when it is in place, is the knowing of that which is just true. The relativist says that even if there is 'a way that things and happenings are', when any knowing that that is how they are is in place, it is always and necessarily the knowing of a situated knower, one always and necessarily situated differently from all other knowers, and thus always and necessarily the knowing of that which is only relatively true. It is my contention that the problem with both positions, the problem which means that they both come up with that which does not capture what is actually the case, is that both are working from within the I AM paradigm. It is that which causes the realists to fail to acknowledge the relevance of the situatedness of knowers. It is that which causes the relativists to fail to see that such situatedness, being contingent rather than necessary, is only contingently rather than necessarily inimical to the making of truth. And it is my argument that, in order to come up with that which does capture what is actually the case, we have to function instead from within the I IS paradigm.

From within that paradigm, there is indeed 'a way that both things and happenings are'. But among the things and happenings that *are* one way and are *not* another way is the thing or happening called 'I'. Instantiated in the nervous system of a *Homo sapiens* as a linguistically structured state of that system, a state which *is* this way and is *not* another way, is the knowing I. And it is because this knowing I is not an extra-reality thing or happening, is not a thing or happening *outside* 'the way things and happenings are' but is itself *one of the things and happenings that are*, that both realism and relativism as characterised above are inadequate. Let me try to spell out what I take to be their problem.

Reality—all that there is—is (definitionally) a closed system: in my terms, the system universe/biosphere/world is closed in the sense that there is nothing outside it to be over-against it; nothing outside it to know it from nowhere. In that part of it which first emerged, in the inorganic universe, there is no awareness; in that part of it which second emerged, the organic biosphere, there is sentient awareness; in that part of it which most recently emerged, the personic world, there is conscious awareness. It has no 'outside'. Because every thing in the system is structurally and causally related to every other thing in the system, it is not possible for one part of it to change in such a way as to leave other parts just the same. And one of the parts of the system, a world-part, is the thing or happening called 'I': the knowing thing or happening called 'I'. So how can such a thing or happening, such a knowing thing or happening, function in such a system?

Clearly not as a thing or happening which freely discovers: that is, clearly not as a thing or happening which, different from all other of the things and happenings which are, discovers, autonomously and uninfluenced by them, the way that those other things and happenings are, and thereby discovers *the* (absolute) truth. But clearly not either as a thing or happening which freely invents: that is, clearly not either as a thing or happening which, different from all other of the things and happenings which are, invents, autonomously and uninfluenced by them, a way that those other things are, and thereby invents *a* (relative) truth. For just as surely as the conservation of mass/energy rules out the former state of affairs, so also the conservation of mass/energy rules out the latter state of affairs also. How then can what actually *is* the case be formulated without that (rather serious!) contradiction?

Only from within the I IS paradigm: that is, only from within a paradigm within which the knowing 'I', also and necessarily, is one of the things or happenings that are; is one of the parts of the closed system; is one of the things or happenings which itself *is* one way and is *not* another way, which 'way' necessarily has characteristics which position it *within* a system which enables and constrains it rather than *over-against* a system which does neither. So: *within* rather than *over-against* the universe-biosphere-world system, the question we have to answer becomes a different question. It is not that we have to ask the question 'what AM I as a knowing thing or happening?': a question which posits the 'I', illegitimately, as a thing outside the system and uniquely independent of it. It is rather that we have to ask the question 'what IS I as a knowing thing or happening?': a question which posits the 'I', legitimately, as a thing or happening inside the system, and, along with all the other things and happenings which constitute the system, dependent on it. And the candidate answer which I am proposing to this the second and legitimate question is that the I is not a thing but a happening: and a happening consisting in a sequence of states of the nervous system of a *Homo sapiens* structured linguistically.

As such a sequential series of states, it is a sequential series of functions of all previous states plus current input. It is thus a happening which *itself* is necessarily one way rather than another way. Within a system by which he was both enabled and constrained, the sequential series of states of a linguistically structured nervous system called 'Socrates' was just such a happening. It's not that he *did not* formulate the second law of thermodynamics (because he just did not think of it, perhaps), but that he *could not* formulate of the second law of thermodynamics (because it could not have been available to the 'I' that such a sequence 'was').

The forgoing can be summarised like this. The I AM realist wrongly construes the individual knowing subject as *discovering absolute truth*: and doing so somehow free of the relative constraints supplied by the nature of its own knowing. The I AM relativist wrongly construes the individual knowing subject as *inventing relative truth*: and doing so somehow free of the realist constraints supplied by the nature of what is to be known. By contrast with both, the I IS knower, as here instantiated neither realist nor relativist, rightly construes communities of knowers as *making ever more valid truth*: and doing so as, via the making of equality, the constraints supplied by the relativities of individual knowings edge developmentally into alignment with the constraints supplied by the nature of what is to be known.

2. Developing a variety of ways of challenging arguments against equality

In philosophical conversation, arguments against equality are many. Some come from those who are unequivocally opposed to it, some from those who, while they can't quite get their heads around outright opposition, nonetheless feel uncomfortable with some of the implications of an equality agenda. Among the former—the hard nosed, perhaps—it is two arguments which usually predominate: first, that equality destroys variety and is therefore bad (we'd all be similar); second, that equality involves identity and is therefore worse (we'd all be the same). And among the latter—the soft centred maybe—it is two others that are more often used: first, that some increase in equality is perhaps alright, but only if it is achieved by 'levelling up' rather than 'levelling down'; second, that desirable as it might be, complete equality is anyway impossible because it runs counter to something called 'human nature'. These are the sorts of arguments, each of which I will now briefly consider, which a movement in ideas such as the one I'm championing would need both to elaborate (for the sake of clarity) and counter (for the sake of truth).

Beginning with the case made by the hard-nosed, their first argument is that equality destroys variety. This one, or so it seems to me, offends against one of the fundamental principles of ethics: the principle that persons are to be treated as ends in themselves rather than as means to the ends of others. If another person becoming more like me is in some way better *for that other*, it is just immoral for me to regret such becoming as a diminution of variety. When some are illiterate and some can read, there is more variety in the world than when all can read. When some starve in garrets and some feast in palaces, there is more variety in the world than when all live in modest but adequate houses. And so on and so on. For my money, down with such 'variety'! Other persons cannot be made the means to my search for it. I will not search for it.

But there is another 'variety' point to consider as well: one which works completely counter to the usual case. It is that far from fostering variety, a world of many different small-scale societies actually fosters sameness. And that what would foster variety would be a world of one very large-scale human community. For to belong to, and identify with, any grouping short of the grouping of all, of the whole human community, is to belong to that which requires that I should be (or try to make myself) the same as the others in my pack. How can 'we' exclude 'them'—and such groupings are at least as much about excluding them as they are about including us—if we don't accentuate our samenesses over against theirs. It is, after all, our samenesses which make us the insiders: just as it is their samenesses which make them the outsiders. Only when there is neither an inside nor an outside, when there is just one human community, can group samenesses give way fully to personic difference. And by far the most important thing necessary to the creation of one human community is the creation of equality.

This of course is a stance which raises the whole vexed question of the creation of what has come to be called 'the global village'. Because villages are traditionally intensely hierarchical places which, belying their image of the cosy, are just as likely to be backward, punitive and sometimes vicious places, I am not entirely comfortable with the 'village' word here. I prefer the term 'globalisation': indeed, I like the term 'globalisation'. And since this is a liking which I know will raise the eyebrows of many a potential co-egalitarian, I must defend it.

Though not (as some have been led wryly to comment) in quite the way envisaged by Marx with his notion of 'the withering away of the state', everywhere there is talk of the break-up of nation states. International corporate economic power is escaping their control, making the governments of all but the most dominant of them, so it is said, increasingly just irrelevant because increasingly just powerless. This is a process marked by the globalisation of both capital and labour; by massive privatisation; and by deepening inequalities. The world is coming to be characterised by the existence of large-scale multi-national companies rather than small-scale uni-national states. And there is a deal of hand-wringing over that. Anti-globalisation protests proliferate. And given that organisations and movements which fundamentally challenge the economic and political status quo are, at the moment, much more conspicuous by their absence than their presence, there is a very marked and very understandable tendency on what used to be called 'the left' to identify with such protest. Indeed, it sometimes seems to be the case that such identification takes place on no better grounds than that it is better to be doing something rather than nothing. But that is a stance which, if one is not careful, can result in identification with whichever dissidence is currently most visible and/or most noisy. I think that identification with anti-globalisation protests is frequently of this sort. And that as such, it is a stance which, for three main reasons, is at least highly questionable, at most thoroughly counterproductive.

Firstly, it frequently seems to involve countering global capitalism with small-scale and 'local' capitalism. And that means fostering precisely that form of the beast which historically has been its most reactionary, most undemocratic, and *in extremis* most vicious variety. The corner shop is the corner shop and not Marks and Spencers because it has not ('yet', hope its owners) made it. The owner of the small family run business is such an owner and not John Sainsbury for the same reason. It is of the very logic of economies based on the private ownership of the means of production for profit that you just *must* boom (or bust), that you just *must* grow (or go). To defend small-scale (read 'as yet relatively unsuccessful') capitalist enterprise, particularly your own, against large-scale (read 'relatively successful') capitalist enterprise, particularly someone else's, can look more than somewhat like a particularly blatant piece of special pleading. By contrast with which, principled anti-capitalism means anti-capitalism: big, small, or middle-sized.

Secondly, it just as frequently seems to involve being insufficiently open to what might well turn out to be new political realities. When thinking about making non-capitalist worlds, the traditional Marxist approach saw the gradual workers-revolution-by-workers-revolution weakening *within* each nation-state *of* monopoly capitalism. What we appear to be getting instead is the gradual capitalist-merger-by-capitalist-merger weakening *of* each nation-state *by* monopoly capitalism. The culmination of the one process was dreamed to be a global revolution of all workers: the culmination of the other seems set to be a global merger of all capitalists! But as far as our dreams of equality are concerned, it seems to me that it was as wrong then to see the first as their beginning as it would be wrong now to see the second as their end. To the extent to which such revolutionary change was ever envisaged as being centrally about equality—which was never to any great extent—the task then looked like being the transformation of many local systems of inequality into many local systems of equality. The task now could look like being the transformation of one global system of inequality into one global system of equality. It's not obvious that the second is in any sense the harder of the two. Let me suggest why not with a story.

In October 1993, the Moscow police seized three political activists and took them to the local jail. In an attempt to get them to confess to the murder of two policemen, some of the victims' fellow officers gave then a severe beating. But when next day their families found out what had happened, they sought out a friend with access to email. He at once posted details of what had happened on several international bulletin boards, and rounded off his account by giving the phone number of the police station. The first protest call came within the hour from Japan. It was followed by a steady stream from all over the globe. And according to the Institute for Global Communications, the three were released within three days, and all charges against them dropped. Now I relay this story using my computer, and with the aid of Microsoft software. As I do so, I know that of the 500 million or so personal computers in use in the world right now, a huge majority of them are also using Microsoft software. And while that part of me (which is all of me) which hates not capitalists but capitalism finds the wealth of Bill Gates an obscenity, that part of me (which is all of me) which loves equality finds such a story a delight. Why should it not be at least as likely that capitalism will be ended by the 'one-off' public acquisition of that which has already become large-scale and global as that it will be ended by the piecemeal public acquisition of that which is still small-scale and local? Indeed, why should it not be not only at least as likely, but also perhaps more desirable? It was, after all, a version of the piecemeal —'socialism in one country', as it was called—that had such catastrophic consequences in the tragedy which was (and still is) the former Soviet Union and the former Eastern Europe. This is *not*, repeat *not*, a recipe: still less a forecast. Who from now can possibly know the detail of just how the 'only natural' economic system which is capitalism will give way before the advance of the 'justly moral' economic system which will be based on equality? Certainly not me.

Thirdly, it involves a cop-out on loving. The only grounds of principle on which to counter global capitalism are the grounds of a global alternative: with, for example, a movement *for* global governance rather than *against* global trade. And this is not just because protest movements excoriating that which we are against need to give way to project movements mounting the case for what we are for, important though that undoubtedly is. It is more fundamentally because the human rights and human freedoms which give us human justice just cannot exist as the possession of one group, one nation, one community, one civilisation. If they are not global rights and global freedoms giving us global justice, they are not the rights and freedoms of justice at all. The large-scale all-inclusiveness of global enterprise is morally beautiful. It is the (only current) form of its ownership, the capitalist form of its ownership, which needs addressing.

The second argument mounted by the hard-nosed opponents of equality is that it requires not merely the diminution of variety but actually the creation of identity. And it is my claim that just as surely as the previous argument offends against one of the fundamental principles of ethics, so this one offends against what my hypothesis anyway would take to be one of the fundamental principles of ontology. One way of making this point would run something like this: that while we can say 'the even prime is equal to the square root of four' precisely because the two concepts are *not* identical, we cannot say 'the even prime is equal to the even prime' or 'the square root of four is equal to the square root of four' precisely because the two concepts *are* identical. In other words, far from equality *entailing* identity it actually *precludes* identity. A different approach, but one with the same result, is that of Wittgenstein. With his argument that 'to say of two things that they are identical is nonsense, and to say of one thing that it is identical with itself is to say nothing at all', what he does—and usefully—is to problematise the very concept of identity. Clever stuff. But because it is one which I take to be both more compelling and more directly related to relations between persons, I am rather more taken by an argument which arises directly from the I IS hypothesis.

In its terms, while no two material things can be in the same place (whether here or there), an infinite number of happenings can occur at the same time (whether now or then). And in its terms, individual members of the species *Homo sapiens* are material things in space: they are here or there. But persons are happenings in time: they are now or then. For two members of the species *Homo sapiens* to be identical, they would need to stand in the same place—which is impossible. But for two persons to be equal, they would need to understand from the same moment of comprehension (stage of development)—which most certainly is not. Such an approach is making the claim that it is just a category mistake to think that equality between persons is of the same order as identity between members of the species *Homo sapiens*. And a category mistake, what's more, which is sourced—as so often they are—in the reduction of the one to the other.

When arguing the case against equality, both of these hard-nosed arguments—and others like them, for they do not always present so discretely distinct as I have presented them here—are very widely used. Indeed, frequently when arguing a case to the contrary of either of them, I over and over again find myself trying to sort out the same confusion. While I am trying to consider the question as to whether (a) persons (b) should be or will be (c) equal, my opponent seems to think I am trying to consider the question as to whether (a) organisms (b) are (c) identical. If we are to save ourselves from fruitless circularities, all three of the differences here identified need to be understood as differences. In other words, it needs to be understood (a) that a person is not an organism; (b) that normative statements are not positive statements, and that the future is not the present; and (c) that equality is not identity. And a movement in thought such as I am suggesting we need would have to have high on its agenda the tackling of such confusions.

Turning now to the doubts about equality coming from persons more equivocal, I suggested that they also were most commonly two: the first concerning the notorious 'levelling down or levelling up' debate; the second concerning the impossibility of achieving that which runs totally counter to human nature.

On the first: it is a commonplace among liberal philosophers and/or those of what might be called the centre left that while 'levelling up' (adopting policies which give more resources to the have-nots) might sometimes be appropriate, 'levelling down' (adopting policies which give fewer resources to the haves) is almost never so. To a disinterested observer, two things just might occur. It just might occur that this could be a position not totally unconnected with the fact that many of such folk might be expected to be among those who would need to be 'levelled down': that is, that this is I AM interest speaking. And it just might occur that Person A having more-than-before in a situation where Person B has the-same-as-before actually *is* a situation where Person B's position relative to that of Person A has effectively been 'levelled down': that is, that this is incoherence speaking. In the heat of the argument, those defending such a position have even been known to hurl the dreaded 'politics of envy' accusation at those they accuse of being down-levellers. This, to say the least, is odd. In a piece headed 'Do liberty and equality conflict?'—in *Living as Equals* (1996)—Ronald Dworkin suggests, I would have thought uncontroversially, that someone 'envies the resources of another when he would prefer those resources, and the pattern of work and consumption that produces them, to his own resources and choices.' But the egalitarian, of course, has no such preference. What he or she is after is precisely an *equal* distribution. But although it is valid as far as it goes, the above is really just an immediate response to what is an irritating daftness. In terms of the I IS hypothesis, in the context of the new paradigm I am suggesting that we need, there is something much more fundamental that needs to be registered about levelling up and levelling down. For it is my claim that such concepts can only work at all, however incoherently, for I AM persons enmeshed in 'the having ideology'. For I IS persons, levelling up and levelling down take on completely new meanings. For them, since being 'up' and being 'down' are both dis-beneficial vis-à-vis being equal, then just as surely as one is levelled 'up' from having less to having more, so one is levelled 'up' from having more to having less. A win-win situation indeed!

It is important here yet again to register my complete opposition to moralism. Yes, the loving mode, the one which recognises the invalidity of the 'levelling up versus levelling down' argument *is* being presented as the more fully human mode, the higher mode. But no, those who do not so recognise, those who are still working within the having ideology, are not being blamed for moral inadequacy. I take moralism to be that process whereby an a-historical and apparently moral judgement stands in for a thoroughly historical and actually moral explanation. Such moralism is a very present phenomenon. In the words of Chantal Mouffe, the current lack of any 'project of real political transformation' leaves the door wide open to 'moralising liberalism'. But working within the I IS paradigm, it is never appropriate to moralise or blame, only appropriate to try to make conditions of experience such that things will be able to be different. The having ideology will not fade and eventually die because people decide to give up sin. It will fade and eventually die as the result of an inevitable development, albeit uneven, across the selving process. And later I will have more to say about that inevitability.

On the second: I need briefly to revisit an old debate which went on amongst Marxist economists. Where traditional economics sees scarcity as a trans-historical feature of the human condition, Marx thought it could be overcome by the creation of 'abundance'. The debate was about just what was to be understood by abundance. And in such discussion, I was always on the side which insisted that it was something which, as well as being manipulable from the supply side (increasing production) was also manipulable from the demand side (creating different sorts of demanders). This, I now see, was hardly bold enough. I have considerable sympathy with the viewpoint—expressed here in the words of Peter Dews (*Radical Philosophy*, issue 111)—that 'the mere standard of living of the Western democracies may be a violation of the categorical imperative'. And in such a world, it becomes increasingly obvious that while increases in provision are manipulable from the supply side, abundance is actually *only* manipulable from the demand side: it's creation is *all* about creating different sorts of demanders. But that's no problem. Far from being impossible because the nature of humans is for ever fixed, such creation is actually inevitable because the morality of persons is for ever developing. But once again, the anti-moralism clause. The having ideology will not fade and eventually die because people decide to give up sin. It will fade and eventually die as the result of an inevitable development, albeit uneven, across the selving process. And again: later in this chapter I will have more to say about that inevitability.

3. Pointing up the error of confusing poverty with inequality

This confusion is just ubiquitous. Try to engage someone in a discussion about inequality, and there is an almost knee-jerk reaction that what you want to engage in is a discussion about poverty. I don't think it would be an exaggeration to say that almost every book, every journal article, every newspaper column, every television programme on the problem of inequality that I have ever read or seen finishes up—and often starts!—as a book, article, column, programme on the problem of poverty. This is the fish-in-water syndrome again. Just as fish are those least likely to be aware of water as an issue (or at all), so 'having-more-thans'—from amongst whom, for the most part, come those who write the books and make the programmes—are those least likely to be aware of 'having-more-than' as an issue (or at all).

The analogy with feminism is telling here. The having-less-than phenomenon 'woman' was identified as 'problem' or 'question' long before the having-more-than phenomenon 'man' was recognised as equally problematic. Similarly, psychological and sociological studies of femininity-as-problem preceded by many years psychological and sociological studies of masculinity-as-ditto. And even when the latter did start to emerge, two less than helpful confusions emerged with them. First came a truly awful piece of self-deception which could go under the levelling-up-levelling-down head again: for example, women (the problem) needed to be levelled 'up' in the world of paid work, but that did not that mean that men (the non-problem) needed to be levelled 'down' in the world of child-care or domestic chores. And second came the idea—expressed here in the language of Harry Christian from his 1994 book *The Making of Anti-sexist Men*—that 'men's anti-sexism involves a contradiction' because 'it's a political movement for giving up power—or at least 50% of it'. The first is directly analogous to the version of such nonsense already discussed on inequality more generally, where the argument was that the poor (the problem) need to be levelled 'up' to have more, but that does not mean that the rich (the non-problem) have to be levelled 'down' to have less: an argument the utter incoherence of which does not seem to prevent some of the most unlikely of people from embracing it—just like the fish. And the second just ignores what the I IS hypothesis anyway would claim are crucial distinctions. To the I AM animal-human in us, the giving up of powerlessness (if we are secondary) is no less a problem than the giving up of power (if we are primary): for both are bewildering. And to the I IS human-human in us, the giving up of power (if we are primary) is no more a problem than the giving up of powerlessness (if we are secondary): for both are liberating.

Starting from some understanding of poverty, and trying to shift to an understanding of inequality, one of the difficulties has been historical. For earlier generations, the main problem was even to see poverty as it really was: that is, not as a fact of life, but a problem to be solved. In a letter to Boswell, Samuel Johnson wittily suggested that 'the great deal of arguing that we hear to represent poverty as no evil' only serves to show that it is 'evidently a great one': because 'you never knew people labouring to convince you that you might live very happily upon a plentiful fortune.' What I am suggesting is that for us, now, the difficulty is a different one. We are used to accepting that Beveridge's 'five giants on the road to reconstruction'—disease, idleness, ignorance, squalor and want—are problems alright. But while most of us think of them as fundamentally problems of poverty, I want to suggest that we need to learn to see them as problems of inequality; if you like, to look at Johnson's comment rather less as clever quip, rather more as crucial insight—however unintended. It is very hard for I AM consciousnesses—those of normal persons living in the having mode—to conceive of the possibility that 'more' might not be synonymous with 'better'. And it is that difficulty which means that the real problem—not the problem of poverty, not even the problem of wealth, but actually the problem of inequality—is all too rarely understood. In terms of Dr. Johnson's remark, the reason why 'you never knew people labouring to convince you that you might live very happily upon a plentiful fortune' is because it is just taken for granted, almost universally, that you can. What I am saying is that whether or not 'you can' depends on who 'you' is. If 'you' is an as yet only normal person, it's true: if 'you' has become an already moral person, it's not.

Over and over again wires have been crossed and then re-crossed on these issues. Equality has been thought about, written about, dreamed about, in the context of a consideration of poverty: hailed by those who have wanted passionately to end the evils of poverty; denounced by those who have seen it as necessary to accept that, human nature being what it is, the poor will always and necessarily be with us. What I am saying is that the movement in thought I am suggesting we need is one which would aim to break that limiting association. We need to stop thinking of equality and poverty in the same breath, in the same context, as part of a discourse from which riches are excluded. For no less significantly than it addresses itself to the 'problem of poverty', any campaign for greater equality must address itself also to the reciprocal 'problem of riches'. Equality is not about ending poverty. Equality is about ending the poor/rich differential. And ending it for the benefit of everybody: for the benefit of those currently with more just as significantly as for the benefit of those currently with less. Christianity's 'bias to the poor' is a bias to us all. Equality is also 'unto this first'.

And just occasionally studies are done which throw the real problem, the inequality problem, into some sort of relief. Such was the case, for example, with the work of Richard Wilkinson. As presented in his book *Unhealthy Societies* (1996), he was able to show that 'in the developed world, it is not the richest countries which have the best health, but the most egalitarian'; that 'regardless of the fact that health differences within societies remain so closely related to socio-economic status, once a country has passed the threshold level of income associated with the epidemiological transition, its whole population can be more than twice as rich as another without being any healthier'; and that 'the scale of income differences in a society is one of the most powerful determinants of health standards in different countries, and that it influences health through its impact on social cohesion'. In defence of this last point, a letter he wrote to the *New York Review of Books* (8.10.98) pointed out that 'there is a close relation between greater income inequality and low levels of trust, higher hostility, and—as some 34 studies attest—higher rates of homicide and violent crime'. As such a rare attempt to tackle what I anyway take to be the real problem—the problem not of poverty (or even of wealth) but of inequality—this is work which, despite what I suspect are some conceptual problems with it, can only be hugely welcomed. As a particular example illustrative of the fact that it is inequality rather than poverty (or wealth) which is the real problem, it is invaluable. And a movement in thought such as the one I am envisioning would in part consist in encouragement of and support for such research: that is, research into the problems presented by inequality *in its own right*.

4. Developing an understanding of the necessity of determinism

In much of his extensive writing on equality, the philosopher Richard Norman regularly insists that equality be defended as that distributive principle which best expresses a commitment to cooperation. His argument is that when people cooperate on less than equal terms, the cooperation they are involved in is necessarily less than complete. The supposedly cooperative nature of their enterprise is necessarily in conflict with the actually inegalitarian nature of their relationships. And the conclusion he draws is that therefore equality *ought* to obtain. Of course I go with the spirit of what he is saying. But my own position is nonetheless not insignificantly different. I am suggesting not that such conflict is the ground for saying that equality *ought to* obtain (a normative statement), but that such conflict is the ground for saying that equality *will* obtain (a positive statement). I am claiming that because the inegalitarian nature of their relationships belie the supposedly cooperative nature of their enterprise, there is thereby intrinsic to the situation a conflict which means that those relationships are intrinsically unstable, and thus inherently the pressure points for change. Carefully, for this is nothing if not difficult territory, let me borrow and adapt language from two other sources, from biology and from mathematics, to try to convey what I mean.

As I have already mentioned—in the section on 'justice-*and*-benevolence' in Chapter 7—the biologist John Maynard Smith developed the concept of 'an evolutionarily stable strategy' (ESS) as one which, because it maximises life chances—life being of the essence for things organic— could not be invaded by an alternative. Such strategies can be thought of in terms of what among mathematicians are called 'attractors', the settling of a system into one of which is called settling into 'a basin of attraction'. By analogy, I am trying to develop the concept of 'a developmentally stable strategy' as one which, because it maximises love chances—love being of the essence for happenings personic—could not be invaded by an alternative. And suggesting that for personic relationships, it is because it is the equality strategy that has that 'attractor' status that they are inevitably pulled into it as 'a basin of attraction'. My claim is that equality relationships have such status because of the defining nature of what I am calling the personic. The consciousness which is language use brings into existence not only self but also, and concomitantly, Other. At first, each self, newly emergent from its animal ancestry, is only a first-person I AM self: one which, although it comes to recognise the selves of others, indeed makes its own self precisely out of differentiation from the selves of others, nonetheless sees its own self—the self it has, or 'owns'—uniquely as first-person self: with some (few) others as second person selves; and some (most) others as third person selves. But as the steadily accumulating *and never lost* linguistically structured experience which constitutes the persons of the world creates the steadily accumulating *and never lost* consciousness of the less-than-desirable consequences of such thinking, the pull of the 'basin of attraction' represented by equality becomes ever stronger. Put another way, it's not just that, with experience, knowledge of the nature of the self and the nature of the other grows: though it does. It's not just that, with experience, knowledge of the nature of inter-self relationship accumulates: though it does. And it's not even just that, with experience, knowledge of the instability of primary-secondary relationships and of the alternative stability of equal relationships builds: though that happens too. It's two other things. First, it's that experience being not that which a person *has* but that which a person *is*, it's not something which can be in any sense lost. All the experience of all the years of all the persons who have ever existed is present experience. It is the experience which constitutes present persons. And second, it's that experience being thus constitutive of persons and accumulative, present persons will be both continuous with those others but also different from those others. And not only different from them: also more experienced than them. Or as T. S. Eliot puts a similar point: 'We know more than they did. And they are what we know'.

What I am saying is that experience is continuous, irreversible, a one way street. It's not possible for a person to 'un-experience' the experience of which it is constituted. It is not possible for a person to 'unknow' the knowing of which it is made. What I am also saying is that just as, at any one time, there is *a* way (not *several* ways) that the universe is, and *a* way (not *several* ways) that the biosphere is, so at any one time there is *a* way (not *several* ways) that the world is. And that as far as the world is concerned, at each of its succeeding 'any one times', the way that it is is a way constituted by less of the experience (fewer of the persons) *less* aware that if people co-operate on other than equal terms their relations with one another are less than fully co-operative and thus less than fully loving, and more of the experience (more of the persons) *more* aware that if people co-operate on other than equal terms, their relations with one another are less than fully co-operative and thus less than fully loving. In short: that the world consists in a steadily growing recognition of the necessity of equality to full co-operation. That which pulls towards such a 'basin of attraction' is simply experience.

And it's a pull which becomes that much stronger when the knowledge provided by such experience becomes scientifically systematised. 'The object of all science, whether natural science or psychology, is to co-ordinate our experiences and to bring them into a logical system'. This is

Einstein—again. And what I am claiming is that when the linguistically structured experience which we are is experience of contradiction—in this case, of contradiction between inequalities and purported co-operation—then the co-ordination and bringing-into-a-logical-system which he is suggesting is 'the object of all science' is thwarted, and thus the pressure to 'unthwart' inevitably with us. Indeed, it would seem to me that we could very usefully describe the whole of the history of the 40,000 or so years since language use ushered in the personic as a process of shift from many 'us-and-them' societies of persons co-operating normally (in various primary/secondary relations) towards a single one-and-all' community of persons co-operating morally (in a single equality relation).

And a spectacularly rapid shift at that. My answer to the (much asked!) question 'so if the lower level of consciousness represented by the linguistically structured experience of I AM selves in primary/secondary societies will *inevitably* generate the pressure to shift to the higher level of consciousness represented by the linguistically structured experience of I IS selves in equal communities, why on earth is it taking so long?' is that it isn't. True, communities of equality are currently conspicuous primarily by their absence. But get that into appropriate perspective. Consider this.

If we think about the level of the organic, first the emergence of the double-helical DNA form of evolution-by-natural-selection from among what have been estimated to have been its probably thousands of precursors took probably a good few million years. At the beginnings of life, the single-celled heterotrophs, those unable to make a food supply and thus cannibalising each other via a fermentation process, would 'very quickly' (in palaeontology-speak, within about 500 million years) have run into a food crisis if the evolutionary pressure to make your own supply had not resulted, after about 300 million years or so, in the various autotrophs who could do just that. The autotrophs of all autotrophs were the photoautotrophs. They were the ones which evolved the capacity (called photosynthesis) to use sunlight to synthesise their nutrition from carbon dioxide and water, and then spent another 1000 million years or so making a living that way. In the process, they created enough free oxygen for other forms of life, eventually those called animals, to emerge some 800 million years or so ago. It was out from amongst these last that, in time, the very recent arrival *Homo sapiens* joined the crowd, some mere half million or so years ago. Getting hold of even just a rough and ready outline such as this can perhaps help us to realise over just how lengthy a time the process which was the emergence of human living took place. Another way to do it would be to chart the millions-of-years-long shift from prokaryotic to eukaryotic cells, for example, or get some hold on the time taken to get from the 570,000 million year old 'shelly fossils' to the (by comparison) tiny number of vertebrate species.

By almost mind-boggling comparison, the speed with which, across no more than perhaps 50,000 years or so, the making by moral creation of the I IS form of the personic out of its single I AM precursor has taken place almost beggars belief. It is only the quite spectacular egocentrism of I AM thought which by so crazily over-estimating the significance of its own few score years—there are those who will confidently announce that something or other will never happen, and when challenged say, without a blush, 'well anyway not in my lifetime'—fails to be able to see the quite breathtaking speed with which the emergence of the justly moral from the merely natural is taking place. Where the evolution of human living took hundreds of millions of years, the development of human loving is probably going to take no more than a few thousand. Indeed, in the mutuality of 'philia' (a relationship becoming ever more common across the last 3000 years or so); in the democracy of science; in the steady individuation of the social; in the beginning of the end of deference: in such things and many, many more, goodly chunks of what will become the I IS personic are already around.

My claim is that it was as inevitable that the personic, once existent, would develop as it was inevitable that the organic, once existent, would evolve: that is, that while it is *wrong* to think of a person as being determined evolutionarily via genetic structuring (that is what is true of member of the species *Homo sapiens*), it is nonetheless *right* to think of a person as being determined developmentally via linguistic structuring. That both should be determined—though so spectacularly differently—is hardly surprising. After all, the hypothesis that the personic is a *state* of a *part* of the organic (of that part of the organic which is a human nervous system) is one which mirrors pretty closely the theory that the organic is a *state* of a *part* of the inorganic (of that part which is 'a collection of carbon-based matter which is stable, open, dissipative, yet capable of maintaining a stable configuration far from thermodynamic equilibrium'). That the second is a theory and the first only an hypothesis reflects the state of our current knowledge in the two areas. When Crick and Watson worked out the structure of deoxyribonucleic acid, they discovered the mechanism whereby the information necessary to the making of an organic entity (a body) is carried. And it is a prediction of the I IS hypothesis that it will be when we work out the structure of some as yet unrecognised characteristic of language that we will also discover the mechanism whereby the information necessary to the making of a personic entity (a mind) is carried. Until then, all we can do is speculate.

We can speculate, for example, on what light, if any, the work of Steve Freeland and Lawrence Hurst ('The genetic code is one in a million', *Journal of Molecular Evolution*, 1998) can throw on the issues. They have been studying the emergence of the DNA-based genetic code, the one which became standard over 3,000 million years ago and is now used by almost all living organisms on the planet, from among the 10^{20} estimated as possible. Although in a very small number of places—in the evolution of mitochondria, for example—unconventional codes do exist, the strong selective pressure for a universal code, they think, would have been considerable. This would have been because most organisms exchange DNA, as for example when a virus infecting two different species transfers some from one to the other. And our speculation might be to wonder whether there are any parallels here with the way in which there is pressure for one language to lift us out of the Babel represented by the fact of there having been thousands of them.

We can speculate that while, at the level of the organic, the organism is the genes' way of replicating the information they carry genetically, it might be that at the level of the personic the person is the (?'s) way of replicating the information it carries linguistically. We can further speculate—about the nature of that (?)—that just as half of the genes strung along the chromosomes of the cells of a *Homo sapiens* are replicated to become half of the genes strung along the chromosomes of the cells of its genetic offspring (of which process we can therefore say, and without contradiction, that the genes, chromosomes and cells are those of both the one *Homo sapiens* and the other), so some number of the phonemes strung along the morphemes of the sentences of one person are replicated to become some number of the phonemes strung along the morphemes of the sentences of its linguistic other (of which process we can therefore say, and without contradiction, that the phonemes, morphemes and sentences are those of both the one person and the other). And we can speculate on what, if anything, George Williams' 1966 delineation of a gene as a unit of DNA '*visible* to selection' (at the level of the organic) might prompt us to think about concerning linguistic units '*audible* to election' (at the level of the personic). And so on and so on.

But all of this, and much like it, can only be speculation. It will be increased scientific understanding—about the nature of language, about the nature of information, about the nature of nervous systems—which in time will take us further. Meanwhile, the best hypothesis, it seems to me, is that with the steady emergence of mutually aware selves comes, inevitably, the steady development of the demands of loving, and thereby the steady making, as attractor into its basin of attraction, of the pull of the ever more equal. The development, like all development, is uneven. At any one time, the linguistic structuring which is some persons will consist in experience which makes them more loving, more aware of the need for equality, than the linguistic structuring which is other persons. But while to their individual biographies this will be a difference which will make all the difference in the world, to the grand narrative of the personic it will matter hardly at all. For ineluctably, steadily, *that* is a narrative which, with increasing scientific understanding turning it steadily instead into an explanation, just *will* be pulled to that 'basin of attraction' which is fully loving equality.

* * *

These then are my four examples of issues, at different levels, with which a movement in thought such as I envisage would need to engage. They are, of course, in no way supposed to constitute an exhaustive list. There would also, for example, be very practical issues concerning the strategy and tactics of any such movement to consider. But such considerations are for places other than here, and times other than now. Here and now, as I approach the end of what has, somewhat to my surprise, become a rather lengthy text, there are other issues to address. As far as that bulkiness itself is concerned, perhaps the only thing I can say in my defence is that what George Eliot called 'that tempting range of relevancies called the universe' —or as I would put it, 'the universe, the biosphere and the world'—just actually *is* relevant to the very broad and general issues which I have been trying to address. But as far as some other of the characteristics of my text are concerned, specifically some five other characteristics, I want, to conclude, to say more.

First, let me try to offer some sort of explanation as to why I have been able to write it only somewhat late in my life. I think there are two main reasons. The less important reason is that like ninety-nine per cent of the human race, I have needed to spend most of my life working at other things: earning my living, bringing up my growing children, looking after my ageing parents. Only in the last very few years, retired a little early on a small but adequate pension, with children now adults and parents now dead, have I been able to give my almost all to the task. The only people ever in that position earlier in their lives fall into two groups. There are those few who live directly on the labour of others: people who get all or a significant part of the necessary money as private income, and/or all or a significant part of the necessary time from being serviced, traditionally by people called 'servants' and/or 'wives'. And there are those few academics who are actually paid, in both money and time, to think, discourse, research and write. But for reasons which will be obvious from the reading of this book, I think these are privileges which, like all other privileges, are almost always more apparent than real. At the very least, I find it difficult to imagine an hypothesis with the content of this one emerging from such places. But the more important reason is that it has taken that long just because it has taken that long. I have found the I IS hypothesis a very difficult one to come by. The ideas which have fed it have been accumulated unevenly, with now less now more conviction, across almost a lifetime. I have needed to struggle with them: to struggle to conceptualise them, grasp them, formulate them, apply them. Of course I know that there is certainly a sense in which I haven't, even yet, struggled quite hard enough: that much has been left rough and ragged, insufficiently thought through, or just in error. I think, however, that the guts of the I IS hypothesis are both probably right, and if right, very important. And if that *is* the case, then 'it has taken that long just because it has taken that long' is an entirely appropriate explanation.

Second, let me record that it was difficult to find a structure for the book. The main problem was to situate a beginning, a middle and an end, preferably in that order. For after living with the developing ideas for so long, they had come to be such a seamless web in my own head that it was difficult to know how to un-spin them so that they might be offered, as candidate web-building-material, to others. The structure I came up with has I hope enabled me to do that. In Part One and Part Three, I have tried to present what I take to be the generality of my case: in Part Two, some of the detail of its working out. In addition, the 'book-ends' device whereby Part One and Part Three are deliberately some sort of mirror image the one of the other will I hope point up two things: first the way in which the organic is so very different from, sometimes just the very opposite of, the personic; and second the way in which I IS is both derivative from, but also development upon, I AM. And the 'scientific investigation' device of Part Two, whereby the 'observation' and 'experiment' chapters present somewhat catalogue-ish versions of ideas which in the 'hypothesis' and 'theory' chapters appear in what I trust is a more structured fashion, will I hope provide access to the ideas at a series of different levels and to a series of different extents. If this structure and these devices work, then the result should be that the book will be able to speak to many different people at many different levels and with many different concerns. Because we are each to some extent like *all* other persons, to some extent like *some* other persons, and to some extent like *no* other persons, it is entirely possible that, in what I have written, only one thing will speak to *all* readers; a few more things to *some* readers; but all things to only *one* reader—me! Indeed, it has often been suggested that it is when we cannot find the book that we want to read on a subject that we are driven to try to write it for ourselves. This perhaps is what I have done. I have not been trying to write *about* philosophy. I have been trying to *do* philosophy. About the nature of ipseity, I have been trying to offer what I think is at least to some extent an original perspective. And I have been trying to present that perspective in such a way that it might leave some sort of mark—on her or his perception of the world—of more or less anyone who encounters it.

Third, let me register that, on occasion, it was difficult to find some of the vocabulary appropriate to the book. I wanted words as unambiguous as possible, as clear as possible, but sometimes they were just not available. In particular, they were not available in two (related) cases: first, language representing constant millisecond-by-millisecond change was hard to come by; second, language re-presenting place (space) concepts as their more appropriate duration (time) equivalents was often non-existent. These of course are gaps in the language in no way theoretically neutral. How could they be? Our theories of how the world is are made out of language. And language where the *fixed in place* does service for the *developmental in time* is absolutely characteristic of an I AM as opposed to an I IS theoretical stance. For this reason alone—though no doubt also for others—I know that my language has sometimes failed. Right now, all I can do is acknowledge that. And suggest that until the conception of persons as developing entities in time rather than as fixed entities in space is more fully understood, language is likely to continue to jolt and stumble in this area. (And 'area' is a space word!)

Fourthly, let me admit that it was difficult to know how to deal with the experts. Given that my subject matter has been primarily 'philosophical conversation', I felt quite comfortable in turning for reference to relatively easily accessible and popular scientific and philosophical books and magazines: after all, they are very often the sources from which the information used in such conversation is gleaned. But the distinction between academic philosophy and philosophical conversation is, of course, considerably less than absolute. So my problem was: in dealing primarily with the latter, what should be my stance vis-à-vis the former? For its solution, I must thank, at least in part, one Joseph Joubert. 'One must think when one writes that the scholars are there, but it is not to them that one is addressing oneself'. He said that. I know because when I came across it I jotted it down as apposite. And then when I wanted to use it, I didn't know where or when I had found it. I didn't even know who Joseph Joubert was—all of which goes to illustrate the point I want to make here. Mine is a serious book, but not a scholarly book. When I register that 'the scholars are there', it does make me nervous. But then without—I admit—being totally convinced by it, I remember a comment from Einstein to the effect that to do real intellectual work, it is better to avoid academic life: to be a shoe-maker by day, and do one's thinking by night. Be that as it may, while the seriousness of my intent means that my book contains much reference to the words of others, the non-scholarliness of its execution means that I don't always know which they are. Wherever I do recognise the source of such reference, I have mentioned it. But I suspect that there will be many instances where I don't. Let me therefore offer a heart-felt thanks to each and every Joseph Joubert who may lurk in these pages. But also reiterate my earlier claim that while I am not addressing myself, anyway primarily, to those who, 'qua' academic philosophers, teach the subject in universities, I am addressing myself to those very same people who, 'as' persons, take part—as do we all—in philosophical conversation.

Fifthly, and finally, let me offer some explanation of titles: first, of the sub-title I rejected; second, of the title I chose. There are two propositions which I take to be the most basic of the book: the one, that which insists upon the differentiation of the organic biosphere and the personic world; the other, that which differentiates animal-human I AM persons from human-human I IS persons. It is these two propositions which result in the two title decisions.

On the rejected subtitle: I have argued that just as it is one planet (among the millions which make up the inorganic universe) which is host to that living which is the organic, so it is one species (among the millions which make up the organic biosphere) which is host to that loving which is the personic. And since this is a differentiation which, I have suggested, has almost incalculable consequences for our thinking about the human, it is one which I tried to encapsulate in my first attempt at a sub-title. The book argues that the *competition* intrinsic to evolution by natural selection is caused by inequality among organisms (variation) | and not enough of what they need (scarcity). By absolute contrast with that, it also argues that the *cooperation* intrinsic to making by moral election | results in | equality among persons (justice) | and enough of what they want (abundance). It suggests that that the *evolution* | starts out from | already existing | unequal organisms | presenting as species | in conditions of scarcity. And by absolute contrast with that, it also suggests that *making* | ends up with | newly created | equal persons | presenting as selves | in conditions of abundance. The division of my sentences here into their phrasal chunks is in order to draw attention to the way in which, at every stage, the organic and the personic are always fundamentally different, and sometimes diametrically opposed. And it is that difference and opposition which I thought I could demonstrate, if a bit cheekily, by a direct Darwinian allusion. Going more explicitly vertical:

where he had presented	I would have presented
a theory of evolution concerning the speciation of entities which live	*a theory of development concerning the selving of entities which love*
under the title	under the sub-title
On the origin of species by means of natural selection	*On the progress of selves toward ends of moral election*

That I eventually chose not to use such a sub-title was not at all because I did not think it apposite. I think it *is* apposite. It was rather because, relative to the more prosaic 'new materialist theory of the person', I did not think it gave a clear enough guide to potential readers as to what the book is actually about.

On the chosen title: I have argued that, as animal-human persons, we have mad a steadily fading because less coherent I AM God (an ideological idea, *purporting to provide* immunity from a relativising critique) out of our steadily fading because less coherent I AM image of ourselves; but that, as human-human persons, we are making a steadily growing because more coherent I IS God (a scientific idea, *actually providing* immunity from a relativising critique) out of our steadily accumulating more coherent I IS image of ourselves. And that this differentiation also is one which has almost incalculable consequences, this time for our thinking about the good.

For an I AM self, 'that-which-is-good' is that which will enable its first-person self to flourish. For such a self, such a 'that-which-is-good' is not only *available to* the first person whether other persons have it or not, but its unavailability to those others is, as often as not, actually a *condition of* its availability to that one. Thus do I AM persons selve in a mode which rests relatively comfortably in inequality. For an I IS self, by contrast, 'that-which-is-good' is that which will enable all just- person selves to flourish. For such a self, such a 'that-which-is-good' is not only *unavailable to* one person when other persons have it not, but its unavailability to those others is always actually a *condition of* its unavailability to that one. Thus do I IS persons selve in a mode which rests only comfortably in equality. And thus do I think it legitimate simply to define a human-human position as one which always endeavours to make its relationships those of equality: that is, which always tries to build *from* that which is normal (having persons in society) *to* that which is moral (loving persons in community).

The theory of the good which is captured by this approach is one which, anyway among the professionals, would I think be called a 'substantive' theory: that is, it is a theory which offers a criterion by which to judge whether or not an action is a moral action. Utilitarianism, for example, is a substantive theory: with its 'greatest-happiness-of-the-greatest number' criterion, it is a theory which suggests that if an action promotes that greatest happiness it is good, and if it frustrates that greatest happiness it is bad. Well, so too is mine then a 'substantive' theory. With its greater-equality criterion, it is a theory which suggests that if an action results in there being more equality than previously there was, then it is good: and if it results in there being less equality than previously there was, then it is not good. Or to put it succinctly: where utilitarianism uses what has been called 'a felicific calculus', the I IS hypothesis uses what could be called 'an aequalific calculus'. To substantiate my claim that it is this in which the good consists, I have needed to argue that what we are in need of is a completely new paradigm concerning the nature of persons, and thus of the God which they make in their own image. With quiet apologies this time to nobody, it is this which explains my choice of title.

POSTSCRIPT

Once upon a *first* time, we most of us thought that the earth was 'first planet' in a universe which was 'out there' and Other, and which spun around it for its benefit. Then came Copernicus, Galileo, Newton and the rest. And the place of the earth in the universe was decentred for ever. Though of course it made no difference to how things seemed. The sun still 'rose' in the morning and 'set' in the evening, even at the Meteorological Office. And only with the technology of flying, and then of space travel, could we get the common sense which gave us a flat-and-still earth finally to give way, before the evidence to the contrary, to a *first* radically different understanding.

Once upon a *second* time, we most of us thought that humans were 'first creatures' in a biosphere which was 'out there' and Other, and which was made for their succour. Then came Darwin and his ilk. And the place of humans in the biosphere was decentred for ever. Though of course it made no difference to how things seemed. Long after evolution had shown us otherwise, human animals still seemed to be of a totally other order than all the rest. Only with the technology of gene sequencing, could we get the common sense which told us that neither a carrot nor a parrot had anything much in common with a *Homo sapiens* finally to give way, before the evidence to the contrary, to a *second* radically different understanding.

Once upon a *third* time—and the third time is our time—we most of us think that 'I' am 'first person', in a world which is 'out there' and Other, and which is there for me to make my way in as I choose. But now comes—well, let's just call it 'the train'. If you put your ear to the track, you can hear 'the train' coming. It's the *neuroscience* train; the *cognitive science* train; the *psycholinguistics* train. It's the train—or so I contend—which, when it arrives, will carry away both the 'personic exceptionalism' and the 'personic reductionism' which I identified in my preface. And the place of 'I' in the world will be decentred for ever as well. Though of course it will make no difference to how things seem. It will still seem as though 'I' am 'in here'. And that I am either 'in here' looking out upon you who, among other people and things, are 'out there'. Or that I am 'in here' looking in upon 'my thoughts', past or current, which, hidden from you, are 'in here' with me. Only with the technology of – well, who knows what – will we get the common sense which tells us that 'I' am the free-willing originator of 'my' actions finally to give way, before the evidence to the contrary, to what will be a *third* radically different understanding.

The revolutionary nature of the decentring of the earth in the universe meant that it took many centuries to achieve: and was not without some heart-ache (and some burning flesh!) for those who lived (and died) through it. The perhaps even more revolutionary nature of the decentring of the human in the biosphere took two or three centuries to achieve: and in some places—for example, among creationists and in the resurrection of 'the argument from design'—is not yet achieved, and continues to create considerable angst for those who still cannot understand and accept it. It is my contention that the almost certainly even more revolutionary nature of the decentring of the 'I' in the world will be the hardest of all to accommodate. It will probably take place more quickly than either of its two predecessors: scientific progress does seem to be exponential. But for those who live through it, for us and our immediate descendants, it will probably be the hardest of all for us to get our heads around. We will need all the compassion and tenderness we can muster for each other in our joint attempts to understand it.

But for them: for those who, in the not-too-distant future, will start to live comfortably in the full knowledge that it is so, it will, quite literally, change their world. When as a consequence of their equality, they can at last replace moral philosophy with moral science, then they will understand why Doris Lessing, at the end of one of her futuristic novels, had one of her characters look back to us, in compassionate anguish, and exclaim: 'oh you poor animal humans!'

APPENDIX A

Complete list of 'ands' from The Oxford Companion to Philosophy

(01) acquaintance and description
(02) acts and omissions
(03) affirmative and negative propositions
(04) analytic and synthetic statements
(05) appearance and reality
(06) a priori and a posteriori
(07) art and morality
(08) atheism and agnosticism
(09) autonomy and heteronomy
(10) base and superstructure
(11) boo and hoorah theory (actually entered as 'boo-hoorah theory')
(12) bourgeoisie and proletariat
(13) choosing and deciding
(14) compatibilism and incompatibilism
(15) conservatism and Romanticism
(16) contingent and necessary statements
(17) cosmology and religious belief
(18) denotation and connotation
(19) de re and de dicto
(20) determinables and determinates
(21) dictionaries and encyclopaedias of philosophy
(22) Dionysian and Apollonian
(23) egoism and altruism
(24) embraced and reluctant desires
(25) emotion and feeling
(26) emotive and descriptive meaning
(27) ends and means
(28) fact and value (actually entered as 'fact-value distinction')
(29) for-itself and in-itself
(30) formal and material mode
(31) from and matter
(32) freedom and determinism
(33) genus and species
(34) God and the philosophers

(35) heterological and homological
(36) I and thou
(37) individualism and anti-individualism
(38) is and ought
(39) justice and benevolence
(40) knowledge and science
(41) law and continental philosophy
(42) law and morals
(43) liberty and equality
(44) life and science
(45) literature and philosophy
(46) macrocosm and microcosm
(47) master and slave
(48) maximin and minimax
(49) methodological holism and individualism
(50) mind and body (actually entered as 'mind-body problem')
(51) monism and pluralism
(52) motives and motivation
(53) national and regional philosophies
(54) necessary and contingent existence
(55) necessary and sufficient conditions
(56) negation and double negation
(57) neustic and phrastic
(58) non-being and nothing
(59) objectivism and subjectivism
(60) objectivism and subjectivism, ethical
(61) open and closed thought
(62) ordinary language and philosophy
(63) particulars and non-particulars
(64) pessimism and optimism
(65) phenomena and noumena
(66) philosophy and science
(67) politics and determinism
(68) politics and the philosophers
(69) primary and secondary qualities
(70) psychology and philosophy
(71) public and private (actually entered as 'public-private distinction')
(72) pushpin and poetry
(73) quantity and quality
(74) realism and anti-realism
(75) reasons and causes
(76) religion and morality
(77) religiousness A and B

(78) saying and showing
(79) self-regarding and other-regarding
(80) sense and reference
(81) sign and symbol
(82) spontaneity and indifference
(83) statements and sentences
(84) stories and explanations
(85) structure, deep and surface
(86) subject and predicate
(87) substance and attribute
(88) teaching and indoctrinating
(89) tender and tough minded
(90) theology and philosophy
(91) unconscious and subconscious mind
(92) war and philosophy

APPENDIX B

From complete list of 'ands', those chosen

(01) acts and omissions (*)
(02) appearance and reality (*)
(03) arts and sciences
(04) autonomy and heteronomy (*)
(05) chance and necessity
(06) egoism and altruism (*)
(07) ends and means (*)
(08) fact and value (*)
(09) freedom and determinism (*)
(10) good and evil
(11) I and thou (*)
(12) identity and relationship
(13) ideology and science
(14) individual and society
(15) individualism and anti-individualism (*)
(16) is and ought (*)
(17) justice and benevolence (*)
(18) knowledge and science (*)
(19) liberty and equality (*)
(20) master and slave (*)
(21) mind and body (*)
(22) natural and cultural
(23) natural and moral
(24) natural and social
(25) needs and wants
(26) objectivism and subjectivism (*)
(27) praise and blame
(28) public and private (*)
(29) quantity and quality (*)
(30) realism and anti-realism (*)
(31) reason and emotion
(32) reasons and causes (*)
(33) right and wrong
(34) stories and explanations (*)
(35) time and space (*)

(*) indicates those taken from *The Oxford Companion to Philosophy* (1995)

BIBLIOGRAPHY

Note: Publishing details relate where possible to a currently available edition.

Appleyard, Brian, *Understanding the Present: Science and the Soul of Modern Man* (Tauris Parke, 2003)
Aristotle, *Categories* (Clarendon P., 1975)
Armstrong, Karen, *A History of God* (Vintage, 1999)
Arnhart, Larry, *Darwinian Natural Right: The Biological Ethics of Human Nature* (State Univ. of N.Y. Press, 1998)
Austin, J.L., *How to Do Things With Words* (O.U.P., 1976)
Bhaskar, Roy, *Reclaiming Reality* (Verso, 1989)
Becker, Ernest, *The Birth and Death of Meaning* (Penguin, 1972)
Bermudez, Luis, *Philosophy of Psychology* (Routledge, 2006)
Bernal, J.D., *Science in History* (Pelican, 1969)
Bickerton, Derek, *Language and Human Behaviour* (Routledge, 1996)
Bowden, John, *Jesus: The Unanswered Questions* (S.C.M., 1988)
Brown, Andrew, *The Darwin Wars* (Simon & Schuster, 1999)
Buber, Martin, *I and Thou* (Continuum International, 2004)
Bultmann, Rudolph, *Jesus and the Word* (Prentice Hall, 1982)
Cassirer, Ernst, *Essay on Man* (Yale U.P., 1977)
Chalmers, A.F., *What Is This Thing Called Science?* (Open Univ. P., 1999)
Christian, Harry, *The Making of Anti-Sexist Men* (Routledge, 1994)
Churchland, Patricia, *Brain-Wise* (Bradford Books, 2002)
Churchland, Patricia, *Neurophilosophy: Towards a Unified Understanding of the Mind-Brain* (M.I.T. Press, 1989)
Churchland, Patricia & Paul, *The Chuchlands and their Critics* (Blackwell, 1996)
Churchland, Paul, *Matter and Consciousness* (M.I.T. Press, 1988)
Clark, Andy, *Mindware* (O.U.P. Inc. U.S.A., 2001)
Cohen, Gerry, *Self-ownership, Freedom and Equality* (C.U.P., 1995)
Comte, Auguste, *Cours de Philosophie Positive* (Adamant Media Corp., 2001)
Crossley, Nick, *Intersubjectivity* (Sage Pubns., 1996)
Cupitt, Don, *After God* (Weidenfeld & Nicolson, 1997)
Cupitt, Don, *Taking Leave of God* (S.C.M., 2001)
Daly, Mary, *Beyond God the Father* (Beacon, 1973)
Dante, *The Divine Comedy* (O.U.P., 1998)
Darwin, Charles, *The Descent of Man* (Penguin, 2003)
Darwin, Charles, *The Origin of Species* (Penguin, 1982)

Dawkins, Richard, *River out of Eden* (Phoenix, 2001)
Dawkins, Richard, *The Selfish Gene* (O.U.P., 2006)
Deacon, Terrence W., *The Symbolic Species: the Co-evolution of Languages and the Human Brain* (W.W. Norton, 1998)
Dennett, Daniel, *Consciousness Explained* (Penguin, 1993)
Dennett, Daniel, Darwin's *Dangerous Idea* (Penguin, 1996)
Dennett, Daniel, *Kinds of Minds* (Basic Books Inc., 1996)
Descartes, Rene, *Discourse on Method* (Penguin, 1973)
Deutsch, David, *The Fabric of Reality* (Penguin, 1998)
Dostoevsky, Fyodor, *The Brothers Karamazov* (Penguin, 2003)
Dworkin, Ronald, *Living as Equals* (O.U.P., 1999)
Eagleton, Terry, *The Illusions of Postmodernism* (Blackwell, 1996)
Eliot, T.S., *Tradition and the Individual Talent* (In Poems etc Alexis Gregory, N.Y., 1971)
Epstein, Joshua & Axtell, Robert Growing, *Artificial Societies: Social Science from the Bottom Up* (M.I.T. Press, 1996)
Erikson, Erik, *Childhood and Society* (Vintage, 1995)
Flanagan, Owen, *Consciousness Reconsidered* (M.I.T. Press, 1994)
Fletcher, Richard, *The Conversion of Europe: from Paganism to Christianity, 371-1386 AD* (Harper Collins, 1997)
Fortey, Richard, *Life: An Unauthorised Biography* (Flamingo, 1998)
Freud, Sigmund, *Civilisation and its Discontents* (Penguin, 2004)
Fromm, Erich, *The Art of Loving* (HarperCollins, 1995)
Fromm, Erich, *Beyond the Chains of Illusion* (Simon & Schuster, 1985)
Fukuyama, Francis, *The End of History* (Penguin, 1993)
Giddens, Anthony, *Central Problems in Social Theory* (Palgrave Macmillan, 1979)
Glynn, Ian, *An Anatomy of Thought* (O.U.P., 2003)
Gadamer, Hans-Georg, *Truth and Method* (Sheed and Ward, 1975)
Goldman, Alan H., *Moral Knowledge* (Routledge, 1990)
Goodman, Nelson, *Starmaking: Realism, Anti-Realism and Irrealism* (M.I.T.Press, 1996)
Gore, Charles, *The Philosophy of the Good Life* (Murray, 1934)
Gould, Stephen Jay, *Wonderful Life* (Vintage, 2000)
Gramsci, Antonio, *Prison Notebooks* (Lawrence and Wishart, 1973)
Habermas, Jurgen, *The Theory of Communicative Action* (Polity P., 1986)
Halsey, A.H., *No Discouragement* (Palgrave Macmillan, 1996)
Hampson, Daphne, *Theology and Feminism* (Blackwell, 1990)
Hawking, Steven, *A Brief History of Time* (Bantam, 1995)
Hegel, G.W.F., *Phenomenology* (O.U.P., 1979)
Hegel, G.W.F., *The Philosophy of Right* (Dover Pubns. Inc., 2005)
Hessen, B., *On the Social and Economic Roots of Newton's Principia* (Current Book Distributors, Sydney, 1946)
Hick, John, *An Interpretation of Religion* (Yale U. P., 2005)
Hick, John (ed), *The Myth of God Incarnate* (S.C.M., 1993)
Holbrook, David, *Lost Bearings in English Poetry* (Vision P., 1977)
Honderich, Ted (ed), *The Oxford Companion to Philosophy* (O.U.P., 1995)
Horgan, John, *The End of Science* (Perseus Bks., 1996)
Horgan, John, *The Undiscovered Mind* (Free P., 2000)
Howe, Michael, *Genius Explained* (C.U.P., 2001)

Jaspers, Karl, *The Origin and Goal of History* (Greenwood P., 1977)
Kant, Immanuel, *Critique of Judgement* (Dover, 2005)
Kant, Immanuel, *Critique of Practical Reason* (Hackett Pub. Co. Inc., 2002)
Kant, Immanuel, *Critique of Pure Reason* (Dover Pubs., 2004)
Kilminster, Richard, *The Sociological Revolution: from the Enlightenment to the Global Age* (Routledge, 1998)
Koslowski, Barbara, *Theory and Evidence: The Development of Scientific Reasoning* (M.I.T.Press, 1996)
Kuhn, T.S., *The Structure of Scientific Revolutions* (Univ. of Chicago P., 1996)
Lampe, G.W.H., *God as Spirit* (Clarendon P., 1977)
Leakey, Richard, *The Origin of Humankind* (Basic Books Inc., 1996)
Lewin, Roger, *Complexity* (Phoenix, 2001)
Lewis,C.S., *The Four Loves* (Fount, 2002)
Lockwood, Michael, *Mind, Brain and the Quantum* (Blackwell, 1989)
MacIntyre, Alasdair, *Three Rival Versions of Moral Enquiry* (Univ. of Notre Dame P., 1992)
Macmurray, John, *Conditions of Freedom* (Prometheus Bks., 1993)
Macmurray, John, *Persons in Relation* (Humanity Books, 1998)
Macmurray, John, *Reason and Emotion* (Prometheus Books, 1999)
Marx, Karl, *Manifesto of the Communist Party* (Foreign Languages P., 1970)
Marx, Karl, *Towards a Critique of the Hegelian Philosophy of Right* (C.U.P., 1970)
Matthew, Patrick, *Naval Timber and Arboriculture* (Longman, 1831)
McNeill, David, *Hand and Mind: What Gestures Reveal about Thought* (Univ. of Chicago P., 1996)
Mill, John Stuart, *The Subjection of Women* (Penguin, 2006)
Monod, Jacques, *Chance and Necessity* (Collins, 1972)
Montefiore, C.G., *Some Elements of the Religious Teaching of Jesus according to the Synoptic Gospels* (Arno P. 1973)
Moore, G.E., *Principia Ethica* (C.U.P., 1993)
Morris, Desmond, *The Naked Ape* (Vintage, 2005)
Munz, Peter, *Philosophical Darwinism: On the Origin of Knowledge by means of Natural Selection* (Routledge, 1993)
Oatley, Keith, *Best Laid Schemes: the Psychology of Emotions* (C.U.P., 1992)
O'Hear, Anthony, *Beyond Evolution* (Clarendon P., 1999)
Ong, Walter J., *Orality and Literacy* (Routledge, 2002)
Passmore, John, *Man's Responsibility for Nature* (Duckworth, 1980)
Pattison, George, *The End of Theology* (S.C.M., 1998)
Pinker, Steven, *The Language Instinct* (Penguin, 1995)
Pinker, Steven, *Words and Rules* (Phoenix, 2000)
Plato, *The Republic* (Penguin, 2003)
Popper, Karl, *The Logic of Scientific Discovery* (Routledge, 2002)
Popper, Karl, *The Philosophy of Karl Popper* (Open Court, 1974)
Rawls, John, *A Theory of Justice* (Harvard U.P., 2005)
Reich, Wilhelm, *The Mass Psychology of Fascism* (Souvenir P., 1997)
Reuther, Rosmary Radford (ed), *Woman Guides: Readings Towards a Feminist Theology* (Beacon Press, 1983)
Richards, I.A., *Poetries and Sciences* (Routledge 1970)
Richards, Janet R., *The Sceptical Feminist* (Penguin, 1994)
Robinson, Richard, *An Atheist's Values* (Clarendon P., 1964)

Rorty, Richard, *Contingency, Irony and Solidarity* (C.U.P., 1989)
Rorty, Richard, *Objectivity, Relativism and Truth* (C.U.P., 1990)
Rosen, Michael, *On Voluntary Servitude* (Polity P., 1996)
Russell, Bertrand, *Why I am not a Christian* (Routledge, 2004)
Ryle, Gilbert, *The Concept of Mind* (Penguin, 2000)
Sanders, E.P., *The Historical Figure of Jesus* (Penguin, 1995)
Sanders, E.P., *Jesus and Judaism* (S.C.M., 1985)
Schelling, F.W.J.V., *Philosophical Enquiries into the Nature of Human Freedom* (Open Court, 2003)
Schrodinger, Erwin, *What is Life?* (C.U.P., 1992)
Schweitzer, Albert, *The Quest for the Historical Jesus* (S.C.M., 2000)
Scruton, Roger, *An Intelligent Person's Guide to Philosophy* (Penguin, 1999)
Scruton, Roger, *Modern Philosophy* (Routledge, 2001)
Searle, John, *The Construction of Social Reality* (Penguin, 1996)
Sennett, R. & Cobb, J., *The Hidden Injuries of Class* (Random House, 1988)
Shea, Victor and Whitla, William (ed), *Essays and Reviews: The 1860 Text and Its Reading* (University of Virginia Press, 2000)
Skinner, B.F., *Beyond Freedom and Dignity* (Hackett Pub. Co. Inc., 2002)
Sloterdijk, Peter, *The Critique of Cynical Reason* (Univ. of Minnesota P., 1988)
Smith, Adam, *The Theory of Moral Sentiments* (Dover Pubns., 2006)
Smith, Adam, *The Wealth of Nations* (O.U.P., 1998)
Smolin, Lee, *The Life of the Cosmos* (O.U.P., 1999)
Soper, Kate, *What is Nature?* (Blackwell, 1995)
Spinoza, B., *Ethics* (Penguin, 1996)
Stevens, Wallace, *Collected Poems* (Faber, 2006)
Troeltsch, Ernst, 'On the Historical and the Dogmatic Method in Theology' (1913), in *Religion in History* (Fortress Press, 1997)
Vermes, Geza, *Jesus the Jew* (S.C.M., 2001)
Ward, Keith, *The Concept of God* (Fount, 1977)
Wherwell, William, *The Philosophy of the Inductive Sciences* (Frank Cass, 1967)
Wiles, Maurice, *Working Papers on Doctrine* (S.C.M., 1976)
Wilkinson, Richard, *Unhealthy Societies* (Routledge, 1996)
Williams, Bernard, *Ethics and the Limits of Philosophy* (Routledge, 2006)
Williams, George, *Adaptation and Natural Selection* (Princeton U.P., 1966)
Williams, Raymond, *Keywords* (Fontana, 1988)
Wilson, Edward O., *Consilience* (Abacus, 2003)
Wilson, Edward O., *Sociobiology* (Harvard U.P., 2000)
Wilson, R.A., *The Miraculous Birth of Language* (Guild Bks., 1941)
Wolpert, Lewis, *The Unnatural Nature of Science* (Harvard U.P., 1998)

NAME INDEX

SUBJECT INDEX